Technical Biochemistry

Oliver Kayser · Nils J. H. Averesch

Technical Biochemistry

The Biochemistry and Industrial Use of Natural Products

Oliver Kayser
Fakultät Bio- und Chemieingenieurwesen
TU Dortmund University
Dortmund, Nordrhein-Westfalen, Germany

Nils J. H. Averesch
Microbiology and Cell Science
University of Florida
Gainesville, FL, USA

ISBN 978-3-658-47120-0   ISBN 978-3-658-47121-7 (eBook)
https://doi.org/10.1007/978-3-658-47121-7

Translation from the German language edition: "Technische Biochemie" by Oliver Kayser and Nils J. H. Averesch, © Der/die Herausgeber bzw. der/die Autor(en), exklusiv lizenziert an Springer Fachmedien Wiesbaden GmbH, ein Teil von Springer Nature 2024. Published by Springer Fachmedien Wiesbaden. All Rights Reserved.

This book is a translation of the original German edition "Technische Biochemie," 2nd edition, by Oliver Kayser and Nils J. H. Averesch, published by Springer Fachmedien Wiesbaden GmbH in 2024. The translation was done with the help of an artificial intelligence machine translation tool. A subsequent human revision was done primarily in terms of content, so that the book will read stylistically differently from a conventional translation. Springer Nature works continuously to further the development of tools for the production of books and on the related technologies to support the authors.

© The Editor(s) (if applicable) and The Author(s), under exclusive license to Springer Fachmedien Wiesbaden GmbH, part of Springer Nature 2025.

This work is subject to copyright. All rights are solely and exclusively licensed by the Publisher, whether the whole or part of the material is concerned, specifically the rights of translation, reprinting, reuse of illustrations, recitation, broadcasting, reproduction on microfilms or in any other physical way, and transmission or information storage and retrieval, electronic adaptation, computer software, or by similar or dissimilar methodology now known or hereafter developed.
The use of general descriptive names, registered names, trademarks, service marks, etc. in this publication does not imply, even in the absence of a specific statement, that such names are exempt from the relevant protective laws and regulations and therefore free for general use.
The publisher, the authors and the editors are safe to assume that the advice and information in this book are believed to be true and accurate at the date of publication. Neither the publisher nor the authors or the editors give a warranty, expressed or implied, with respect to the material contained herein or for any errors or omissions that may have been made. The publisher remains neutral with regard to jurisdictional claims in published maps and institutional affiliations.

This Springer imprint is published by the registered company Springer Fachmedien Wiesbaden GmbH, part of Springer Nature.
The registered company address is: Abraham-Lincoln-Str. 46, 65189 Wiesbaden, Germany

If disposing of this product, please recycle the paper.

Preface to the 1st Edition

Biology for engineers may seem like a contradiction at first glance. Upon closer examination, this contradiction dissolves, as with typical subjects such as bioengineering, biotechnology, and food technology, engineers have rightfully claimed their place in biological questions. Biology and its related disciplines genetics, molecular biology, and bioinformatics are no longer conceivable without the canon of biologically oriented engineering education, and without the high technical expertise, biotechnical processes would probably hardly be feasible.

The biological education of prospective bio-process engineers and biotechnologists is strongly based on the classical university education of biologists. Many important fundamental biological, biochemical, and genetic laws and processes are taught, without fully meeting the needs of engineering education. From our own experiences with bioengineering students, they receive a very solid education that enables them to understand biological concepts and strategies, but hardly allows them to transfer this knowledge to engineering problems. This book aims to bridge this gap. Unlike well-known and very well-established textbooks on biology, biochemistry, and genetics, biological concepts will be presented and linked with technical and engineering problems.

The aim of the book is to illuminate chemical and biochemical principles in natural product biosynthesis and to explain their biotechnological and bio-process engineering production pathways. The book consists of three parts: Firstly, the necessary teaching material is conveyed, which should provide a basic understanding of biochemistry, natural substances, and their chemistry and biological effects. Secondly, "infoboxes" provide information on current developments and trends. Thirdly, further literature references and weblinks are suggested for the text and the boxes, allowing the reader to undertake in-depth self-study. Here, there are also self-study questions to check and repeat the understanding of the content conveyed in each chapter.

Why this book and why in this form? Traditionally, books are written by experienced scientists who try to convey the content of their field from the perspective of research and their lectures. After long discussions, Nils and I dared to attempt to see and write a textbook for students from the perspective of the students and lecturers. In addition to pure knowledge transfer, the goal was also to integrate direct access to further information via modern media such as the internet directly into the text. Therefore, behind the QR codes, you will find numerous hyperlinks to Wikipedia, KEGG, PubMed, and many homepages of distinguished colleagues.

Oliver Kayser
Nils J. H. Averesch
Dortmund/Brisbane
January 2015

Preface to the 2nd Edition

Dear readers,
Ten years after the first edition of the book *Technical Biochemistry*, a new edition is all the more urgent to incorporate many biochemical developments of recent years. Over the last decade, new biochemical and genetic engineering techniques, processes, and also new products have been discovered in research or have moved from research to the market or into industrial application.

You are holding this second edition of *Technical Biochemistry* in your hands. We have completely redesigned and rewritten the book because we have learned that science in biochemistry has brought new research areas such as synthetic biology and the computer-aided design of new metabolic pathways into the laboratories. An example of this is natural product biochemistry, which until a few years ago was characterized by the characterization of biosynthetic enzymes and the cloning of their genes in microorganisms. What was laborious molecular biological work in the last decade, such as the elucidation of the function of individual enzymes with the help of mutants, can now largely be determined with the aid of computers. For these various reasons, we would like to include updates in the areas of protein modeling, metabolic engineering, and natural product biotechnology.

The learning objectives of this current second edition are:
- Conveyance of applied biochemical fundamentals for the in-depth understanding of industrially relevant natural substances and their manufacturing processes
- Guidance for the critical evaluation of new biotechnical processes and their biochemical foundations against the backdrop of Synthetic Biology and Genomic Sciences

Upon studying the book, it quickly becomes clear that the classical world of simple natural substances and metabolites of primary metabolism no longer defines the image of modern technical biochemistry. It has been shown in recent years and is now standard that biotechnological synthesis of highly complex natural substances such as paclitaxel, artemisinin, or polypeptide antibiotics define the current state of-the-art. The original goal of knowing biochemical metabolic pathways is no longer sufficient. Knowledge of regulatory mechanisms, molecular biological basics, and genetics not only round off biochemistry but make it a modern cross-disciplinary subject in the life sciences.

We would like to provide an introduction to the basics of biochemistry for readers who are studying biochemistry, biotechnology, or bio-process engineering. This book is not intended to replace the proven and very comprehensive textbooks on biochemistry. Rather, it is meant to be a supplement to general biochemistry, which

Preface to the 2nd Edition

will illuminate the technical questions in the further course of study. We hope to be able to show with this book how deeply biochemistry is rooted in biotechnological professions and in our everyday life.

Oliver Kayser
Nils J. H. Averesch
Dortmund/Stanford
September 2024

Contents

1	**Biology and Technology: An Introduction**	1
1.1	History of Biochemical Process Engineering	5
1.2	Technical Biochemistry Today	8
2	**Fundamentals of Technical Biochemistry**	9
3	**The Basis of Biochemical Reactions—Primary Metabolism**	15
4	**Bioorganic Reactions and Building Blocks of Natural Products**	21
4.1	From the Building Blocks to the Structure	24
5	**Technical Enzymes**	31
5.1	Technical Applications of Enzymes	33
5.2	Extremophile Organisms and Extremozymes	34
6	**Vitamins**	37
6.1	Vitamin B_2	38
6.2	Vitamin B_5	38
6.3	Vitamin B_{12}	39
6.4	Vitamin C	40
6.5	Vitamin D	43
6.6	Vitamin E	44
7	**Photosynthesis—Basics and Application**	47
7.1	Photosynthesis—The Beginning of Everything	48
7.2	Dark Reaction	50
7.3	The Somewhat Different Photosynthesis	52
7.4	The Artificial Photosynthesis	52
7.5	Photolysis for the Production of Biohydrogen	53
7.6	Photosynthesizing Microorganisms as New Producers	54
8	**Sugar Metabolism**	57
8.1	Glycolysis	58
8.2	Pentose Phosphate Pathway	59
8.3	Citric Acid Cycle	62
8.4	Technical Acids	63
8.5	Oxidative Phosphorylation	67
8.6	Alcoholic Fermentation	70
8.7	Lactic Acid Fermentation	72
9	**Amino Acid Metabolism**	77
9.1	Amino Acids as Building Blocks for Proteins	78

9.2	Chemistry of Amino Acids	79
9.3	Classification of Amino Acids	81
9.4	Technical and Economic Importance	82
9.5	Biosynthesis of Aliphatic Amino Acids	82
9.6	Biosynthesis of Aromatic Amino Acids	93
10	**Fatty Acid Biosynthesis and ABE Metabolism**	**97**
10.1	Fats and Fatty Acid Biosynthesis	98
10.2	ABE Metabolism	101
10.3	Technical Fats and Oils	103
11	**Secondary Metabolism and Biochemical Pathways of Significance**	**107**
11.1	Polysaccharides	109
11.2	Heterogeneous Polysaccharides	116
11.3	Aminopolysaccharides	118
12	**Phenolic Natural Products**	**127**
12.1	Simple Phenols and Phenylpropanes	128
12.2	Biosynthetic Classification	132
12.3	Lignans	133
12.4	Lignins	134
12.5	Coumarins	136
12.6	Flavonoids	137
12.7	Styrylpyrones and Stilbenes	141
12.8	Tannins (Tannins or Polyphenols)	143
13	**Terpenes**	**147**
13.1	Terpene Metabolism	148
13.2	Biosynthesis of Terpenoids	149
13.3	Monoterpenes and Essential Oils	151
Infobox 13.1: Steam Distillation		155
13.4	Sesquiterpenes	156
13.5	Diterpenes	160
13.6	Triterpenes	163
Infobox 13.2: Is Licorice Dangerous?		168
13.7	Tetraterpenes	169
13.8	Polyterpenes: Rubber	171
14	**Alkaloids**	**175**
14.1	Definition	177
14.2	Chemistry	179
14.3	Technical Importance	179
14.4	Biosynthesis	180
14.5	Functions in the Plant	181
14.6	Alkaloids Derived from Phenylalanine and Tyrosine	183
14.7	Alkaloids derived from Tryptophan	191
14.8	Quinoline Alkaloids	201
14.9	Alkaloids Derived from Ornithine	205

14.10	Alkaloids Derived from Histidine	210
14.11	Purine Alkaloids	211
14.12	Alkaloids Derived from Arginine	212

15	**Antibiotics**	217
15.1	History	218
15.2	Non-ribosomal Peptide Biosynthesis	220
15.3	Penicillins and Cephalosporins	222
15.4	Polyketide Antibiotics	227

16	**Environmental Biochemistry**	235
16.1	Biochemistry of Wastewater Treatment	236
16.2	Biogas and Methanogenesis	237

17	**The Future of Bioprocesses Engineering**	245
17.1	Artificial Photosynthesis	247
17.2	Natural Substance Biotechnology	248
17.3	Bioinformatics and Artificial Intelligence	249
17.4	Bioelectricity and Biological Fuel Cells	252

Supplementary Information

Answers to the Self-Control Questions	254
Important Databases	267
Glossary	269
References	271

About the Authors

Oliver Kayser

was born in 1967 in Recklinghausen. He went to the Westfälische Wilhelms-University Münster to study pharmacy from 1986 to 1991. After a stay abroad at the University of Florida, he began his doctorate at the Free University of Berlin in the field of Pharmaceutical Biology in 1992, switched to industry, and received his postgraduate qualification (in German Habilitation) in 2003 in the subjects of Pharmaceutical Technology and Pharmaceutical Biotechnology. From 2004, he was an Assistant Professor and later Adjunct Professor for Natural Product Biosynthesis at the University of Groningen, NL, where he dealt with combined biosynthesis and plant cell cultures for the production of biologically active secondary metabolites in heterologous cell systems. Since 2010, he holds the Chair of Technical Biochemistry at the Technical University of Dortmund and is trying to forge new paths in teaching as a dean of studies.

Nils J. H. Averesch

was born in Essen in 1984. In 2011, he completed his Bioengineering studies at the Technical University of Dortmund and took up a doctoral position at the University of Queensland (Australia) with a focus on Metabolic Engineering in 2012, completing his PhD in 2016. From 2017 to 2018, Nils was a research associate at the NASA Ames Research Center in California, conducting research in the field of synthetic biology. In 2019, he moved to Stanford University, initially as a postdoc (2019 to 2021), and later on (until 2024) as an academic staff member and head of a research group in the field of microbial biotechnology. Currently (since 2024), Dr. Averesch is an Assistant Professor at the University of Florida where his research comprises the development of microbial cell factories that use greenhouse gases and other waste-derived feedstocks as substrates. The goal is to create a sustainable chemical industry on Earth, while simultaneously developing circular biological production platforms for life-support on long space research missions.

Frequently Recurring Abbreviations and Symbols

a	Year	IPP	Isopentenyl diphosphate
aKG	alpha-Ketoglutarate (= 2-Oxoglutarate)	k	Kilo
Abb.	Figure	KEGG	Kyoto Encyclopedia of Genes and Genomes
ATP	Adenosine triphosphate		
ca.	approximately	L	Liter
CoA	Coenzyme A	LC	Liquid Chromatography
COR	Codeine reductase	LSD	Lysergic acid diethylamide
COX	Cyclooxygenase		
		MEP-Weg	Methylerythritol phosphate pathway
Da	Dalton (1 Da = 1 g/mol)		
DHAP	Dihydroxyacetone phosphate	m	milli
		min	Minute
DMAPP	Dimethylallyl diphosphate	mol	Mole
DMT	Dimethyltryptamine	MS	Mass Spectrometry
DNA	Deoxyribonucleic acid	MVA-Weg	Mevalonate pathway
DXP-Weg	Deoxyxylulose phosphate pathway	µ	micro
		NAD	Nicotinamide adenine dinucleotide
F16BP	Fructose-1,6-bisphosphate		
F6P	Fructose-6-phosphate	NADP	Nicotinamide adenine dinucleotide phosphate
FAD	Flavin adenine dinucleotide		
FMN	Flavin mononucleotide	n	nano
FPP	Farnesyl diphosphate		
		PEP	Phosphoenolpyruvate
g	Gram	PET	Polyethylene Terephthalate
GABA	γ-Aminobutyric acid	PG	Prostaglandin
GAP	Glyceraldehyde 3-phosphate	PLA	Polylactic acid
GC	Gas Chromatography	PP-Weg	Pentose phosphate pathway
GMP	Good Manufacturing Practice	PYR	Pyruvate
GPP	Geranyl diphosphate	RNA	Ribonucleic acid
GTP	Guanosine triphosphate	RuBisCO	Ribulose-1,5-bisphosphate carboxylase/oxygenase
HMG-CoA	3-Hydroxy-3-methyl-glutaryl-CoA		
		SAM	S-Adenosylmethionine
HPLC	High-Performance Liquid Chromatography	sec	Second

t	Ton	UTP	Uridine triphosphate
Tab.	Table	UV	Ultraviolet
THC	Tetrahydrocannabinol		
THF	Tetrahydrofolate	w/w	Weight/Weight
TPP	Thiamine diphosphate		
TX	Thromboxane	ZNS	Central Nervous System
Q10	Ubiquinone-10		

Biology and Technology: An Introduction

Contents

1.1　History of Biochemical Process Engineering – 6

1.2　Technical Biochemistry Today – 8

© The Author(s), under exclusive license to Springer Fachmedien Wiesbaden GmbH, part of Springer Nature 2025
O. Kayser and N. J. H. Averesch, *Technical Biochemistry*, https://doi.org/10.1007/978-3-658-47121-7_1

Chapter 1 · Biology and Technology: An Introduction

> **Learning Objectives**
> - General definitions
> - History of biology technology
>   - Fermentation and brewing
>   - Industrial evolution of biotechnology and bio-process engineering
> - Natural substances
>   - Significance
>   - History

Biology and technology may seem to be a contradiction at first glance, as the often non-exact and unpredictable world of biology is difficult to describe with the mathematical and absolute world of engineering. We have often read about bionics, which show us that many technical inventions such as dirt-repellent surfaces or construction principles for towers or skyscrapers can be traced back to biological phenomena such as the lotus effect or the statics of the grass stalk. Bionics with clear references to biochemistry are rare, which is why we must strive for comparisons with biotechnology.

> **Definition**
>
> **Technical Biochemistry** is an application-oriented science of chemical changes in the metabolism of organisms, which includes methods and techniques for the production, development, and testing of chemical compounds ranging from small molecules to natural substances and proteins.

> **Definition**
>
> **Biotechnology** is an interdisciplinary science for the use of cells, organisms, or parts thereof for the production of chemical compounds or diagnostic applications.

Fermentation in bioreactors is an attempt to mimic the biochemistry of a production organism's cell under artificial conditions (Fig. 1.1). How difficult and demanding this technical implementation is, is described in this book. Engineering

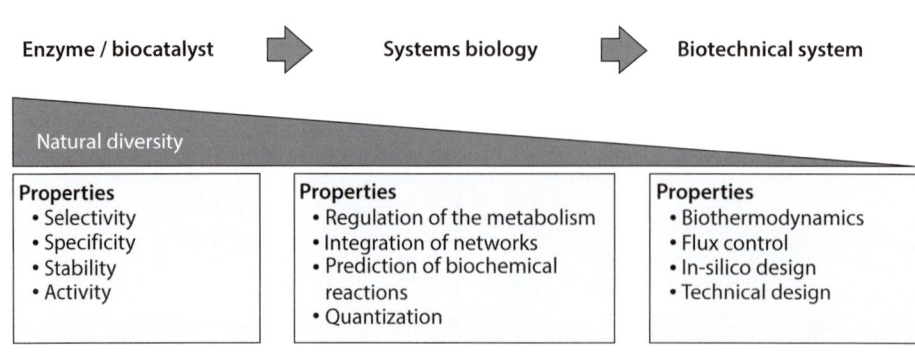

Fig. 1.1 From biochemical diversity to standard models in Technical Biochemistry

and mathematics in biological application are not the only ones, but they rightly penetrate the field of biology, as shown by biotechnology and recently nanobiotechnology. In an increasingly environmentally conscious society, it is necessary for the chemical industry to think about sustainable production processes, often referred to as **Green Chemistry.** Modern production processes are operated biotechnologically as far as possible, if the production quantities and costs allow it. Of course, we cannot expect all our products to be biologically and sustainably produced in the near future. The amount of renewable natural resources alone would not be available, because in a world with a high population and increasing demands on nutrition and mobility, this endeavor is not feasible with our current knowledge and level of technology. If biology and biochemistry contribute to minimizing environmentally hazardous processes in which heavy metals are used as catalysts or organic solvents, a great progress would already be achieved today.

> **Definition**
>
> **Bio-process engineering** is an application-oriented science that uses methods of natural sciences and engineering to develop biochemical and biotechnical processes for production.

What is meant by Technical Biochemistry? Technical Biochemistry is a subfield of biochemistry that deals with the application of biochemical methods and findings in technical practice (◘ Fig. 1.2). Important topics include the development of processes for the production of products such as food, pharmaceuticals, and biofuels, as well as the development of processes for wastewater treatment and energy production. What technical processes are already established today without which biochemistry would be unthinkable? Immediately, millennia-old techniques such as beer brewing and bread baking come to mind, but in the last 100 years, additional processes have been added. In this book, we learn how a biogas plant technically works, how penicillin and other antibiotics have been produced since the 1930s, and how laundry enzymes have made our lives easier.

The basic idea of technology in biology is the desire to use biological cellular phenomena for biochemical reaction performances to produce chemical substances. However, the big problem for the technical biologist is the high variability of cellular organisms, which needs to be controlled in an engineering manner. Cells and

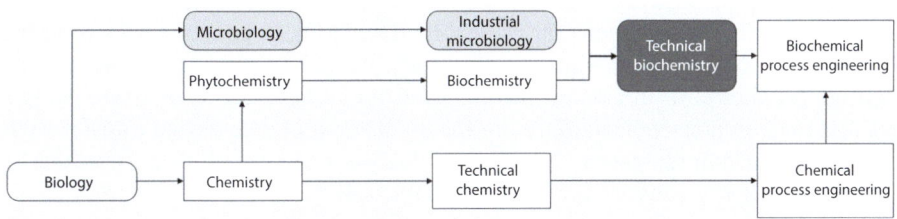

◘ **Fig. 1.2** Industrial Evolution of Technical Biochemistry

organisms are chemically communicating biological units that interact and react with their neighbors and the environment. This reaction is a fundamental property to remain viable and survive. Consequently, high rigidity and reduced adaptation to external factors are not in the biological and one could also say not in the biothermodynamic sense of biology (◘ Fig. 1.1). The cell is a thermodynamically open system that follows the principle of entropy increase. It is completely clear and wrong to claim that biology and biochemistry do not follow the principles of physics.

Even if we could ask the question why this biochemical reaction takes place in the case of glycolysis, despite two non-endergonic enzyme catalysis steps, the physical and thermodynamic principles are important for the development of subjects such as biotechnology or bio-process engineering. While biotechnology still focuses on the exploration of cellular basics and bio-process engineering focuses on mathematical and engineering modeling, both disciplines have a clear justification. Both want to master the biochemical and genetic "black box" cell and make biochemical reactions quantifiable. Of course, it is desirable to know exactly what is happening biochemically in the cell, but it is only sufficient knowledge. Only through the technical work with the biological phenotypes and a sufficient prediction of the biochemical reactions does the existing knowledge suffice to design and control processes.

◘ Figure 1.2 shows the development of the various subjects and disciplines in bio-based technology. The disciplines of biology and chemistry were the scientific basics from which all other disciplines have developed over the last almost 200 years. We will see that one of the first very primitive biochemical techniques was the extraction of plants. It is important to understand that with the progress of technology, from the bucket for simple fermentation approaches to modern computer-controlled fermenters, technical knowledge has increasingly penetrated the biological world. Technical Biochemistry is not at the end of development, but is a discipline in the midst of the canon of biotechnology and bio-process engineering. Other disciplines not shown in ◘ Fig. 1.2, which also have a major influence on the technization of biology, are bioinformatics, applied genetics, and the upcoming nanobiotechnology.

In the recognized canon of subjects, which is reflected in university courses, there is currently no study of Technical Biochemistry. However, in other courses such as Biotechnology and Bio-process Engineering, Technical Biochemistry is included as a focus. It may be surprising that there are two subjects like Biotechnology and Bio-process Engineering, which sound very similar at least by name and where one can also assume similar study contents. If you look at the definitions in the boxes above, you will see that Biotechnology wants to convey very strong biochemical and cellular phenomena in the study, while Bio-process Engineering has its origin in chemical process engineering and puts questions of quantitative biology and mathematical consideration in the foreground of knowledge transfer. Here, it is about understanding and designing production processes that are of interest to industry in order to produce quantities on a ton scale.

In the future, it will not be easy to recognize biology in bio-based engineering sciences when production organisms are reduced to their genome in databases. Enzymes as catalysts are optimized with the help of artificial intelligence (AI), and the design of genes and metabolic pathways is created on the computer and outsourced

1.1 History of Biochemical Process Engineering

to gene tech companies that clone this information precisely into the desired organism and deliver it functional. With synthetic organisms, own fermentations are carried out, the course and success of which are monitored based on the transcriptome.

1.1 History of Biochemical Process Engineering

The origin of a biochemical process engineering cannot be determined with certainty. Was the deliberate fermentation to produce beer (◘ Fig. 1.3) and vinegar a beginning, or were it the very simple processes in the Middle Ages that required basic knowledge? Here we think of the medicine behind the monastery walls, which was characterized by the cultivation of medicinal plants and their later extraction or distillation. With pure natural substances, it becomes even more difficult, as it is impossible to give an exact date, because even the question of the isolation of natural substances must be critically questioned whether it is biochemistry in the strict

◘ **Fig. 1.3** Brewer in Nuremberg around 1425

sense. Nevertheless, let's try to take a look at the beginnings of the extraction and the purification of natural substances. We can state that plant-based biochemistry, also known as phytochemistry, had its beginnings in the last third of the 17th century in France. At the Société de Pharmacie in Paris in 1666, a broad program was set up to extract plants, albeit with rainwater, by pressing or distillation. The first demonstrably isolated natural substance was sucrose, which the pharmacist Marggraf extracted from the sugar beet in 1747. Carl Wilhelm Scheele isolated pure citric, wine, oxalic and malic acid, which were of no great interest in the economy at the time. The real beginnings of Technical Biochemistry and the pharmaceutical industry as a whole started with the chemistry of quinine. Here, the first isolation by the Frenchman Desrone with rainwater and alcohol is certainly worth mentioning, industrial extraction processes were later developed by Friedrich Koch, which are still valid in their basic features today. Morphine, cocaine and strychnine followed in the 19th century, to name just a few secondary natural substances.

The heyday of biochemistry began at the beginning of the 20th century with the merger of medical physiology and chemistry. Noteworthy are the isolation of adrenaline from the adrenal glands by the Japanese Takamine (1854–1922) in 1906, the discovery of hormones in the 1920s and of vitamin D by A. Butenandt (1903–1995) (◘ Table 1.1).

However, many biochemical phenomena remained unclear, and the elucidation of biochemical metabolic processes such as glycolysis, citric acid cycle, urea cycle, and gluconeogenesis, the last three of which were clarified by Adolf Krebs, was groundbreaking. This was followed by groundbreaking discoveries to this day, whose discoverers were often awarded the Nobel Prize for their long-term research.

If one applies a narrower definition of Technical Biochemistry, one has to wait until the 1940s until the first substances were actually produced, whose biochemical background was recognized and understood. Noteworthy is the discovery of penicillins by Alexander Fleming in 1928, whose fermentation and production was greatly advanced by World War II and the high number of wounded. Less known, but equally important and also strongly promoted during World War II were the glucocorticoids. They were considered the second miracle drug, as their anti-inflammatory and partially analgesic effect was of great importance for war casualties. The glucocorticoids, like the other steroids, have great technical significance, as they marked the beginning of **Industrial Microbiology.** Russell Marker may not have received the Nobel Prize, but without his understanding of plant steroids as a starting material for the biochemical semisynthesis of steroids, for example, the "pill" would not have been cost-effectively producible. For the production of penicillin, *Penicillium* producers were used and the product was subsequently chemically modified, while fungi such as *Rhizopus nigricans* chemically altered steroidal basic bodies after feeding and partially replaced the organic chemist (Kayser 2024).

After World War II and with the beginning of the age of genetic engineering in the mid-1960s, the importance of Technical Biochemistry in industry also grew. With new techniques such as genome shuffling, recombinant genetic engineering in the 1990s, and the highly modern CRISP/Cas technology in the 2010s, metabolic pathways and their regulation could be changed initially inaccurately and later tailor-made. New

Table 1.1 Historical classification of the isolation of technically relevant natural substances (selection)

Year	Natural Substance	Organism	Scientist as historic person in Technical Biochemistry	Effect/Benefit
1747	Sucrose	*Beta vulgaris*	Marggraf, S.	Sweetener
1806	Morphine	*Papaver somniferum*	Sertürner, F.	Analgesic
1817	Emetine	*Ipecacuanha Root*	Pelletier, P.J. & Magendie, F.	Antiparasitic
1818	Strychnine	*Strychnos nux vomica*	Pellentier, P.J. & Caventou, J.B.	Cardiac tonic
1819	Atropine Scopolamine	*Atropa belladonna*	Runge, F.	Mydriatic
	Colchicine	*Colchicum autumnale*	Pelletier, P.J. & Caventou, J.B.	Gout remedy
	Piperine	*Piper nigrum*	Ørsted, H.C.	Spice
1820	Quinine	*Cinchona pubescens*	Pelletier, P.J. & Caventou, J.B.	Malaria remedy
1821	Caffeine	*Coffea arabica*	Runge, F.	Neural stimulant
1827	Coniine	*Conium maculatum*	Gieseke, A.L.	Neurotoxin
1828	Nicotine	*Nicotiana tabacum*	Posselt, H., C.W. & Reimann, K.L.	Insecticide
1832	Thebaine	*Papaver somniferum*	Pelletier, P.J.	Precursor for opiate synthesis
1839	Lobeline	*Lobelia inflata*	Reinsch, H.	No application
1928	Penicillin	*Penicillium* spp.	Fleming, A.	Antibiotic
1848	Papaverine	*Papaver somniferum*	Merck, G.	Muscle relaxant
1860	Cocaine	*Erythroxylon coca*	Niemann, A.	Local anesthetic
1900	Paclitaxel	*Taxus* spp.	N.N.	Cancer drug
1885	Ephedrine	*Ephedra sinica*	Nagai, N.	Cardiac tonic Asthma
2004	Artemisinin	*Artemisia annua*	Tu, Y.	Malaria therapeutic

fields such as **Metabolic Engineering,** which has evolved into **Synthetic Biology**, suggest that in principle there are no real bioorganic, genetic or biochemical boundaries anymore. Whether this will be a curse or a blessing for us humans will be shown in our moral responsibility in the coming years, but the benefit for medicine and technology in a demographically aging world must be clearly recognized.

1.2 Technical Biochemistry Today

The current achievements of Technical Biochemistry are impressive, as the field strives to provide solutions to pressing issues of our society. Climate-friendly biofuels and "green" solvents such as butanol and acetone are produced through the optimized Acetone-Butanol-Ethanol (ABE) metabolism. Improved enzymes for the food industry, optimized proteins as recombinant therapeutics, and new techniques for the biosynthesis of plant natural substances like vanillin and resveratrol in microorganisms are just a few examples. It is not unlikely that biotechnological processes to green hydrogen and better biodegradable plastics will also be developed. We need to provide answers today for future climate-neutral energy for transport and chemical industry (with or without genetic engineering) by advancing Technical Biochemistry. We also need to explain how to produce food and medicines with genetically modified microorganisms, plants, and perhaps even animals like insects. It is no longer the "if", but the "how" that needs to be discussed in a global context. Technical Biochemistry will support the development of modern biotechnological processes and will certainly cross scientific boundaries multiple times, forcing the discussion of ethical dilemmas as we venture into the future.

Fundamentals of Technical Biochemistry

> **Learning Objectives**
> - Delineation of Biotechnology and Biochemical Engineering
> - Basic knowledge of fermentation
>   - Downstream/Upstream
>   - Bioreactor control
>   - Process optimization

Biochemistry appears to most students as a subject that seems too complicated to fully grasp its beauty and excitement. This is unfortunate, because the biochemical processes in all living things (biology) are characterized by processes (biochemical reactions) that follow the principles of chemistry and physics. To prevent particularly interested technicians from being deterred from further reading, it should be noted at this point that this book aims to arouse a basic interest and demonstrate the great importance of biochemistry for applied technology, but also for our cultural development as humanity. The aim of this book is not to explain the basic metabolic pathways, but to illustrate biochemistry using examples from technical and medical practice. In doing so, we have to delve into some biosynthesis pathways and explain biochemical reactions. However, the goal is to explain the basic principles and concepts in such a way that it becomes clear where technology and biology come together, and that these two sciences do not contradict each other. If interest is aroused after reading, we recommend standard works for deeper insights that satisfy the thirst for knowledge at any depth and complexity (see appendix for further reading).

Let's recall the definition mentioned at the beginning for the subject of Technical Biochemistry. The words application, method, and technique are important. In reference to general biochemistry, we find many parallels concerning the research and development of biochemical technical processes. Very common goals of biochemical work are:

- Investigations into information exchange within an organism or between cells
- Investigations of metabolism and its biocatalysts
- Physical and analytical methods and techniques for qualitative and quantitative recording of biochemical phenomena

In contrast to classical biochemistry, which deals with the study of metabolism in humans (**Medical Biochemistry or in plants** (Plant Biochemistry), Technical Biochemistry has its roots in microbiology. In almost all technically relevant biochemical processes, microorganisms such as bacteria and fungi are used for production. The use of animals like goats or rabbits in Pharmaceutical Biotechnology for the extraction of blood clotting factors or vaccines is (still) unlikely from today's perspective. A brief overview of production organisms and technically relevant chemical substances is shown in ◘ Table 2.1.

What makes Technical Biochemistry different? For example, it aims to produce a chemical compound in the highest possible yield. To achieve this goal, it seeks a suitable production organism, which is genetically modified to find the best possible

Fundamentals of Technical Biochemistry

Table 2.1 Important production organisms and their products in Technical Biochemistry

Production Organism	Substance	Application	Literature
Bacterium			
Escherichia coli	Recombinant Proteins	Pharmaceuticals Detergent Enzymes	Bang et al. (2021); Gruchattka et al. (2013); Smith and Liao (2011)
Corynebacterium	Glutamate Biofuels Butanol	Flavoring Technology Solvent	Baritugo et al. (2018); Atsumi et al. (2008)
Fungus			
Yeast	Ethanol Vanillin Recombinant Proteins	Fuel Flavoring Pharmaceuticals	Chen et al. (2013); Rahmat and Kang (2020); Xu and Li (2020)

production conditions in an artificial medium of a fermenter. In the genetic development of a production organism, species boundaries do not matter. Enzymes in other species that have better catalytic performance are selected, codon-optimized, and cloned into the new organism as a heterologous gene. We will read that alternatively, species-specific genes are also switched off to reduce or suppress metabolic pathways in the producer that do not optimally allow the flow of substrates to our product. Technical Biochemistry is not about understanding why a microorganism functions as it does, but about expanding its qualitative and quantitative production profile. At the end of the development of a genetically modified producer, the question is not whether it could reproduce and survive in its natural environment, but the factual reduction to the task of producing a specific chemical substance in an ideal way (Fig. 2.1).

To this end, Technical Biochemistry must meet three requirements that work synergistically:
- Offering the best possible conditions for the uptake and utilization of substrates in ideal nutrient media for process**process optimization**,
- Optimization of the **physiology** of the producer. Here, understanding the entire biochemical metabolism and stoichiometry is important.
- Best possible design of enzymes as **biocatalysts** and their expression.

Today's integrated bioprocesses are much more demanding and complex than in the past. As shown in Fig. 2.2, the biological end reaction in the fermenter is not the result of a bioinformatics-based strain development that delivers a systems biotechnological profile with the help of analytics, optimizing the metabolism of the producer and equipping it with synthetic enzymes that deliver a high product yield (Chen and Elowitz 2021).

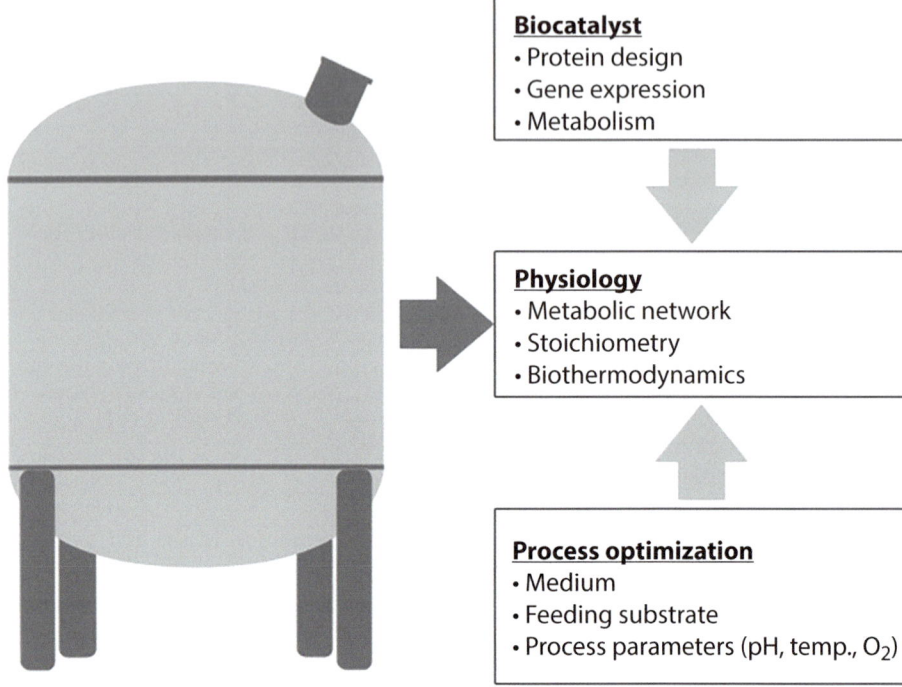

Fig. 2.1 Biochemical process parameters in fermentation

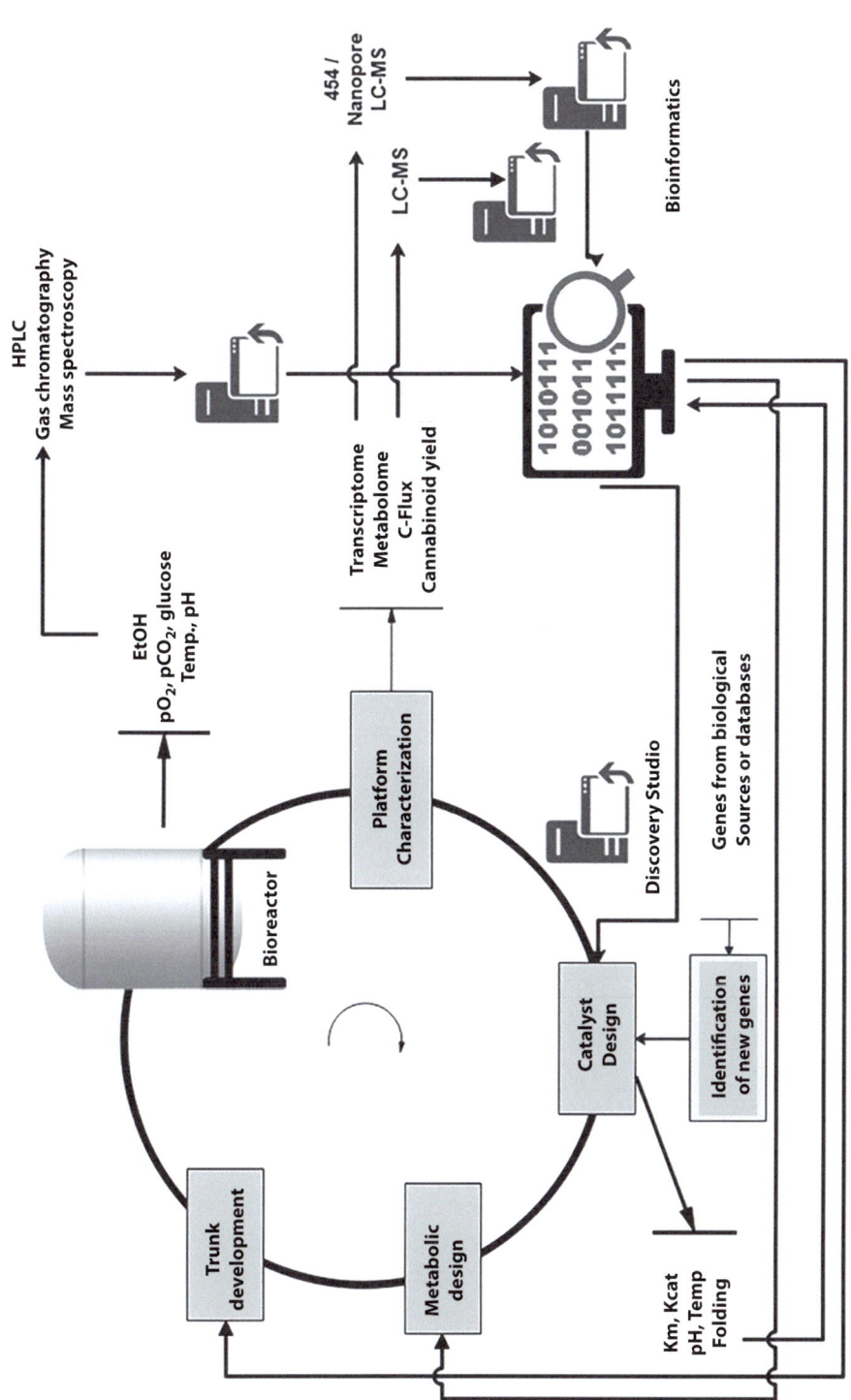

• Fig. 2.2 Process optimization of a genetically modified production organism

The Basis of Biochemical Reactions—Primary Metabolism

> **Learning Objectives**
> - Physiology and biochemistry of the microbiological cell
> - Primary metabolism and secondary metabolism
>   - Differences
>   - Metabolic transitions
>   - Definition of secondary substances
> - Significance of Technical Biochemistry in Biotechnology
>   - Important natural substances
>   - Natural substances with industrial relevance

Where does Technical Biochemistry come in? Basic metabolic pathways occur in a very similar way in all organisms (human, animal, plant, microorganism). Therefore, uniform principles can be identified in the biological cell as the smallest biochemical "factory":
- All cells use energy-rich phosphate compounds such as ATP, GTP, and diphosphates as activated biochemical intermediates.
- The biosynthesis, i.e., the construction (anabolism) and breakdown (catabolism) of similar carbon compounds (amino acids, sugars, fats), occurs in all cells according to similar principles and is metabolized in the same way across species boundaries.
- Metabolic reactions are catalyzed by enzymes, whose catalytic activity depends on the three-dimensional structure of an amino acid chain. Enzymes usually consist of amino acids.
- In most organisms, the nucleotide bases and their coding by the triplet code are the same.
- A number of low molecular weight molecules, on which vital metabolic pathways in primary metabolism depend, are ubiquitous (e.g., cofactors in enzymes, primary metabolites, thiamine, heme), although there are always exceptions to the rule in biology.

The totality of these vital metabolic pathways, which occur in almost all organisms in the same way, is referred to as primary metabolism. This term is generally used because the transitions to secondary metabolism are not always clearly defined. Typical examples of primary metabolism are glycolysis, the citric acid cycle, and fatty acid biosynthesis. In contrast to primary metabolism is secondary metabolism, which varies greatly from species to species. From a biosynthetic perspective, this explains the chemodiversity within species, with metabolic pathways such as alkaloid biosynthesis completely absent in some plant families, while others can have a very complex spectrum of different alkaloid structures.

Natural substances that are biosynthesized in secondary metabolism are therefore referred to as secondary metabolites; they are characterized by the following criteria (◘ Fig. 3.1):
- They do not occur ubiquitously, but characterize certain taxa. The basic structure varies structurally between species.
- They have no significance as an energy source for the organism, and the direct benefit to the organism is difficult to recognize. Secondary metabolites proba-

The Basis of Biochemical Reactions—Primary Metabolism

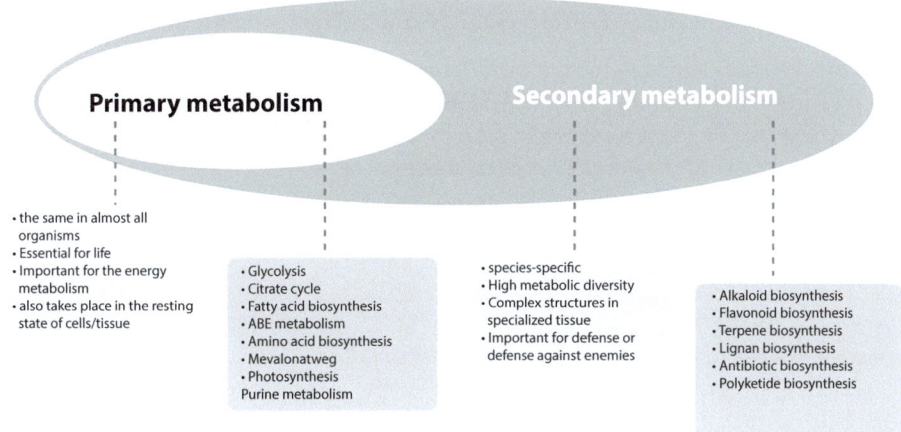

Fig. 3.1 Characteristics of primary and secondary metabolism

bly play an important role in protecting the organism, enabling it to assert itself ecologically. Secondary metabolites often have an antimicrobial effect or repel predators through smell, taste, and toxicity.
— Their biosynthesis is linked to developmental stages of the organism, can occur specifically in certain organs, and is subject to genetic control.
— They are often no longer clearly assignable to metabolic pathways and may have arisen through the combination of several metabolic pathways.

However, secondary metabolites are built up from the same intermediates of primary metabolism, and here too, biosynthesis follows an enzyme-catalyzed metabolic chain. The terms "primary" and "secondary" are generally used, although the distinction is not always clear, so there can be considerable overlaps.

A good example of this dilemma are certain proteins that should actually be strictly assigned to primary metabolism. Some proteins, such as anthrax toxin, botulinum toxin, or the drug cyclosporin, are peptides or proteins with very strong biological activity that should be counted as secondary metabolites. On the other hand, secondary metabolites such as flavonoids, sterols, and phenylpropanes are also known to have primary metabolic functions. Since it is already clear at this point that a strict separation of structure and function is not possible within the scope of this book, this attempt is not made and the use of both terms is largely avoided.

The first glance at the metabolic pathways according to KEGG (Kyoto Encyclopedia of Genes and Genomes) reminds the layman more of a pattern for trousers or the subway map of Tokyo than a biochemical overview with the interfaces of the most important biosynthesis pathways (Fig. 3.2). If one compares the detailed metabolic pathways, one can rightly say that it is even a greatly simplified representation of the links of biochemical conversions, which only partially reflects the real relationships in a cell. However, within the scope of this book, it is important to keep an overview and to deal in depth with individual metabolic pathways in the following chapters. Even here, the high complexity of biochemistry becomes clear.

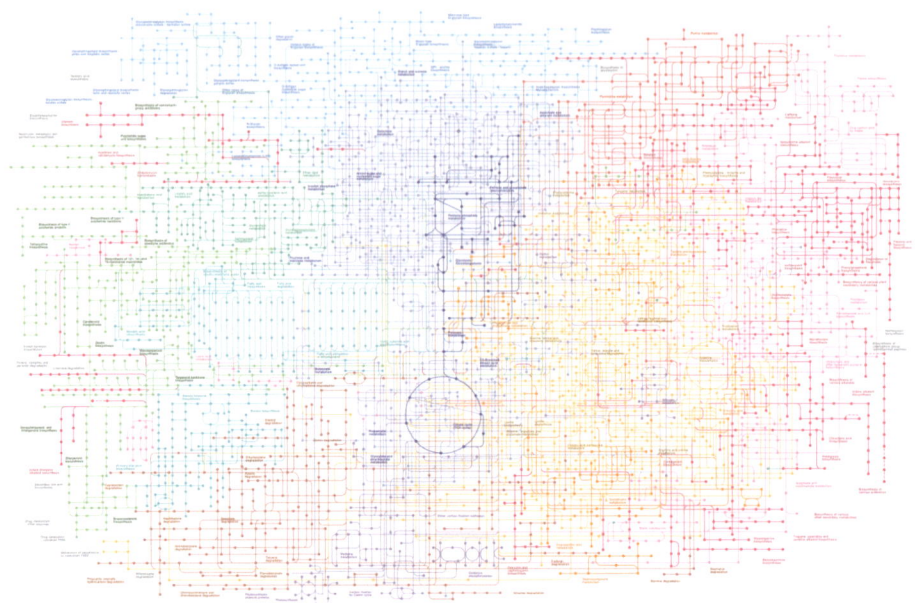

Fig. 3.2 Complex overview of the biochemical metabolic pathways in the representation according to KEGG

Those who want to be even more impressed and excited by the interplay of the individual biosynthesis pathways are recommended to visit KEGG or the Biochemical Pathways at ExPASy. About 60 important metabolic pathways are known, which differ by their intermediates and metabolically relevant end products. Which of the many important metabolic pathways should an engineer know in order to have a solid basis for understanding basic biochemistry and metabolic physiology? In the following chapters, not only the above-mentioned biochemical pathways are discussed (◘ Fig. 3.2), but also the relationships between them are shown and analyzed in order to convey biochemistry as a systematic science in which biosyntheses are never isolated and decoupled from each other. ◘ Figure 3.3 exemplarily shows the network of important biosynthesis pathways with interesting technical products, which can be seen as a red thread. What this book unfortunately cannot do is the virtual journey through the world of enzymes, metabolic intermediates and genes, to which more and more detailed information is given with each mouse click. We have already mentioned an introduction with KEGG above; other sources for concentrated information about enzymes and their specific functions in biochemical reactions are, for example, Brenda or UniProt. Also recommended are the famous "Biochemical Pathways" from Roche Applied Science, which hang as posters at the authors' and in many student dorms.

? Self-check questions
1. Explain the terms ethnomedicine, biochemistry, phytochemistry, and physiology.
2. Explain the difference between primary and secondary metabolism and give two examples (chemical substances) for both areas.

The Basis of Biochemical Reactions—Primary Metabolism

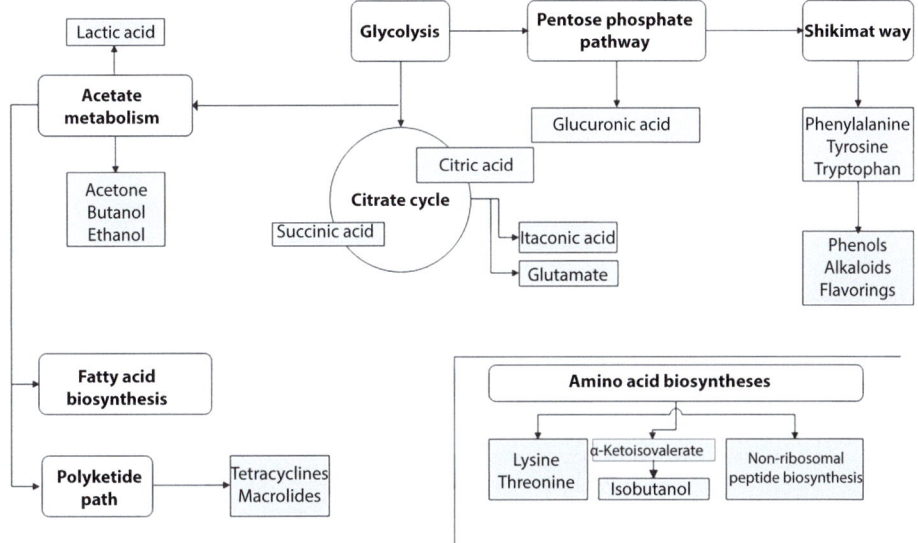

◘ **Fig. 3.3** Overview of the most important technical-biochemical metabolic pathways

3. Provide five criteria that can characterize the term "life".
4. Do viruses meet these criteria to be able to call them alive?
5. Name four technical processes that are characterized by biochemical conversion.
6. Research how many base differences in the DNA exist between you, a chimpanzee, and a banana.

Bioorganic Reactions and Building Blocks of Natural Products

Contents

4.1 From the Building Blocks to the Structure – 24
4.1.1 Alkylating Reactions and Nucleophilic Substitution – 24
4.1.2 Alkylating Reactions and Electrophilic Substitution – 25
4.1.3 Wagner-Meerwein Rearrangement – 26
4.1.4 Aldol Reaction and Claisen Condensation – 27
4.1.5 Schiff Bases and Mannich Reactions – 28
4.1.6 Transaminations – 28
4.1.7 Decarboxylations – 29

References – 30

© The Author(s), under exclusive license to Springer Fachmedien Wiesbaden GmbH, part of Springer Nature 2025
O. Kayser and N. J. H. Averesch, *Technical Biochemistry*, https://doi.org/10.1007/978-3-658-47121-7_4

> **Learning Objectives**
> - The biochemical "Lego" principle
> - Basic bioorganic reactions in biochemistry
>   - Condensations
>   - Rearrangements
>   - Alkylations
>   - Transaminations

Almost all metabolites and secondary natural substances* are constructed from uniform biosynthetic building blocks; one is almost tempted to say that they follow a "Lego principle".

> *Secondary natural product substances are usually small biomolecules that are formed by organisms and fulfill a biological function. However, this term does not refer to natural products such as feathers or cotton, nor does it include macromolecules such as DNA or proteins or inorganic substances.

The fact that this is not the case and that there is a very high structural diversity in nature has to do with the "promiscuity" of enzymes and the combination of biosynthetic routes that are not always strictly compartmentalized. This combination of chemistry and biology, and certainly also chance, led to the chemotaxonomic diversity that has already been mentioned. This becomes clear when natural substances of marine organisms are examined. Because chloride and bromide are freely available dissolved in water, it is not surprising that halogenated natural substances, which carry bromine or chlorine as a substituent, are found more frequently in marine organisms. In contrast to the marine chemical diversity, halogenated natural substances are a real rarity in land plants. In all organisms that practice biochemistry, many biochemical reactions can be seen as bioorganic reactions that follow the same chemical rules and principles. It is important that these building blocks have a certain uniformity in order to be recognized and converted by enzymes as biocatalysts. This is done by enzymes with a tolerance for the substrate, which is described as substrate promiscuity. So how can the biosynthetic building blocks be classified? The natural product chemist Dewick suggests a classification according to the number and type of carbon structures. This sensible biochemical taxonomy for explaining biosynthesis is adopted in the context of the book (Dewick 2009). Even if the formalized representation of metabolites and natural substances as biochemical "Lego blocks" takes some getting used to, this description and representation is important to the two authors. In the further course of the book, the bridge to synthetic biology and genetics is repeatedly made, the principle of which is the synthesis of building blocks by foreign enzymes in a production organism. To promote this understanding, it is important to also view technical biochemistry in a modular way, which meets the engineering science requirement.

C1: The simplest and smallest building block contains only one carbon atom (C1). This is often incorporated as a methyl group. L-methionine (L-Met) is often used as a donor.

C2: This building block consists of two carbon atoms and is supplied in the form of acetyl-CoA as "activated acetic acid". This C2 building block can be used for simple acetylations of hydroxyl groups, but the overriding importance of acetyl-CoA lies in the construction of fatty acids, in the mevalonate pathway to terpenes or in highly complex structures such as polyketides (▶ Chap. 6). In the mentioned natural substance groups, activated acetates react with each other multiple times to form long chains, which are then brought into their three-dimensional structure by specific enzymes.

C5: Many C5 building blocks are isoprenes, which are biosynthesized via the mevalonate pathway (MVA pathway) or the deoxyxylulose diphosphate pathway (DXP pathway). If the biosynthesis takes place via the MVA pathway, C2 building blocks are again involved, as three acetates are converted into the C5 building block by decarboxylation (removal of CO_2). These C5 building blocks can be linked almost arbitrarily with each other and form, for example, the natural substance group of monoterpenes as dimers. A variety of monoterpeneoids form the diterpenes ($2\times$), triterpenes ($3\times x$), tetraterpenes ($4\times$), which are often also referred to as carotenoids. The polymers of isoprene (C5) are rubber or gutta-percha, which are of great technical or medical importance. Terpenes are formed by many plants and microorganisms and represent the numerically largest group of natural substances in nature.

C6C1: As the name suggests, shikimic acid is formed via the shikimate pathway by coupling phosphoenolpyruvate (PEP) from glycolysis and erythrose-4-phosphate from the pentose phosphate pathway. The derivatives or the shikimic acid itself (C6C1) can be oxidized or linked with other biosynthetic pathways to form complex structures such as tannins, flavonoids, and aromatic amino acids, which are completely or partially essential for the mammalian organism.

C6C3: This phenylpropane originally comes from the shikimate pathway and is built up to a C6C3 building block by side chain extension. Important representatives are the two aromatic amino acids L-phenylalanine (Phe) and L-tyrosine (Tyr), which are precursors in natural product biosynthesis, but through modifications are also important neurotransmitters in the neuronal and hormonal regulation system of mammalian organisms (e.g., DOPA, dopamine, adrenaline). By deamination of the amino acids, the actual C6C3 building blocks are formed. The chain can be shortened to C6C2 and C6C1 building blocks by oxidation.

C6C2N: This building block also derives from the aromatic amino acids phenylalanine and tyrosine, however, no deamination takes place here, and the nitrogen remains as an amino group.

Indol C2N: Here, the aromatic amino acid tryptophan (Trp) is the basic structure, which carries an indole ring (N-heterocycle). This indole skeleton is very common in alkaloids (lysergic acids), which have a strong biological effect.

C4N: The C4N building block is another nitrogen-containing heterocycle, which occurs in pyrrolizidine alkaloids. The biosynthetic starting material is the non-proteinogenic amino acid ornithine (Orn) from arginine or the urea cycle. Typical representatives from the group of natural substances are cocaine or hyoscyamine.

C5N: This building block is formed from the amino acid L-lysine (Lys). Decarboxylation produces a diamine, from which piperidine develops. A typical example is nicotine, which is built from lysine.

These nine building blocks form the basis of many natural substances. Unfortunately, it is not always possible to recognize one or more of these building blocks in the complex structure of a natural substance at first glance. The reason for this lies in biocatalysis, which leads to a great variety of structures through different unfoldings and rearrangements in the same molecule. In the following chapters, an attempt will nevertheless be made to provide an overview of structure and biosynthesis. A taste is given by ◘ Fig. 13.13 with the complex terpene structure of paclitaxel. Without going into too much detail, it should be revealed here that this natural substance consists of a diterpene, three phenylpropanes, and two acetate building blocks. Try to find these building blocks, but don't be disappointed if you don't succeed at first (the solution can be found on page 246).

4.1 From the Building Blocks to the Structure

As already mentioned, nine building blocks can be defined that explain the structure and biosynthesis of natural substances. It is clear that every biosynthesis always takes place in the living cell and is essentially bound to enzymes, which form the biocatalytic basis for most chemical reactions in both secondary and primary metabolism. Many who deal with questions of biosynthesis for the first time stumble over the names of enzymes that seem to be able to catalyze only a very specific reaction. It may give the impression that all enzymes are highly specific, which is true in many cases. However, it should be noted that not all enzymes possess this supposedly high substrate specificity. Often, similar substrates can also be converted. This is particularly useful for enzymes involved in degradation reactions for detoxification and can degrade physiologically foreign and dangerous substances due to their promiscuity. We will see in the course of this chapter that there are enzymes that accept a variety of substrates more or less specifically and can indeed biotransform different products from one substrate. At this point, some enzymatic reactions that are either generally important for natural product biosynthesis or of interest due to their specificity will be explained. Enzyme kinetics, which deals with the theory of enzymatic catalysis, will not be covered here. Further reading is referred to.

4.1.1 Alkylating Reactions and Nucleophilic Substitution

The coupling of a methyl group (CH_3) to a molecule is catalyzed by a so-called methyltransferase. Chemically speaking, this is a nucleophilic substitution, which can only take place when the methyl group is coupled as a strong leaving group to a carrier (also called a donor). In biosynthesis, two donors are often found at central biochemical key positions. On the one hand, there is S-adenosylmethionine (SAM), and on the other hand, tetrahydrofolate (THF). SAM plays a special role in general methylations of hydroxyl or amino groups in a molecule. As shown in ◘ Fig. 4.1, the methyl group is bound to a sulfur atom and can be transferred in an S_N2-type reaction to a donor such as an amino (-NH_2) or hydroxyl (-OH) group. In the same way, longer carbon groups can also be transferred, as we will see later in the

4.1 · From the Building Blocks to the Structure

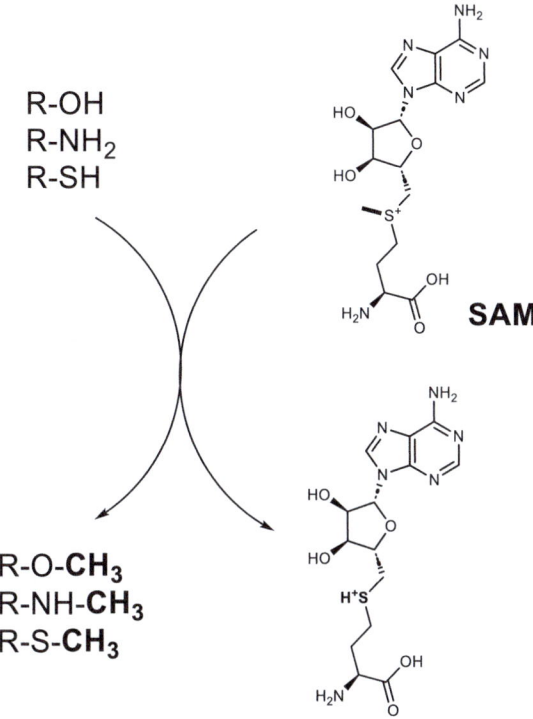

◘ **Fig. 4.1** SAM Methylations

chapter on terpenes for dimethylallyl pyrophosphate (DMAPP). DMAPP plays an important role in biochemistry as a C5 building block (terpene biosynthesis, steroid biosynthesis), as here too the C5 molecule is activated as a good leaving group. The good transferability comes at a price: diphosphates are very unstable, hydrolyze quickly and cannot be stored in biological cells. They are therefore produced as needed. Methylation with tetrahydrofolate (THF) as a donor and acceptor is more complicated, as tetrahydrofolate can appear as a cofactor in various biosynthesis pathways. The chemical mechanism of uptake and transfer is, for example, illustrated in biomethane synthesis (◘ Fig. 16.5). Functionally, the physiological effect can be summarized as a universal donor for C1 building blocks of different oxidation levels -C-, -C=, -CHO, which are also important in methionine biosynthesis (◘ Fig. 16.4). Under the influence of NADP or NADPH, reduction steps can also be introduced, leading to interconversions (example biomethane synthesis).

4.1.2 Alkylating Reactions and Electrophilic Substitution

The linkage of two carbon atoms can often be characterized as electrophilic addition. This type of reaction is found, among other things, in the biosynthesis of terpenes with C5 building blocks. These C5 building blocks can be linked together almost arbitrarily, thus forming simple as well as complex to polymeric natural

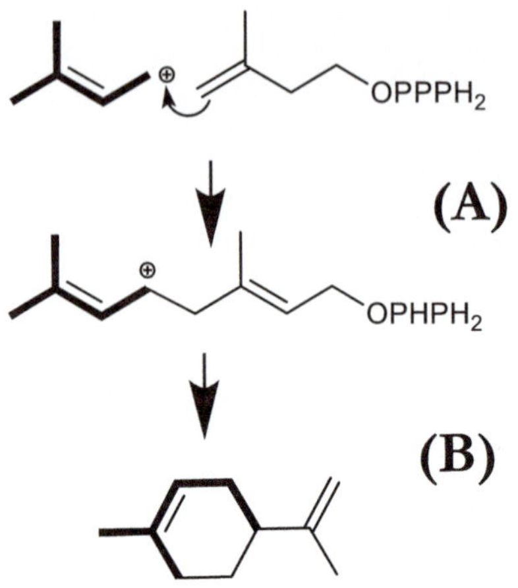

Fig. 4.2 Intramolecular rearrangement

substances. Important enzymes for this biocatalysis are terpene transferases. Chemically, this reaction can be described as a Friedel-Craft alkylation. The linkage of the two carbon atoms is explained by the reaction of a carbocation ($^+CR_3$) with a double bond. Here it becomes clear that the resulting carbocation must lose a proton after the reaction in order to build a double bond and be uncharged again. It does not matter whether this reaction takes place intermolecularly between two substrates (Fig. 4.2a) or intramolecularly within a molecule (Fig. 4.2b). Especially the intramolecular bond formation is of great importance in nature, as it builds ring systems of high structural diversity, which play a major role in terpenes and justify their high number (see also Wagner-Meerwein rearrangement). According to the same chemical mechanism, methyl groups can also be shifted in a double bond chain of alkenes.

4.1.3 Wagner-Meerwein Rearrangement

In natural product biosynthesis, rearrangements often account for the great structural diversity. As already mentioned above, the natural product group of terpenes is a good example. A typical rearrangement is the migration of a carbenium ion (a subgroup of carbocations with triply substituted carbon such as $^+CR_3$) between adjacent carbons. This is also referred to as 1,2-hydride, methyl, and alkyl shift and can occur according to an S_N1 or E1 mechanism, particularly with tertiary carbons. Partially, 1,3-shifts over a greater distance are also observed (C-H, C-C). These shifts are referred to as Wagner-Meerwein rearrangements (Fig. 4.3) and occur in complex biochemical reactions, such as in terpene and steroid biosynthesis. Of course, special enzymes, so-called transferases, are also involved in these reactions, as the intermediate carbenium ion formed is not stable on its own.

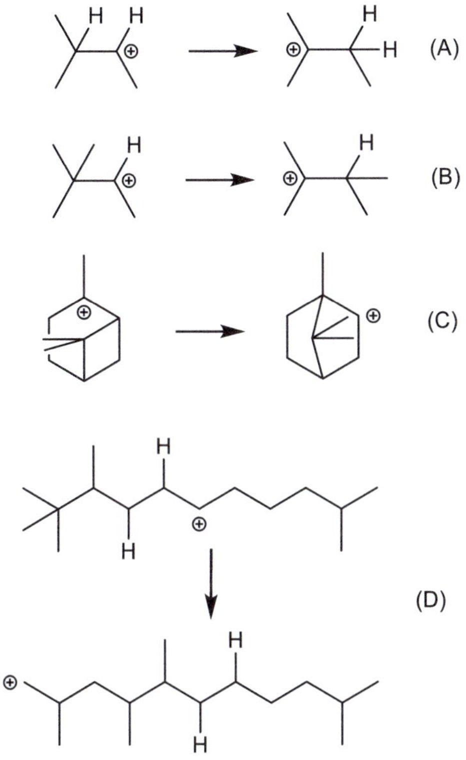

Fig. 4.3 Wagner-Meerwein Rearrangement

4.1.4 Aldol Reaction and Claisen Condensation

A bond between two carbon atoms can also be formed via an aldol reaction or a Claisen condensation. Both are well known in organic chemistry. In both reactions, a proton is first split off, forming a base from a carbonyl system. The resulting enolate anion is stabilized by resonance. Whether a Claisen condensation or an aldol reaction occurs (Fig. 4.4), depends on the type of leaving group: If it is a good leaving group and can be easily split off, a Claisen product is formed. Typical reactions are very often found with esters such as acetyl-CoA. Claisen condensations are found in the biosynthesis of fatty acids and polyketides. In reactions with coenzyme A, it is not always possible to clearly distinguish between the two types, as the cofactor in a Claisen condensation can transfer acetyl groups (acetyl-CoA) or in an aldol reaction malonyl groups (malonyl-CoA). The mentioned Claisen condensation can also proceed in reverse: In β-oxidation, these reactions proceed backwards, i.e., there is a degradation of fatty acids, with acetate groups being released. Details of this reaction will be discussed in the discussion of terpene biosynthesis. If the leaving group is poor and still present in the product, it is an aldol reaction and thus an aldol addition product. Reactions between aldehydes or ketones (crossed aldol condensation/addition) are very typical. Both types of reactions are frequently encountered in biochemistry.

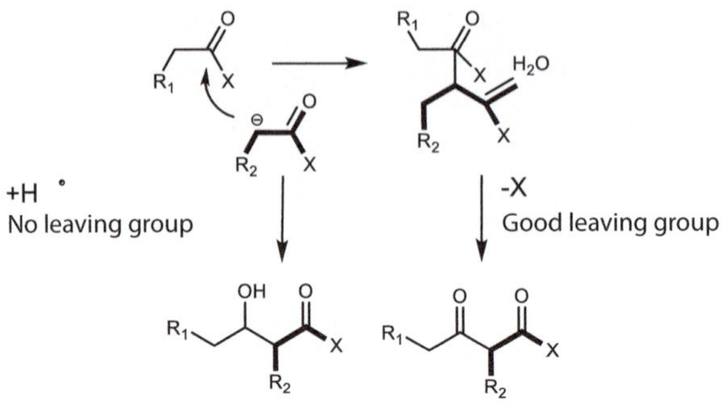

Fig. 4.4 Aldol/Claisen Reaction

+H • No leaving group

−X Good leaving group

Aldol product **Claisen product**

R=X = H, Aldol   R= H, X= OEt

4.1.5 Schiff Bases and Mannich Reactions

The formation of C-N bonds occurs through condensation reactions between amines and aldehydes or ketones. This typical reaction is familiar to everyone from everyday life. For example, cheese turns brown after prolonged storage. This is a typical Mannich reaction, in which the keto group of the sugar reacts with the natural amines of the cheese to form the typical brown Schiff bases. What happens chemically? It is a nucleophilic addition with water elimination to form an imine, which is also referred to as a Schiff base. These Schiff bases are unstable and can also react back. In this case, it is a typical hydrolysis. Schiff bases (also called protonated iminium ion) can act as an electrophilic group in the Mannich reaction. In this case, the iminium ion reacts with a base. The latter can be, for example, an enolate anion or an activated center in an aromatic ring. Mannich reactions play an important role in the biosynthesis of alkaloids, as the ring structures are often built up by linking an amine with an aldehyde or ketone.

4.1.6 Transaminations

Transamination is the transfer of a β-amino group ($-NH_2$) from an amino acid to an α-keto acid (2-oxo acid). In return, the keto group (C=O) is transferred to the amino acid. Thus, the amino acid becomes a keto acid and the keto acid becomes an amino acid. The same system can also be used for the degradation of amines, the deamination. This biochemical reaction relies on glutamate and 2-oxoglutarate as acceptor and donor. The biochemical intermediate 2-oxoglutarate originates from the citric acid cycle and is reduced to the amino acid glutamate. This reaction is depicted in a section of this biosynthesis, for further details, please refer to the citric acid cycle mentioned above. Here too, a coenzyme plays an important role in catalyzing this reaction.

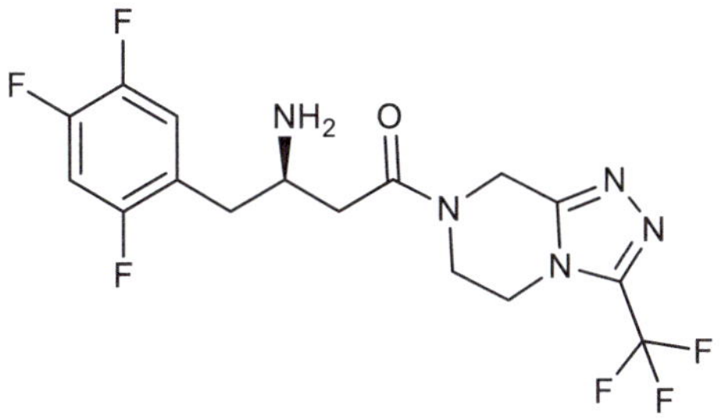

4.1 · From the Building Blocks to the Structure

Fig. 4.5 Sitagliptin

This involves pyridoxal phosphate (PLP), which ensures that the α-proton of the amino acid becomes more acidic and can thus be better cleaved. The negative charge can be captured by resonance stabilization in the pyridine ring. An intermediate product, a ketimine, is formed, which through hydrolysis yields a pyridoxamine and converts the keto acid into an amino acid. This reaction plays an important role in medicine, but also in technology. Typical reactions of this kind occur in the liver, and the doctor determines the amount of transaminases as a measure of possible liver damage. In 2010, transaminases were used in the synthesis of sitagliptin to make organic syntheses more eco-efficient. Sitagliptin (Fig. 4.5), a modern oral antidiabetic, is a good example of the modern use of transaminases as technical enzymes in a complex organic reaction, which makes a significant contribution to making organic chemistry "greener".

Technical Significance For the improved green synthesis of the oral antidiabetic sitagliptin (Fig. 4.5), an optimized transaminase was developed, which enables the stereospecific amination of a ketone (Savile et al. 2010). With this enzyme, a difficult enamine reduction at 250 bar with a rhodium catalyst could be avoided. The chemical problem that MSD Pharma had to solve in the synthesis was the sterically correct amination and the exchange of a ketone with the help of a transaminase, which was not efficient in the wild type. Through directed evolution, protein engineering was carried out over 11 mutation cycles, resulting in a genetic transaminase that chemically enabled a conversion into the (R)-stereoisomer (99.9% yield) with increased solvent tolerance (25–50% DMSO) and a temperature optimum of 40 °C. This protein engineering was so economically and technically interesting that it was awarded the Presidential Green Chemistry Award by then US President Obama in 2010. Other technical enzymes are ketoreductases, glucose dehydrogenases and a haloalkane dehydrogenase in the synthesis of the drugs atorvastatin (Lipitor®) and montelukast (Singulair®).

4.1.7 Decarboxylations

In many biosynthetic routes, natural substances are decarboxylated. Chemically speaking, a C1 building block is removed from the molecule. The decarboxylation of amino acids also occurs through the coenzyme PLP (see transamination). Decarboxylations of β-keto acids are biochemically simple to perform and do not need to be enzymatic if a hydroxyl group is present in the vicinity. A good example is tetrahydrocannabinolic acid, which in *Cannabis sativa* L. is activated as a psychogenic substance only through decarboxylation. The decarboxylation occurs through heating during smoking, as the carboxyl group is not thermally stable. Biochemically, the decarboxylation of β-keto acids at normal pressure and biologically relevant temperatures relies on the coenzyme thiamine diphosphate (TPP). A typical structural feature is a thiazole ring with an acidic hydrogen atom, which provides a carbanion. This can decay into a carbonyl group.

❓ Self-check Questions

1. Explain how biosynthetic building blocks can be classified.
2. Name the main (nine) biosynthetic building blocks, as described by Dewick in 2009.
3. Name a natural substance as an example for a C6C3 and a C6C2N basic body.
4. Explain why the transfer of a methyl group by SAM is a nucleophilic substitution and not an electrophilic addition.
5. Define the term "tautomerism".
6. Research the known metabolic pathways of ethanol for detoxification in the liver.
7. In ◘ Fig. 12.2 a non-natural biosynthesis of vanillin is described. Research which biological species have contributed the four mentioned enzymes in ◘ Fig. 12.2. Explain the purpose of a glycosylation of the finished vanillin in the last step.
8. Explain the difference between glycosylation and glucosylation.

Technical Enzymes

Contents

5.1 Technical Applications of Enzymes – 33

5.2 Extremophile Organisms and Extremozymes – 34

> **Learning Objectives**
> – Definition and significance of technically relevant biocatalysts
> – Technical applications
>   – Principles for discovering new enzymes
>   – Extremophiles and extremozymes
>   – Protein engineering

No biochemical metabolism can fundamentally do without biocatalysts. In biochemistry, these are primarily proteins, which significantly lower the activation energy of a chemical reaction and enable the conversion of a substrate into the desired intermediate or technical product (◘ Fig. 5.1). Enzymes are playing an increasingly important role in a chemical world that wants to become more sustainable, but their use in chemical processes is still complicated.

The basic kinetics of enzymatic catalysis will not be discussed in this book. Here, reference is made to the textbooks of biochemistry. The focus of the chapter is on the technical fields of enzyme catalysis. Besides their use as detergent enzymes or for chemical catalysis, the large application is limited, as the chemical environment in synthesis strongly restricts the catalysis speed. The reason for the limited use is unphysiological synthesis conditions, which often lead to denaturation and inactivation:

Temperature Many chemical reactions do not occur at the typical temperatures of 20–40 °C for ordinary organisms. Enzymes such as DNA polymerases, which are used in PCR reactions, have been known for more than 40 years. New enzymes can be obtained from hyperthermophilic organisms, which allow a shift to high or very low temperatures (psychrophilic organisms). Today, microorganisms are known that die at 122 °C. However, activities above 122 °C are known for isolated proteins.

pH Value The physiological pH value of 7 is a physiological size that is rarely encountered in organic synthesis. Temperature optima for technically relevant enzymes should be set at pH 2–4 or above 8.

Salt Concentration High salt concentrations lower the zeta potential and lead to the agglomeration of proteins. Proteins must be developed that are still active at a salt concentration of 0.9%.

Pressure Little is known about enzymatic reactions in piezotolerant organisms that live under high pressures, as are known from the deep sea. With an average depth of our oceans of 3800 m, hydrostatic pressures of 10 to 40 bar are normal for marine

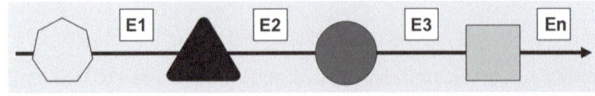

◘ **Fig. 5.1** Basic sequence of a metabolite turnover by enzymes in an imagined metabolic pathway

organisms. Piezotolerant marine organisms are often also halophilic and tolerate high osmotic loads well.

5.1 Technical Applications of Enzymes

Enzymes are highly specific biocatalysts. Due to their exceptional efficiency and specificity, enzymes are used in a variety of technical applications. The following are some technical applications of enzymes:

Food Industry Enzymes such as rennet or chymosin play a major role in the production of cheese and yogurt for the coagulation of milk. These enzymes are important for protein coagulation. Enzymes like amylases are used to break down starch into sugar during the fermentation process, which promotes fermentation and the production of carbon dioxide when bread rises. A final example is fruit juice production. Here, pectinases and cellulases are used to facilitate the extraction of juice from fruits and increase juice yield.

Textile Industry The toxicologically and environmentally questionable process of stone-washing jeans can be carried out in a healthier way. This is achieved by using proteases to create the desired "used" effect on jeans by breaking down and bleaching the surface of the denim fabric. Cellulases are important for the bio-polishing of textiles. They are used to remove unwanted fibers on the surface of textiles, resulting in a softer touch and a smoother appearance.

Detergents Modern detergents are unthinkable without enzymes. Enzymes are used in the formulation of detergents and cleaners to improve effectiveness at lower temperatures and reduce environmental impact. Detergent enzymes play a key role in cleaning clothes, but also as additives in dishwashers. In cleaning and washing clothes, stains are to be removed enzymatically. Because our food is complex, amylases, proteases, and lipases are found in all detergents. Proteases break down proteins, for example, from egg and skin epithelial cells. Amylases help break down polymeric sugars used in foods such as sauces as thickeners. Lipases have the function of breaking down fat stains on our clothes and making them more accessible to detergents. But enzymes are also important in the later treatment of wastewater in the sewage treatment plant (see ▶ Chap. 16). Enzymes like lipases and proteases are used to accelerate the biological degradation of organic contaminants.

Bioenergy Production This field is new and gaining increasing importance. The idea is that waste residues can be better broken down to be fed into the energy-generating primary metabolism. Enzymes such as cellulases and amylases are used in the hydrolysis of biomass to obtain sugar, which is then used for fermentation and subsequent production of bioethanol or biogas.

Pharmaceutical Industry Enzymes serve as biocatalysts in the synthesis of pharmaceutical active ingredients, thereby increasing the efficiency of reactions and minimizing the formation of by-products. Enzymes such as lipases and proteases are used in organic synthesis to produce chiral molecules with high selectivity and purity. To shorten synthesis pathways (e.g., steroid biosynthesis, ▶ Chap. 13) or to avoid toxic catalysts in organic synthesis as shown in the example of sitagliptin in ▶ Chap. 4, technical enzymes are being established today.

Biotechnology and Genetic Engineering Restriction enzymes and ligases have been used in molecular biology for more than 25 years to cut and join DNA molecules, which is essential in the production of recombinant DNA for genetic engineering applications. Many examples are discussed intensively in this book. They also play a major role in protein purification and modification. Enzymes such as proteases and kinases are used in the purification and modification of proteins in biotechnological production to improve their activity and stability.

5.2 Extremophile Organisms and Extremozymes

The physiological limits for the life of extremophile organisms known from the living nature are shown in ◘ Table 5.1 (Stolz 2017). The limits of their life are not necessarily synonymous with the limits of the function of isolated enzymes. In Technical Biochemistry, the aim of science is not to understand extreme habitats and research fields such as astrobiology, but to develop interesting ideas for technical implementations. With the help of genetic engineering, organisms that are already known in technology are again the focus of research. Thermophilic and alkaliphilic enzymes are found in our detergents, and acidophilic producers have long been known in the production of vinegar and citric acid.

Does it make sense to research extremozymes and find new technical applications? This question can be answered affirmatively, as this can lead to robust enzymes in the future that can be used in chemical processes under non-physiological conditions.

Vitamins play a special role for the functioning of biochemical reactions and as cofactors of enzymes, as our body cannot synthesize them itself. They are often supplied to the body as provitamins and activated by biochemical reactions. Vitamins are fat-soluble or water-soluble and do not form a uniform chemical group.

◘ **Table 5.1** Classification of extremophile organisms by their environmental factors

Environmental Factor	Limits of Growth	Type of Organism	Example
Temperature	> 80 °C < 10 °C	hyperthermo- philic psychrotolerant	DNA Polymerase
pH value	< pH 3 > pH 9	acidotolerant alkalitolerant	Citric acid production Detergent enzymes
Salt concentration	> 3.5% NaCl	halotolerant	Glutamate production
Pressure	> 1 bar	piezotolerant	–
Ionizing radiation	increased dose of UV light and γ-radiation	radiation tolerant	–

5.2 Extremophile Organisms and Extremozymes

The physiological effect of the thirteen vitamins will not be discussed here. The technical aspects, such as how these vitamins are biotechnologically produced, are important. The vitamins A, D, E, K are chemically synthesized and are not discussed in this book. Other vitamins such as Vitamin B_1, B_5, B_{12} are biotechnologically produced, and Vitamin C is partially synthesized.

Vitamins

Contents

6.1 Vitamin B_2 – 38

6.2 Vitamin B_5 – 38

6.3 Vitamin B_{12} – 39

6.4 Vitamin C – 40

6.5 Vitamin D – 43

6.6 Vitamin E – 44

© The Author(s), under exclusive license to Springer Fachmedien Wiesbaden GmbH, part of Springer Nature 2025
O. Kayser and N. J. H. Averesch, *Technical Biochemistry*, https://doi.org/10.1007/978-3-658-47121-7_6

Learning Objectives
- Water-soluble and fat-soluble vitamins
- Physiological and biochemical tasks
- Technical processes for the production and extraction of vitamins
  - Semi-biotechnical processes
  - Fermentation

6.1 Vitamin B_2

Riboflavin (Vitamin B_2, ◘ Fig. 6.1) is not only a vitamin in food but also an important yellow dye in the food industry. Biochemically and physiologically, it plays a crucial role as a reduction equivalent of FAM and FAD; it is a building block for the construction of both coenzymes. Its biosynthesis occurs in bacteria, fungi, and plants and begins with the linkage of guanosine triphosphate and ribulose-5-phosphate. Various microorganisms are used for the biotechnological production of riboflavin (also called Vitamin B_2). One of the most commonly used microorganisms for riboflavin production is *Ashbya gossypii*, a filamentous yeast species. *A. gossypii* is known for its ability to produce riboflavin efficiently. This yeast is cultivated in industrial fermentation processes to produce riboflavin in large quantities and then use it in the food, feed, and pharmaceutical industries for various technical applications. The use of *A. gossypii* for riboflavin production has the advantage that this microorganism naturally has a high ability to produce riboflavin and is therefore well suited for commercial purposes.

6.2 Vitamin B_5

Pantothenic Acid (◘ Fig. 6.2) is an important precursor for the biosynthesis of coenzyme A, which we have come to know as a significant activator of fatty acids and acetic acid. Bacteria synthesize pantothenic acid from the amino acid aspartate and a precursor of the amino acid valine. Aspartate is converted into β-alanine.

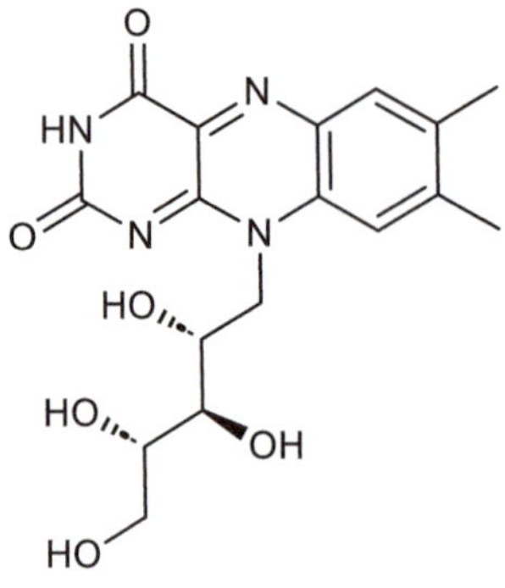

◘ **Fig. 6.1** Riboflavin (Vitamin B_2)

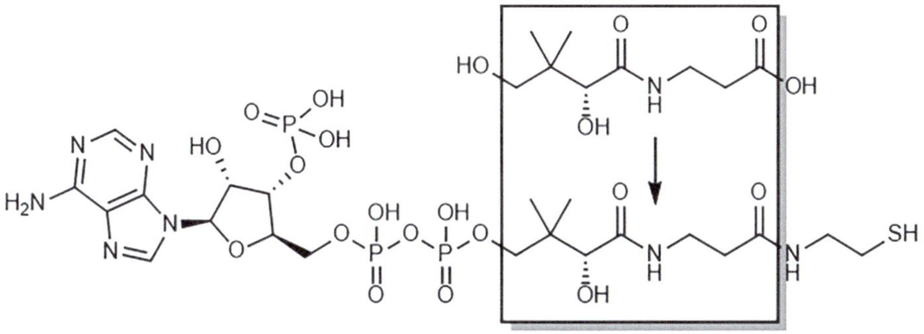

6.3 Vitamin B$_{12}$

◘ **Fig. 6.2** Pantothenic Acid (top) and central position in Coenzyme A (bottom)

The amino group of valine is replaced by a keto group to obtain α-ketoisovalerate, which in turn forms α-ketopantoate after transferring a methyl group and D-pantoate (also known as pantoic acid) after reduction. β-alanine and pantoic acid are then condensed to form pantothenic acid.

6.3 Vitamin B$_{12}$

The structurally highly complex structures are grouped together as cobalamins in the Vitamin B$_{12}$ group (◘ Fig. 6.3). The most important representative is adenosylcobalamin, which is usually referred to as Vitamin B$_{12}$. The structure can be greatly simplified into three units. Characteristic is the porphyrin ring, which chelates a cobalt ion as a central atom. This is stabilized, among other things, by the adenosyl part, which is bound to the porphyrin part (corrin ring) via a phosphate group (◘ Fig. 6.3). 5,6-Dimethylbenzimidazole (DMBI) forms a ligand with the central atom from the lower side (α). Vitamin B$_{12}$ is produced by microorganisms living in ruminants, and we humans mainly ingest this vitamin through meat and fermented dairy products. Interestingly, the highest concentrations of Vitamin B$_{12}$ are found in algae. Adenosylcobalamin is important because it is an essential cofactor in methionine synthesis, which builds S-adenosylmethionine for C1 methylation.

Biosynthesis In ruminants, but also in humans, bacteria have been found that are capable of producing Vitamin B$_{12}$. Until now, it was assumed that the number of producers is small, but according to modern genome analyses, the genetic ability seems to be more widespread than assumed. Biosynthesis can occur anaerobically or aerobically. For industrial production, in addition to anaerobic bacteria (*Propionobacter freudenreichii, Bacillus megaterium*), aerobic bacteria (*P. denitrificans, Rhodobacter capsulatus*) are also important. Although biosynthesis is very complex, two synthesis pathways can be distinguished (◘ Fig. 6.4). On the one hand, the biosynthesis of the corrin ring uroporphyrinogen III from succinate and the adenosyl group.

The biosynthesis of uroporphyrinogen III starts from succinate in the citric acid cycle via four aminolevulinic acid to the linear tetrapyrrole. A fusion provides Uro-

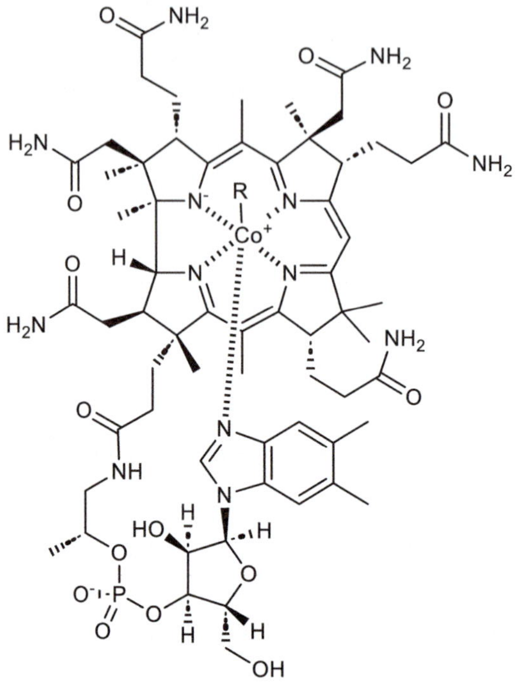

◘ **Fig. 6.3** Adenosylcobalamin (Vitamin B_{12}). R= 5-Deoxyadenosyl, CH_3, OH, CN

porphyrinogen III as a central building block, which can also lead to further biosynthesis of hemoglobin and chlorophyll. With the completion of biosynthesis, further biosynthesis follows.

6.4 Vitamin C

Ascorbic acid (Vitamin C, ◘ Fig. 6.5) is a vinylogous acid that is significant not only in medicine but also in the pharmaceutical and food industries as an antioxidant (E300). Large amounts have been detected in the Acerola cherry (1,500 mg/100 g), the rose hip (1,250 mg/100 g), and the green pepper (139 mg/100 g). Without knowing the chemical structure, the Scottish naval doctor James Lind discovered the physiological effect of lemons in 1747, which he administered to sailors against scurvy (Vitamin C deficiency). The natural and effective Vitamin C is the L-(+)-Ascorbic acid, which has the highest efficacy of the four isomers. For the food industry, the D-Isoascorbic acid is also significant as an antioxidant, but it only has 5% of the effect of Vitamin C.

Ascorbic acid is biochemically formed via the pentose phosphate pathway. In the oxidative phase, glucose is oxidized to glucuronic acid, which is reduced to L-gulonic acid with NADPH. This is lactonized by water elimination (L-gulofuranolactone) and reduced to ascorbic acid by the L-gulofuranolactone oxidase by

6.4 Vitamin C

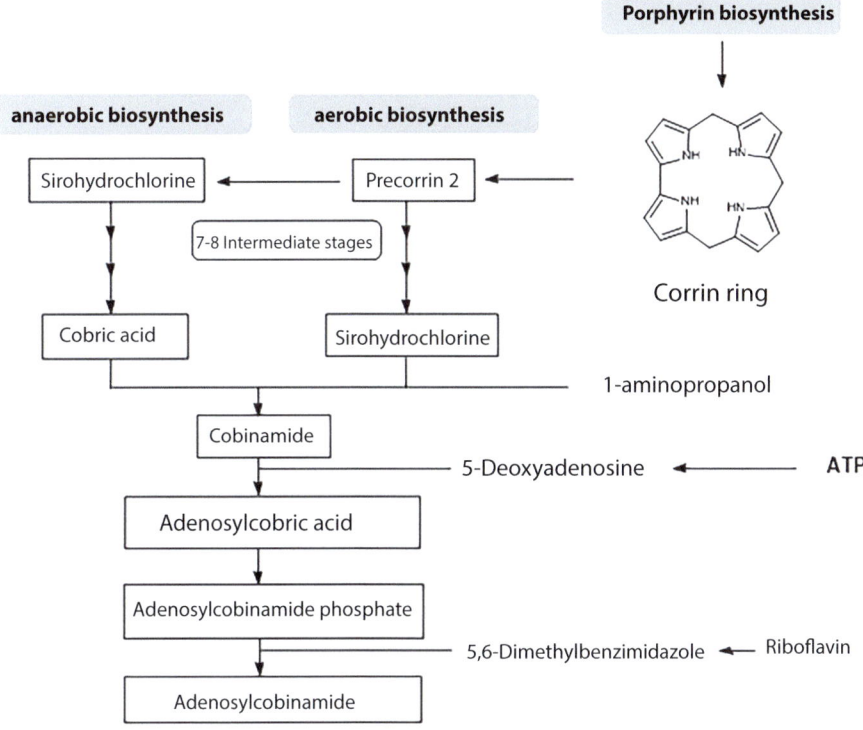

◘ Fig. 6.4 Schematic representation of cobalamin (Vitamin B$_{12}$) biosynthesis

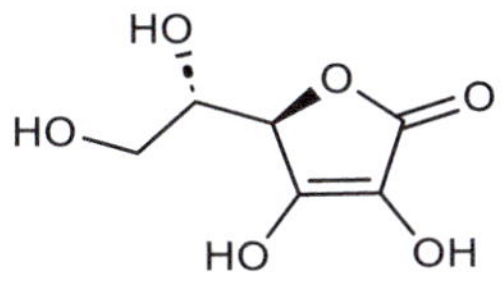

◘ Fig. 6.5 Ascorbic acid (Vitamin C)

incorporating a double bond (◘ Fig. 6.6). In humans, the L-gulofuranolactone oxidase is missing due to a genetic defect.

The original synthesis of L-ascorbic acid by Reichstein was a purely chemical process, which was later converted to a semi-synthetic process for better yields. The synthesis was discovered in 1934 by the chemist Tadeus Reichstein and is still fundamentally used for production. The conversion of sorbitol to L-sorbose is carried out for steric reasons with the help of acetobacter (◘ Fig. 6.7). It consists of microbiological and organic-synthetic steps. The process starts with the oxidation of D-glucose to D-sorbitol. D-glucose is oxidized to D-sorbitol with hydrogen peroxide. This step is catalyzed by a copper-chromium catalyst. In the second stage, the microbiological oxidation of D-sorbitol to L-sorbose takes place. D-sorbitol is oxidized to L-sorbose by the bacterium *Gluconobacter suboxidans*. This step takes

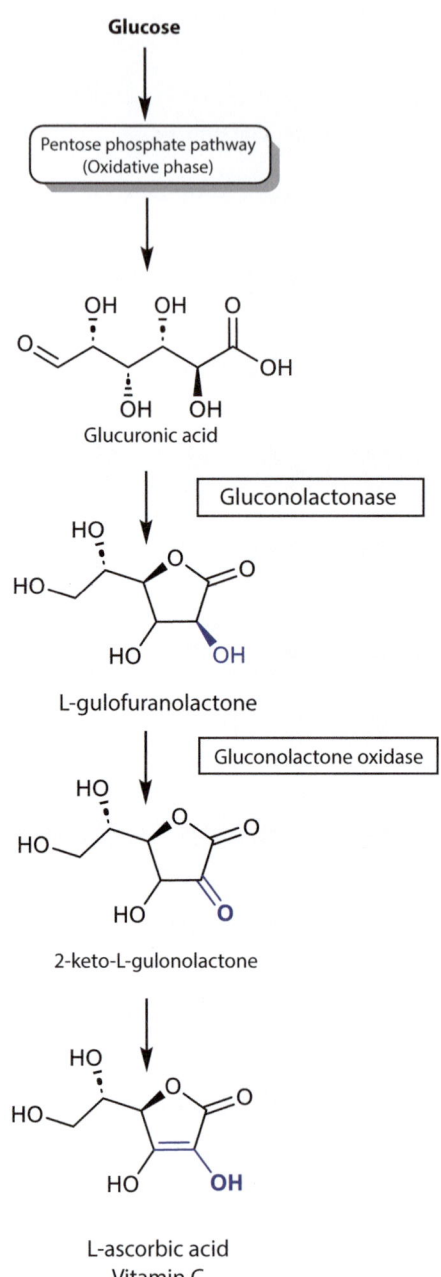

Fig. 6.6 Biosynthesis of ascorbic acid

place at a temperature of 30–35 °C and a pH value of 4–6. For further synthesis, protective groups must be introduced at two hydroxy groups of the L-sorbose. This is done by reaction with acetone and acid. After hydrolysis of the protective groups

6.5 Vitamin D

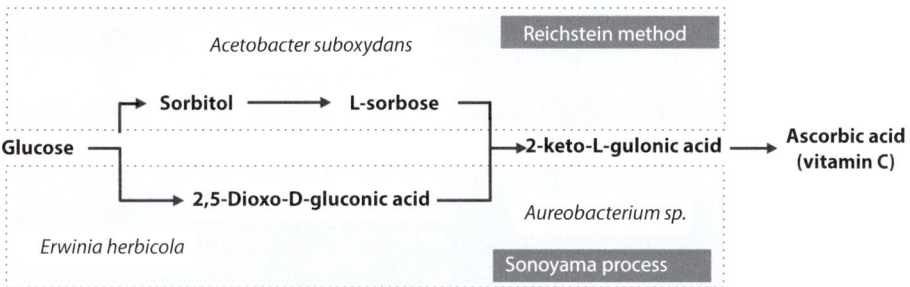

☐ **Fig. 6.7** Technical semisynthesis according to the Reichstein and Sonoyama process

by heating with water, L-ascorbic acid is formed, which leads to L-ascorbic acid through purification.

The Sonoyama process is a purely biotechnologically organized alternative to the biosynthesis of 2-Keto-L-gulonic acid. The process was developed in 1982 by a group of Japanese scientists led by Akira Sonoyama and is now used worldwide for the industrial production of ascorbic acid. With the help of two bacteria, glucose is converted to 2,5-Dioxo-D-gluconic acid. Glucose is first oxidized to 2,5-Dioxo-D-gluconic acid with the help of the bacteria *Pantoea agglomerans* or Erwinia herbicola. This reaction takes place under mild conditions and is very selective. Modern biotechnological processes work with recombinant strains of Glucobacter oxydans, which are fed with sorbitol, which is obtained from glucose by nickel hydrogenation. The Sonoyama process is an efficient and cost-effective process that does not require toxic or environmentally harmful chemicals and selectively delivers ascorbic acid in high purity.

6.5 Vitamin D

The chemical description and synthesis of Vitamin D as an anti-rachitic factor is credited to the chemist Adolf Windaus (1876–1959). He recognized that Vitamin D can be obtained from the precursor cholecalciferol with the help of UV light. The biosynthesis of Vitamin D was an exciting field of research in the 1920s, as for a long time neither the starting compound in the body nor the light-dependent conversion were clear, which was not fully elucidated until 1953. Until its synthesis and market launch as a pure substance under the brand name Vigantol®, Vitamin D was obtained by extraction from fish oil. The source was cod liver oil, i.e., the liver oil of cod, haddock, and other fish. Fish absorb cholecalciferol (Vitamin D_3) via plankton. A deficiency of Vitamin D_3 can cause rickets, which leads to bow legs in children and osteoporosis in older people.

Vitamin D is formed in the skin under the influence of UV from 7-dehydrocholesterol. Under sunlight, this precursor is split into previtamin D_3, which rearranges to cholecalciferol (Vitamin D_3). After transport to the liver and kidneys, there are two oxidation reactions that lead to calcitriol (1,25-dihydroxycholecalciferol), which we generally refer to as Vitamin D. Vitamin D is not produced biotechnologically, although microbiological and recombinantly modified yeasts are known. To-

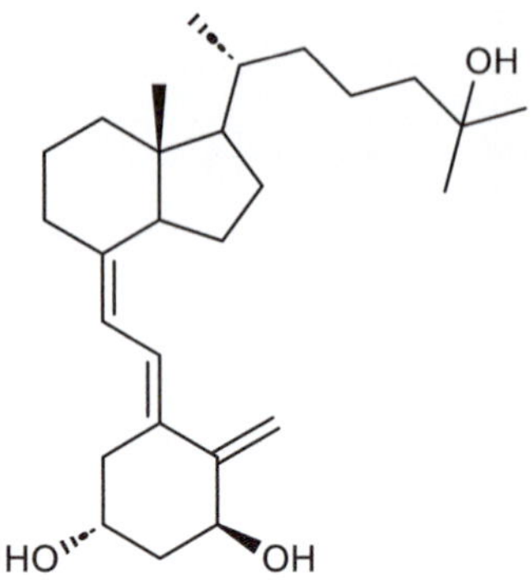

Fig. 6.8 Calcitriol

day's chemical synthesis using the Hoffmann-LaRoche method produces calcitriol (Fig. 6.8) from ergocalciferol with a yield of 22%. The biosynthesis is detailed in Fig. 13.23.

6.6 Vitamin E

Under the designation Vitamin E, there are at least eight Vitamin E derivatives, all of which are fat-soluble and antioxidative and are also referred to as tocopherols. Chemically, they are chromanes with a longer aliphatic side chain. The chemistry is complex, as there are three chirality centers and thus eight different stereoisomers are known. Vitamin E is biosynthesized in plants for the same purpose, i.e., to ensure antioxidative protection of the cell membranes. The aromatic part comes from homogentisic acid (HGA), the side chain from terpene biosynthesis (Fig. 6.9). In the case of the tocopherols, the synthesis of their derivatives is based on the reaction between HGA and phytyl-PP, resulting in 2-methyl-6-phytylhydroquinone. At this point in the synthesis, the 2-methyl-6-phytylhydroquinone can take two different paths. On the first path, the molecule is methylated at C3. The result is 2,3-dimethyl-5-phenylhydroquinone. Subsequently, cyclization via the hydroxyl group at C1 produces the first derivative, δ-tocopherol. After cyclization, a further methylation at C5 of the δ-tocopherol occurs, resulting in α-tocopherol (not shown in the figure). The second path uses the same 2-methyl-6-phytylhydroquinone as a substrate. Here, cyclization via the hydroxyl group at C1 occurs, resulting in δ-tocopherol. Subsequently, methylation at C5 leads to the last derivative, β-tocopherol. The representation of Vitamin E is now done by organic synthesis. The vitamin is found in cereal germs (e.g., wheat germs) as well as in nuts, seeds, vegetable oils, eggs, milk, and butter.

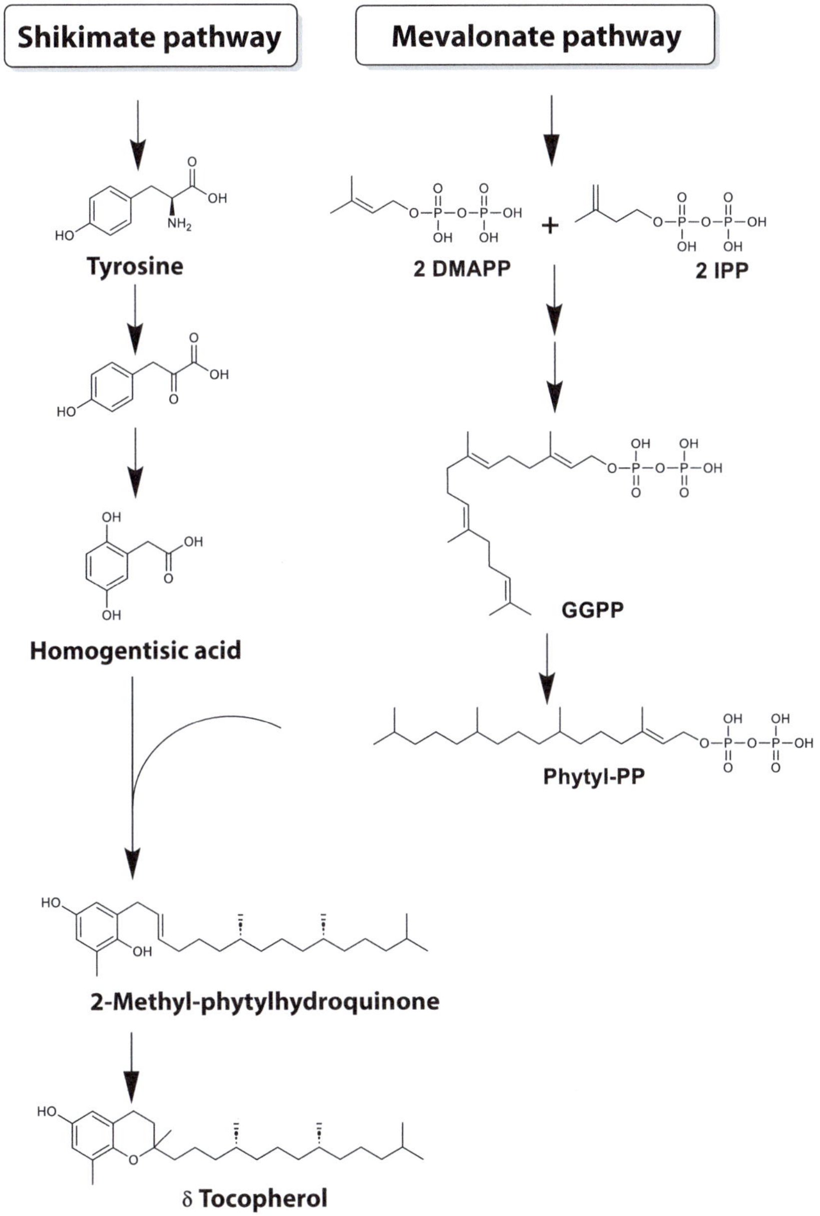

6.6 Vitamin E

Fig. 6.9 Biosynthesis of δ-tocopherol

Photosynthesis—Basics and Application

Contents

7.1 Photosynthesis—The Beginning of Everything – 48
7.1.1 Light Reaction – 49

7.2 Dark Reaction – 50

7.3 The Somewhat Different Photosynthesis – 52

7.4 The Artificial Photosynthesis – 52

7.5 Photolysis for the Production of Biohydrogen – 53

7.6 Photosynthesizing Microorganisms as New Producers – 54

References – 56

© The Author(s), under exclusive license to Springer Fachmedien Wiesbaden GmbH, part of Springer Nature 2025
O. Kayser and N. J. H. Averesch, *Technical Biochemistry*, https://doi.org/10.1007/978-3-658-47121-7_7

Learning Objectives
- Definition and significance of photosynthesis
- Light-dependent reaction and the biochemistry of Photosystems I and II
  - Cytochromes
- Calvin Cycle
  - Stoichiometry
  - Metabolic linkage to glycolysis
- Artificial photosynthesis
- Biochemical production of hydrogen

7.1 Photosynthesis—The Beginning of Everything

Photosynthesis is the biological process in green plants, algae, and even some bacteria that converts carbon dioxide and water into the sugar glucose and the "waste product" oxygen, which is vital for animals and many microorganisms, with the help of light. This occurs in special organelles, such as chloroplasts. ◘ Figure 7.2 shows the main functions (energy and carbon capture and fixation from sunlight and CO_2) and side effects (oxygen production from the photolysis of water) of photosynthesis. The remarkable interplay of individual partial reactions and metabolic pathways is noteworthy. The overall reaction of photosynthesis (Reaction 7.1) can also be considered a redox reaction in which a CO_2 reduction occurs and water acts as an electron donor. The basic equation is (Reaction 7.1):

Reaction 7.1
$$CO_2 + 2\,H_2O + n\cdot h\nu \rightarrow CHOH + H_2O + O_2$$

Why is this equation written so strangely? The representation in the gross equation without ATP and reduction equivalents clearly shows the scientific name for sugar (carbohydrates). Reduced to the simplest stoichiometric ratio, the sugars appear as hydrates (H_2O) of carbon (C). Only in the net equation does it become clear what and how much is actually produced by photosynthesis (Reaction 7.2):

Reaction 7.2
$$6\,CO_2 + 12\,H_2O + n\cdot h\nu \rightarrow C_6H_{12}O_6 + 6\,H_2O + 6\,O$$

The released oxygen does not, as often assumed, come from the bound CO_2, but from the photolysis of water. Therefore, in the following first sum equation (Reaction 7.2)

7.1 · Photosynthesis—The Beginning of Everything

on the left side are 12 water molecules, to get 6 O_2 molecules on the right. Photosynthesis is a complex biochemical reaction that includes a light reaction as well as a dark reaction (Calvin Cycle, ◘ Fig. 7.4).

7.1.1 Light Reaction

Through light, or light energy (h·ν), the charge carrier NADP is reduced to NADPH$_2$ and ATP is provided as an energy carrier. Typical for photosynthesizing organisms is their green color. The reason is the pigment chlorophyll (◘ Fig. 7.1), which absorbs mainly blue and red light with absorption maxima at certain wavelengths (Photosystem I: 700 nm, Photosystem II: 680 nm). It is able to capture the energy of sunlight in the light-harvesting complexes very efficiently.

In ◘ Fig. 7.2, the photosynthesis organisation is shown, which builds up the proton gradient with the help of light energy, which finally enables Reaction 7.1 and 7.2. In these enzyme complexes, which are located in the thylakoid membrane within the chloroplasts in plants, two membrane-integral photosystems are connected in series. Electrons are excited to a higher energy level in two stages via an electron transport chain. This membrane-bound enzyme complex now passes the electrons to the final acceptor NADP⁺, which is thus reduced to NADPH. At the same time, protons are transported across the membrane, which drive an ATP synthase when they flow back, phosphorylating ADP to ATP. Overall, in light-dependent photosynthesis, for every two molecules of water, two molecules of NADPH and about three molecules of ATP are provided. When considering ◘ Figs. 7.1 and 7.2 simultaneously, the energy levels (the standard redox potential is plotted on the

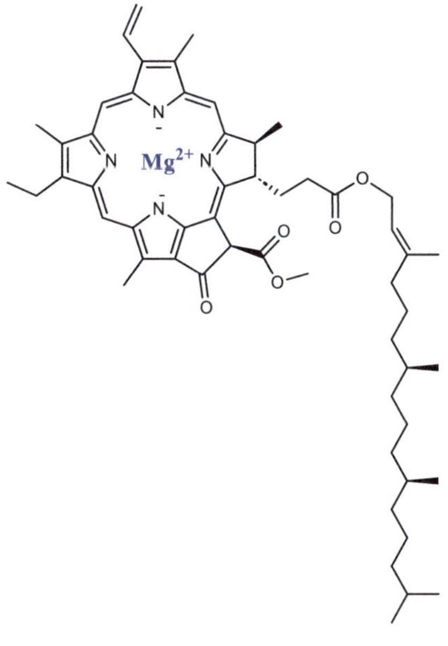

◘ **Fig. 7.1** Chlorophyll

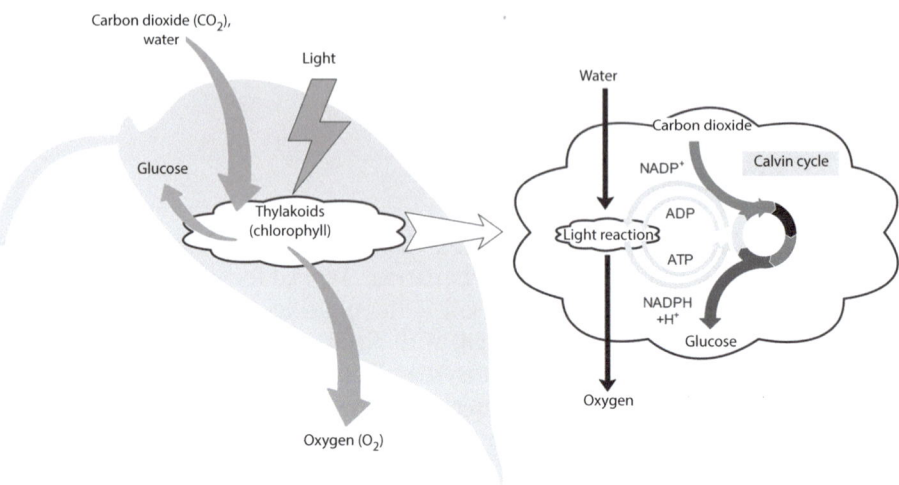

☐ **Fig. 7.2** Overview of the coupling of light reaction and Calvin cycle in photosynthesis

ordinate) during the transfer of electrons between both photosystems (PSI and PSII) and the cytochrome-b6f complex become clear.

7.2 Dark Reaction

Regardless of light, the previously formed coenzymes NADPH and ATP are used for the reduction and fixation of carbon dioxide (☐ Fig. 7.3). Three ATP and two NADPH are required per bound CO_2 molecule. This process also takes place at night and is identical in many plants to the Calvin cycle (☐ Fig. 7.4). The core of the dark reaction is the enzyme RuBisCO (Ribulose-1,5-bisphosphate carboxylase/oxygenase), which catalyzes the fixation of carbon dioxide and thus initiates the Calvin cycle. The key reaction in the Calvin cycle is the conversion of the pentose ribulose-1,5-bisphosphate with carbon dioxide (CO_2) to two molecules of 3-phosphoglycerate by the enzyme ribulose-1,5-bisphosphate carboxylase/oxygenase (RuBisCO). 3-Phosphoglycerate is phosphorylated with the help of ATP and NADPH and thus activated as glyceraldehyde-3-phosphate (GAP) for further biochemical catalyses. Most of this important intermediate is converted to ribulose-1-phosphate through further condensations and rearrangements and after final phosphorylation to ribulose-1,5-bisphosphate. The latter again serves as an acceptor for carbon dioxide, which is chemically bound in a carboxylase reaction.

A small part of the formed GAP can be converted into hexose (fructose-1,6-bisphosphate) in an aldose-ketose equilibrium through an aldolase reaction. The carbohydrates obtained are transported from the chloroplasts into the cytosol and metabolized there, so that the cell can in principle gain energy, but converts the glucose into starch as a storage substance. As already mentioned, assimilation products with the general empirical formula $(CH_2O)_n$ are referred to as carbohydrates. However, a better definition of carbohydrates is to consider them as primary oxidation

7.2 · Dark Reaction

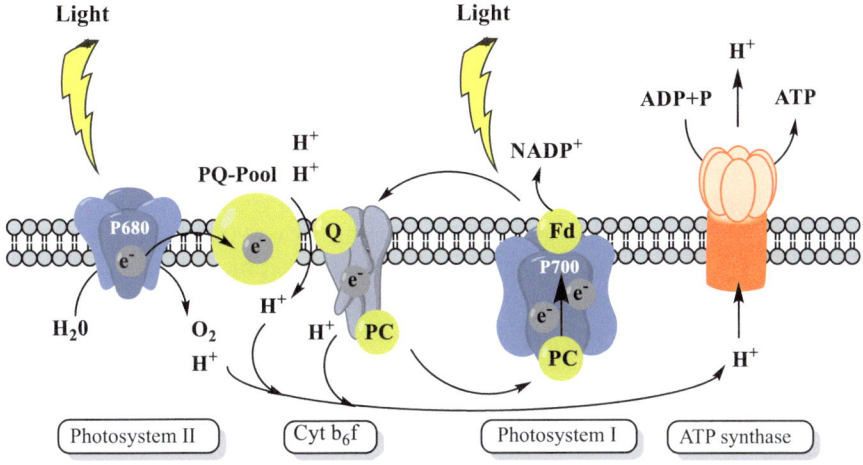

Fig. 7.3 The light reaction, function of the photosynthesis apparatus

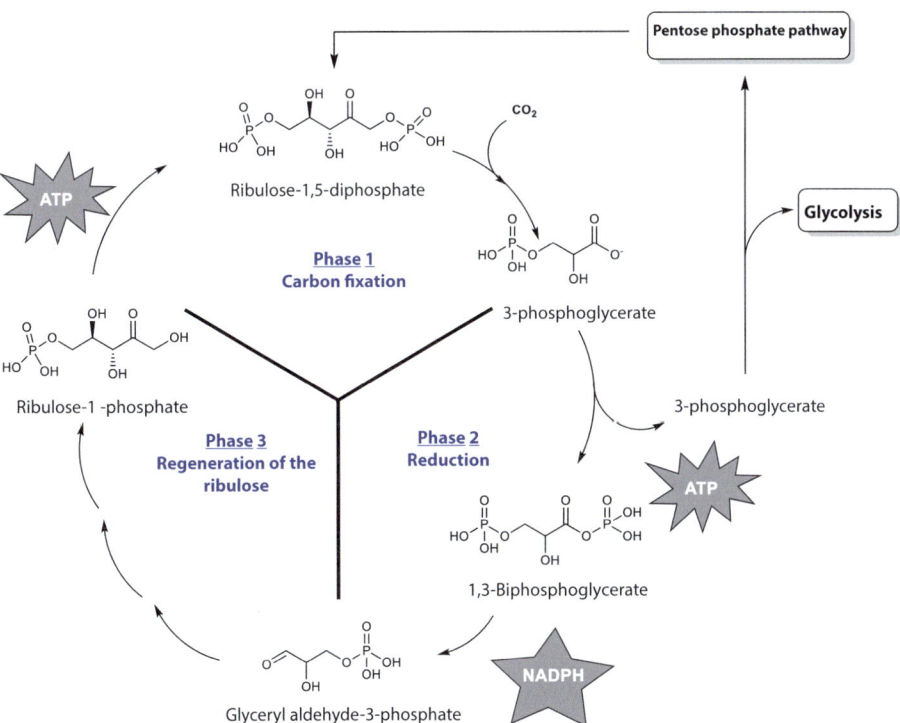

Fig. 7.4 Simplified representation of the Calvin cycle

products of polyhydric alcohols, where a hydroxyl group is oxidized to a keto or aldehyde group. Depending on the position of the carbonyl function, carbohydrates exist as aldoses or ketoses, which are kept in equilibrium by isomerases. In the Calvin-cycle, this equilibrium exists between GAP and 3-dihydroxyacetone phosphate (DHAP) and serves the biosynthesis of fructose-1,6-bisphosphate. It is the first hexose in photosynthesis, but exists in a five-ring system. Phosphorylated fructose serves as a starting metabolite for further sugars and is available via glycolysis and the pentose phosphate pathway as a source for further metabolites and basic building blocks of other biosynthetic pathways. In summary, it can be stated:

- Part of the GAP is returned to the Calvin cycle and metabolized there (carboxylation of ribulose-1,5-bisphosphate).
- Another part is supplied as glucose to glycolysis, broken down to pyruvate and, after decarboxylation, made available via acetyl-CoA as a central building block for further biosynthetic pathways (e.g., citric acid cycle, fatty acid biosynthesis, polyketides).
- Fructose is converted into glucose by isomerases and serves to build up storage substances such as starch and glycogen. In plants, natural substances are also glucosylated, among other things, to increase water solubility. Here too, fructose is involved via glucose.

7.3 The Somewhat Different Photosynthesis

The fundamental work on photosynthesis was carried out on so-called C3 plants. The term C3 plant was introduced later and serves to distinguish from C4 and CAM plants (English: Crassulacean Acid Metabolism). We speak of C3 plants when the first stable product of CO_2 fixation is a C3 body such as phosphoglycerate or glyceraldehyde-3-phosphate. Initial studies in the 1950s showed that important crops such as corn and Chinese reed have a higher photosynthesis performance under arid conditions than the previously known C3 plants. The reason for this is that in C4 plants, CO_2 fixation and the Calvin cycle are spatially separated. This leads to an active enrichment of CO_2 in the endodermis cells of the vascular bundles, which allows a higher photosynthesis rate. Since the pre-fixation of CO_2 to oxaloacetate and storage as malate or aspartate requires energy, C3 plants are more efficient under normal conditions. In CAM plants like pineapple (Ananas comosus), the processes of CO_2 uptake and fixation are temporally separated. This allows the active uptake of CO_2 with energy consumption at night, so that the stomata can remain closed during the day, reducing water loss through transpiration via the stomata. By storing CO_2 as in C4 plants, enough CO_2 is available during the day for the Calvin cycle to metabolize.

7.4 The Artificial Photosynthesis

The light-driven chemical reaction for energy production has long fascinated biologists and biotechnologists. For technical biochemistry, a dream would come true if no educts such as glucose or other sugars had to be added. The educt would be

CO$_2$, which is abundantly available in the air today and would be excellent as a new raw material. This chemical reaction is one of the most demanding tasks of biochemistry and not easy to realize biotechnically. The idea of artificial photosynthesis in a cell of non-plant origin has been a bold venture since the 1960s. The hurdle to overcome was the photolysis of water to hydrogen and oxygen. Unfortunately, the transfer to heterologous organisms such as yeasts or bacteria has not been successful to date. The reason is the complicated electron transport in photolysis, as photolysis involves a complex electron transport across a membrane, involving several electron donors and acceptors. Their assembly and the correct order of transport have not yet been reproduced. In particular, the correct coupling of photosystem II and photosystem I poses a biotechnological challenge.

Nevertheless, two important insights have emerged from this research. On the one hand, photosystem II is a possible biotechnological route for the production of biohydrogen (H$_2$). On the other hand, it appears that in biotechnology it may be easier to genetically manipulate photosynthetic microorganisms such as microalgae or cyanobacteria and optimize their performance (Li et al. 2022; Nagarajan et al. 2016).

7.5 Photolysis for the Production of Biohydrogen

Given that hydrogen is an interesting raw material in the industry of tomorrow (◘ Table 7.1), its production is an exciting challenge. The transition to biosynthesis must be based on two considerations. Firstly, the question of how protons can be obtained through photolysis needs to be clarified, and secondly, how a molecule of hydrogen (H$_2$) can be formed from two protons. There seems to be a solution or a biocatalyst for both reactions.

The cloning of Photosystem II and the decoupling from the rest of photosynthesis seem to have been successful. The second reaction can be catalyzed by a hydrogenase, which allows the reduction to molecular hydrogen (Reaction 7.3):

Reaction 7.3
$$2\,H^+ + 2e^- \rightarrow H_2$$

◘ **Table 7.1** Biotechnically produced quantities of biohydrogen (H$_2$) in the year 2023

Country	Production in tons in the Year 2023
USA	100
China	80
Germany	40
France	30
United Kingdom	20

Reaction 7.3 is only half the truth, because hydrogenases are enzymes that can reversibly form or split hydrogen in both directions. There are different types, with FeFe- and NiFe-hydrogenases being the catalysts of interest for biotechnology. Metals in the active center are crucial for the reaction. For the biotechnological process, the absence of oxygen is also important, as this gas deactivates the catalyst. Hydrogen-producing microorganisms are therefore often found among the anaerobically living archaea, but we humans with our intestinal flora are also active hydrogen producers and release the gas from the intestine via the lungs.

7.6 Photosynthesizing Microorganisms as New Producers

They have been on Earth for millions of years and used light before plants came out of the water onto land and grew upwards. They belong to the bacteria (cyanobacteria) and the single-celled organisms with a nucleus (microalgae), which used light for energy production. Many algae also produce hydrogen, but the yields are too low to have economic significance today. In ◘ Table 7.1, the quantities of biohydrogen produced today in the year 2023 are given.

An example of the study of photosynthetic microorganisms is cyanobacteria. These microorganisms belong to the blue-green algae that are photosynthetically active and use light for energy production. As with plants, carbon dioxide and water are converted into glucose, and glyceraldehyde-3-phosphate is the metabolite that stimulates glycolysis. At the end of glycolysis, acetyl-CoA is converted into the C4 building block acetoacetyl-CoA. Butanol biosynthesis is not genetically established in cyanobacteria, and reprogramming and cloning of genes from clostridia must be carried out using genetic engineering (◘ Fig. 7.5). The basic biosynthesis of butanol is explained in the following chapter. Once the necessary genes are cloned, the reduction of the keto group to 3-hydroxybutyryl-CoA and the saturation, as known from fatty acid biosynthesis, take place. If butyryl-CoA is present, there is no extension with another C2 unit, but the detachment of coenzyme A. Butyraldehyde can easily be reduced to butanol by a butyraldehyde reductase (◘ Fig. 7.5). An example of the study of photosynthetic microorganisms is cyanobacteria. These microorganisms belong to the blue-green algae that are photosynthetically active and use light for energy production. As with plants, carbon dioxide and water are converted into glucose, and glyceraldehyde-3-phosphate is the metabolite that stimulates glycolysis. At the end of glycolysis, acetyl-CoA is converted into the C4 building block acetoacetyl-CoA. Butanol biosynthesis is not genetically established in cyanobacteria, and reprogramming and cloning of genes from clostridia must be carried out using genetic engineering. The basic biosynthesis of butanol is explained in the following chapter. Once the necessary genes are cloned, the reduction of the keto group to 3-hydroxybutyryl-CoA and the saturation, as known from fatty acid biosynthesis, take place. If butyryl-CoA is present, there is no extension with another C2 unit, but the detachment of coenzyme A. Butyraldehyde can easily be reduced to butanol by a butyraldehyde reductase.

The analyses show that in cyanobacteria, the rate-determining step is the formation of crotonaldehyde. For the formation of one molecule, 48 photons of captured light are consumed, which corresponds to the amount for glucose. Comparing

7.6 · Photosynthesizing Microorganisms as New Producers

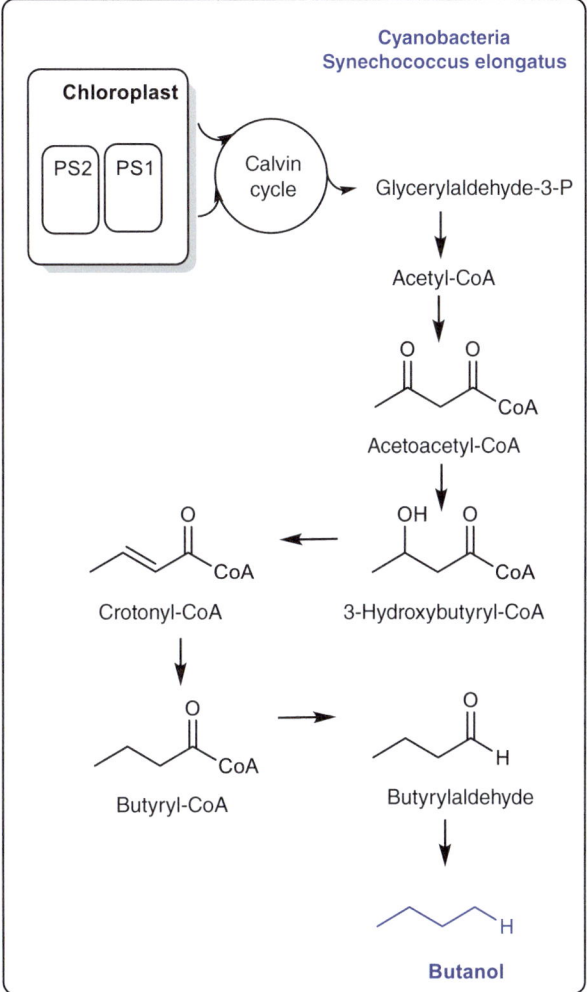

Fig. 7.5 Photobiochemical butanol biosynthesis in *S. elongatus*

photosynthesis between plants and cyanobacteria, it is not as efficient in blue-green algae. Reasons for this are the low provision of carbon sources for the Calvin cycle, a less efficient electron transport, and the lower growth performance of cyanobacteria. An analysis of possible applications of heterologous photosynthesis is discussed in the chapter on the future of biochemistry (Liu et al. 2021).

? Self-check Questions
1. Explain why a spatial separation of the dark and light reaction in the cell is important for photosynthesis.
2. Describe the transport of the photosynthesis end product between leaf and root. Why is this chemical substance useful for transport?

3. Draw a typical representative for an aldose and ketose and explain the chemical differences between the two.
4. Formulate the gross and net equations of photosynthesis.
5. Explain the significance and physiology of the so-called Z-scheme of photosynthesis.
6. Explain the basic purpose of the light-dependent part of photosynthesis. Explain the light-independent cycle comparatively.
7. Name the key reaction in the Calvin cycle and justify your answer.
8. Indicate which redox equivalent for NADH or NADPH has the higher potential, and justify your answer.
9. Distinguish C3, C4, and CAM plants biochemically from each other.

Sugar Metabolism

Contents

8.1 Glycolysis – 58

8.2 Pentose Phosphate Pathway – 59

8.3 Citric Acid Cycle – 62

8.4 Technical Acids – 63
8.4.1 Citric Acid – 63
8.4.2 Succinic Acid – 64
8.4.3 Itaconic Acid – 66

8.5 Oxidative Phosphorylation – 67

8.6 Alcoholic Fermentation – 70
8.6.1 Alcohol as Food – 70
8.6.2 Technical Alcohol – 71

8.7 Lactic Acid Fermentation – 72

References – 75

© The Author(s), under exclusive license to Springer Fachmedien Wiesbaden GmbH, part of Springer Nature 2025
O. Kayser and N. J. H. Averesch, *Technical Biochemistry*, https://doi.org/10.1007/978-3-658-47121-7_8

Learning Objectives
- Basic metabolism of sugar utilization and energy production
- Technically important sugars, acids, and alcohols
  - Ethanol
  - Butanol
- Lactic acid fermentation
- KDPG pathway for ethanol production

The breakdown or chemically the oxidation of sugars is a very prominent catabolic biochemical process for energy provision in (almost) all living beings. Besides energy production, the breakdown of glucose as a central biochemical pathway plays a significant role in the formation of intermediary metabolites, as these serve as the starting point for other metabolic pathways such as amino acid biosynthesis, the biosynthesis of nucleobases for the construction of DNA and RNA, as well as the synthesis of aromatic amino acids and phenols via the pentose phosphate pathway. Since glycolysis and the associated pentose phosphate pathway are of central importance, these two metabolic pathways are explained in more detail.

8.1 Glycolysis

This glucose (glycos) degrading (lytic) metabolic pathway occurs in all cells and is vital. Depending on the tissue and organ, glycolysis is differently active, in humans, for example, in the brain, muscles, and liver. However, it always takes place and is, along with the citric acid cycle, the central pool for the conversion of biochemical basic substances. A look at ◘ Fig. 8.1 shows that glucose is phosphorylated and activated before its utilization. Glucose-6-phosphate is a high-energy compound that is only available in this active form for further metabolism.

In the first biochemical step of glycolysis, glucose-6-phosphate is isomerized to fructose-6-phosphate, which is phosphorylated a second time to fructose-1,5-diphosphate in the second step. The reason for this is that otherwise only one phosphorylated sugar would be available for the splitting of this C6 sugar into two C3 sugars. The resulting trioses glyceraldehyde-3-phosphate (G3P) and dihydroxyacetone phosphate (DHAP) are further central building blocks:
- In the subsequent glycolysis for the conversion to pyruvate as an intermediate for the formation of acetyl-CoA for the citric acid cycle, directly for the ethanolic fermentation and lactic acid biosynthesis
- For the biosynthesis of glycerin for triglyceride biosynthesis
- For the biosynthesis of the amino acids glycine and alanine
- For the biosynthesis of acetone and butanol in technical biochemistry through the ABE metabolism

The technical significance of glycolysis seems to be small, and there are no metabolic products that allow further use. Nevertheless, glycolysis is a very important metabolic pathway for the provision of pyruvate for acetyl-CoA, which can be introduced into other heterologous metabolic pathways. The provision of large quan-

8.2 · Pentose Phosphate Pathway

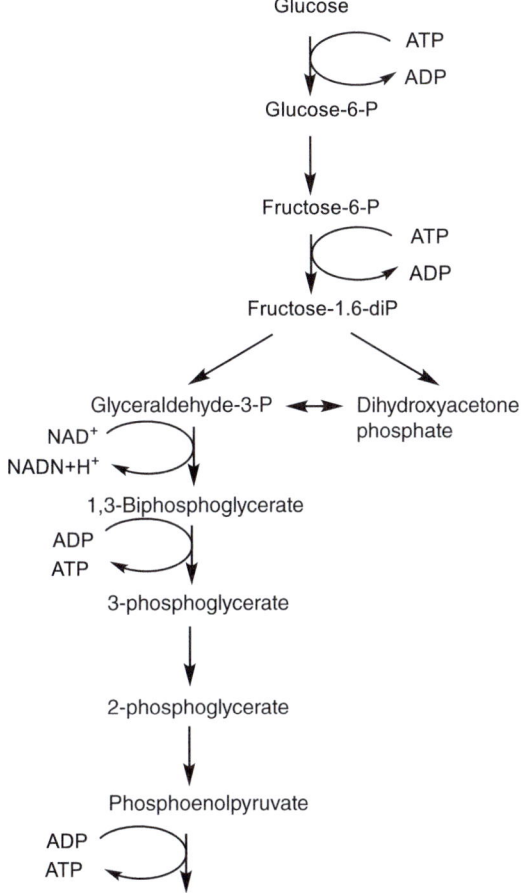

Fig. 8.1 Schematic overview of glycolysis

tities is important in bioprocess engineering to improve the yields of secondary end products. Therefore, glycolysis is being researched to understand the regulatory mechanisms for better decoupling and to find out how more glucose can be introduced into the cell for degradation via glucose transporters in the membrane.

8.2 Pentose Phosphate Pathway

In the pentose phosphate pathway (PP pathway), various C5 sugars, e.g., xylose and ribose, are irreversibly oxidatively or reversibly reciprocally reduced and converted into each other. Therefore, in biochemistry, the PP pathway (Fig. 29) is divided into two phases, with glucuronic acid forming the transition. In the first oxidative phase, glucose-6-phosphate is oxidized to glucuronic acid, which is now converted as a C6 metabolite by cleavage into ribulose-5-phosphate, an important intermediate for the synthesis of DNA, RNA, and many coenzymes such as ATP, coenzyme A, NAD$^+$, and FAD. The PP pathway is, besides the citrate shuttle (not to be confused with the citric acid cycle), the only way to obtain NADPH. In the non-oxidative part, the mutual conversion of trioses, tetroses, and pentoses into each other takes place. Ribulose-5-phosphate is converted into xylulose-5-phosphate, from which two pentose phosphates (so-called C5 sugars) are formed by CO_2 loss and interconversion with glyceraldehyde-3-phosphate (GAP). For technical biochemistry, the biosynthesis of the largest C7 sugar sedoheptulose from erythrose and glyceraldehyde is important, as it is used as a reactant in the shikimate pathway. Aromatic amino acids such as phenylalanine, tyrosine, and indirectly tryptophan are formed via the shikimate pathway.

Through complex molecular shifts and conversion reactions, pentose phosphates can be metabolized to C2 and C3 building blocks, which in turn are involved in the construction of hexose phosphates, thus closing the cycle. These reactions are catalyzed for the C2 building blocks by the enzyme transketolase and for the C3 building blocks by the enzyme transaldolase. The steps of the pentose phosphate pathway can be summarized as a biochemical reaction:

– for the degradation of glucose and fructose to ribulose as a building block for the biosynthesis of nucleotides,
– for the formation of NADPH as a reduction equivalent in anabolic biochemical reactions,
– for the degradation of glucose and fructose to pentoses, as a building block for the biosynthesis of coenzymes such as ATP or FAD,
– for the provision of sedoheptulose as a reactant for the shikimate pathway.

A look at ◘ Fig. 8.2 shows that part of the pentose phosphate pathway corresponds to the Calvin cycle. Photosynthesis follows ribulose-1,5-diphosphate. All reactions proceed via activated, i.e., phosphorylated, intermediates, with one of the most important intermediate and end products being phosphoenolpyruvic acid (synonym: phosphoenolpyruvate, PEP).

The importance of the PP pathway for technical biochemistry is evident in the production of **gluconic acid** (◘ Fig. 8.3) in the beverage industry. The great popularity of biologically produced lemonades has led to a great interest in non-alcoholic fermentation by *Glucobacter oxydans* . The biosynthesis is a third way besides lactic acid and ethanolic fermentation (◘ Fig. 8.4). The chapter on Technical Acids explicitly addresses the production of organic acids, but the biochemical basis of the PP pathway should already be pointed out here. The importance of the PP pathway for technical biochemistry is evident in the production of gluconic acid in the beverage industry. The great popularity of biologically produced lemonades has led to a great interest in non-alcoholic fermentation by Glucobacter oxydans. The biosynthesis is a third way besides lactic acid and ethanolic fermentation (◘ Fig. 8.4). The chapter on Technical Acids explicitly addresses the production of organic acids, but the biochemical basis of the PP pathway should already be pointed out here.

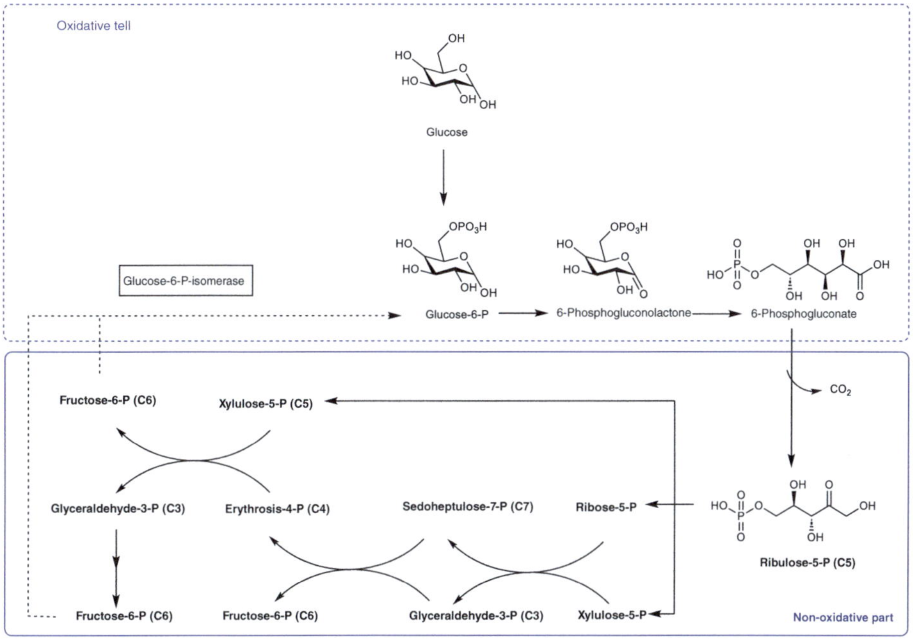

8.2 · Pentose Phosphate Pathway

Fig. 8.2 Schematic overview of the pentose phosphate pathway (PP pathway). With dashed blue lines: oxidative part of the PP pathway; with solid blue line: Non-oxidative part with sugar conversion reactions

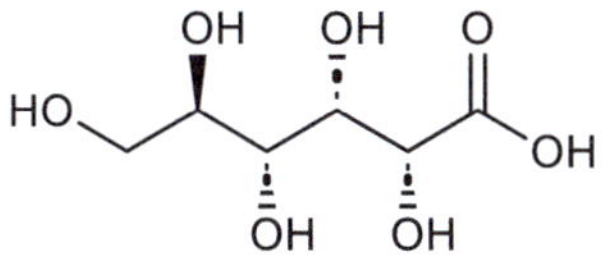

Fig. 8.3 Gluconic acid

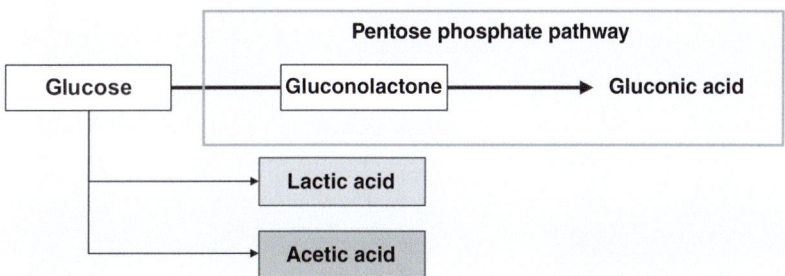

Fig. 8.4 Overview of biosynthetic pathways to glucuronic acid, lactic acid and acetic acid

8.3 Citric Acid Cycle

This biochemical cycle can be described as the central hub in the cell, where the most important intermediates or their degradation products flow in and out, where energy production takes place, and from where many anabolic metabolic processes obtain their starting materials. Chemically, the citric acid cycle is an amphibolic cycle, important for the provision of phosphorylated and reduced substances, and is located in the mitochondria (in prokaryotes in the cytosol) as the powerhouses of the cell. Under **Amphibolism** we understand the combination of catabolic and anabolic metabolic processes. Upon closer inspection, the cycle consists of an anabolic part, characterized by the synthesis of C4 carboxylic acids, and a catabolic part, determined by the generation of reducing equivalents (NADH), GTP, and the degradation of dicarboxylic acids. The cycle ends with the transfer of the acetate unit by coenzyme A to oxaloacetate. A simplified overview is shown in ◘ Fig. 8.5, a more complex representation can be found in the download with the panels of biochemical metabolic pathways. However, at this point, a formal discussion of this metabolic pathway should suffice. Activated acetic acid (acetyl-CoA) is transferred to oxaloacetate in the first step. In the subsequent circulation, the activated acetic acid is "burned" (chemically oxidized) to two moles of CO_2, and eight protons are transferred to three NAD^+ molecules and one FAD molecule. The energy gain is positive, as ADP is phosphorylated to ATP. However, the citric acid cycle is not the central ATP pool, as the majority is gained through oxidative phosphorylation in the respiratory chain. Of great importance is the synthesis of GTP by transferring phosphate groups to GDP. This energy carrier is known not for providing energy in the cell, but for activating certain signal transduction pathways (GDP receptors).

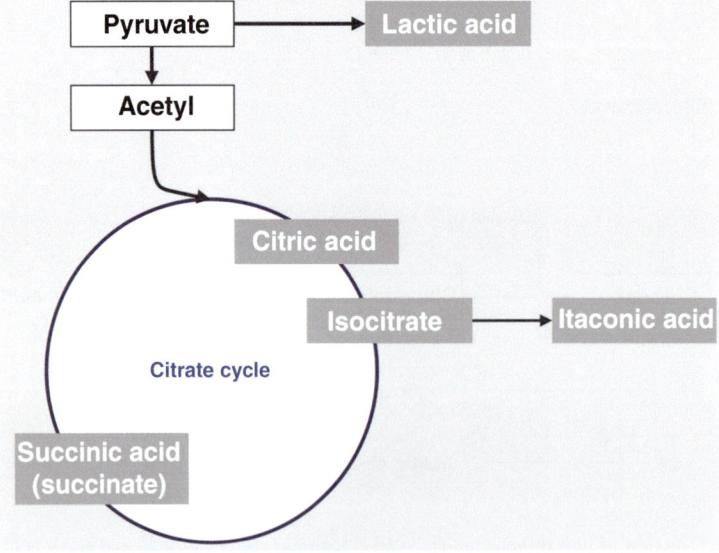

◘ **Fig. 8.5** Technical acids, derived from the citric acid cycle

8.4 · Technical Acids

Considering the importance of the citric acid cycle from a technical perspective, it becomes clear that **citric acid**, **succinic acid** and indirectly **itaconic acid** are three important organic acids provided. This chapter shows how a genetic mutation has led to the accumulation of citric acid in *Aspergillus* spp. The biosynthesis of succinic acid as an important platform chemical with 2,000 to 3,000 tons per year is carried out in *Actinobacillus succinogenes* and *Anaerobiospirillum succinicipro- ducens*. Starting from citric acid, itaconic acid can be considered a byproduct of the citric acid cycle. Dicarboxylic acid is formed in *Aspergillus terreus* via *cis*-aconitic acid.

8.4 Technical Acids

8.4.1 Citric Acid

The metabolite that gave this important biochemical pathway its name is the tricarboxylic acid citric acid. Long before science recognized the metabolic function of this acid, the acid contained in lemon extracts was valued and used in foods. Until the First World War, lemons were used as a raw material for the extraction of citric acid until genetically modified Aspergilli were discovered, which accumulated citric acid due to a mutation in the aconitase. The use of the *A. niger* mutants led to a collapse of lemon cultivation, especially in southern Italy, and to an increase in biotechnological production.

Biosynthesis Presenting the biosynthesis of citric acid here seems superfluous, as it occurs in one step by adding the acetyl group to oxaloacetate at the beginning of the citrate cycle. Of biotechnological interest is the question of intracellular regulation, as this strong acid quickly lowers the pH value and the vitality of the cell. For control, the citrate-malate shuttle is used biotechnologically. Both substances (citrate and malate) have the common metabolite oxaloacetate. Oxaloacetate has a special metabolic function, as it can be converted to citrate with the help of acetyl-CoA or formed by oxidation from malate to be efficiently excreted (◨ Fig. 8.6).

Considering the citrate-malate shuttle (◨ Fig. 8.7) in its entirety, the reverse reaction from citrate to malate must be taken into account. To enable this, malate is exchanged for citrate on the one hand. Since the desired product citrate is available in large quantities, more pyruvate must be transported from the cytosol into

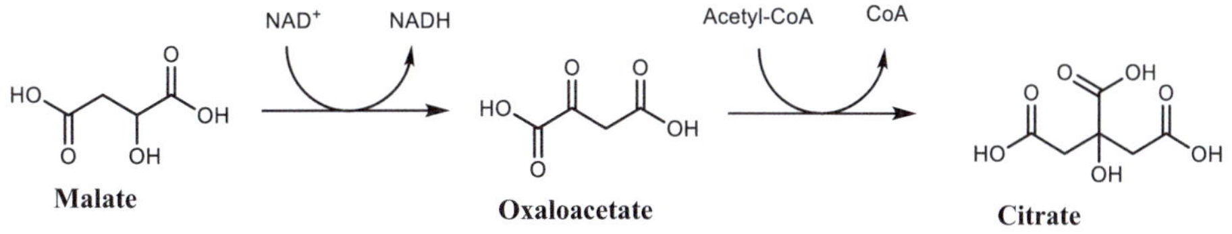

◨ **Fig. 8.6** Oxaloacetate as a metabolite in the citrate-malate shuttle

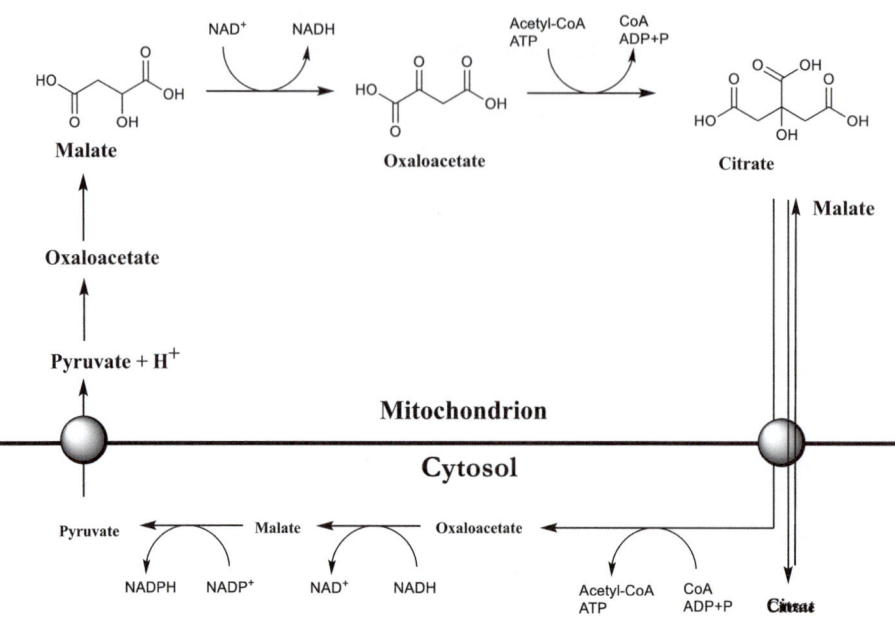

◘ **Fig. 8.7** Citrate-Malate Shuttle

the mitochondria. This exchange is facilitated by transporters that exchange citrate, malate, and protons in the membrane. The role of protons (H$^+$) has been reconsidered in the evaluation of transport performance. The pH value is also important for later production and should not exceed pH 2, otherwise the by-products oxaloacetate and glucuronic acid are formed too strongly.

The production of citric acid is a highly aerobic process that requires balanced process control. If the oxygen concentration is too high due to strong aeration, the CO_2 concentration decreases, which is disadvantageous for biosynthesis, as CO_2 is a mandatory substrate of pyruvate carboxylase for the biosynthesis of citric acid from malate via oxalate. The transport from the mitochondria not only takes place via the citrate-malate shuttle, but also via tricarboxylic acid transporters, which can directly pump out the citric acid. Current production volumes are up to 500 m^3 with a yield of 140 g/L. The annual production of citric acid is 36,000 t (as of 2020) and is used as an acidifier, flavor enhancer, food preservative, and as an additive in household cleaners; the annual production is 36,000 t (as of 2020).

8.4.2 Succinic Acid

Succinic acid is a dicarboxylic acid, which is a central component of the citric acid cycle. Due to the two carboxyl groups, succinic acid is an interesting platform chemical for polymers.

Biosynthesis The biosynthesis direction (◘ Fig. 8.8) from oxaloacetate via fumarate and malate to succinate (succinic acid) seems unusual, but is initiated by the

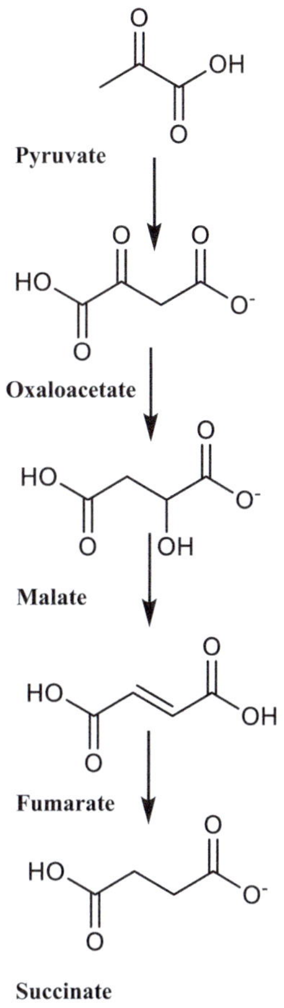

8.4 · Technical Acids

Fig. 8.8 Biosynthesis of succinic acid

carboxylation of pyruvate to oxaloacetate. With the provision of oxaloacetate, the part of the citric acid cycle can also run in reverse. To achieve a high yield, natural producers such as *Mannheimia succiniciproducens, Basfia succiniciproducens* and *Anaerobiospirillum succiniciproducens* utilize the parallel conversion of pyruvate to lactic acid, formate and the replenishment of the acetyl-CoA pool by knock-down mutations of the gene Δpyruvate formate lyase (formate, acetyl-CoA) and Δlactate dehydrogenase to D-lactate. Since cultivation requires expensive complex media, *Escherichia coli* and baker's yeast have been switched to as alternative producers. Here too, the two genes mentioned above were deleted and the pyruvate carboxylase genes from *Rhizobium etli* were cloned into the heterologous producers. A problem with all productions are the substantial amounts of acids that unfavorably lower the intracellular pH value. In addition, succinic acid induces a feedback inhibition in *E. coli*, which is why a two-stage process with aerobic and anaerobic phases is necessary (Lee et al. 2008).

8.4.3 Itaconic Acid

Itaconic acid is obtained from citric acid and leaves the cycle through decarboxylation of *cis*-aconitic acid. At the end of this simple biosynthesis is the unsaturated dicarboxylic acid, which is converted into a cyclic anhydride by dehydration. The acid was first detected in *Aspergillus itaconicus*, but the technical representation is carried out in *A. terreus*. The technical importance lies in the use for plastics, fibers, paints and surfactants with quantities of about 30,000 tons per year (as of 2023).

Biosynthesis The biosynthesis seems short and simple (◘ Fig. 8.9), yet the question must be answered how the physiology of the fungus can be maintained when *cis*-aconitate, an important metabolite, leaves the citric acid cycle. Bentley and Thiessen were able to show as early as 1957 that *cis*-aconitate is exchanged in the citrate-malate shuttle between mitochondria and cytosol for malate, which refills the cycle. The biosynthesis of itaconic acid takes place in the cytosol, where the aconitate decarboxylase (cadA) is present. Biotechnologically, the biosynthesis rate is increased when the enzyme 6-phosphofructo-1-kinase (pfKA) is altered. PfKA is a regulatory enzyme that is inhibited by citrate when an excess of glucose provides too much acetyl-CoA for the citric acid cycle. In the absence of feedback regulation, a lot of citric acid is formed, which is converted into itaconic acid. The problem is the pH sensitivity of *A. terreus* to too high concentrations of organic acids. For this reason, the *A. niger* strain AB 1.13 was genetically modified and the cadA gene with its promoter was cloned into the organism. The current production titers are at 80 g/L (Steiger et al. 2013).

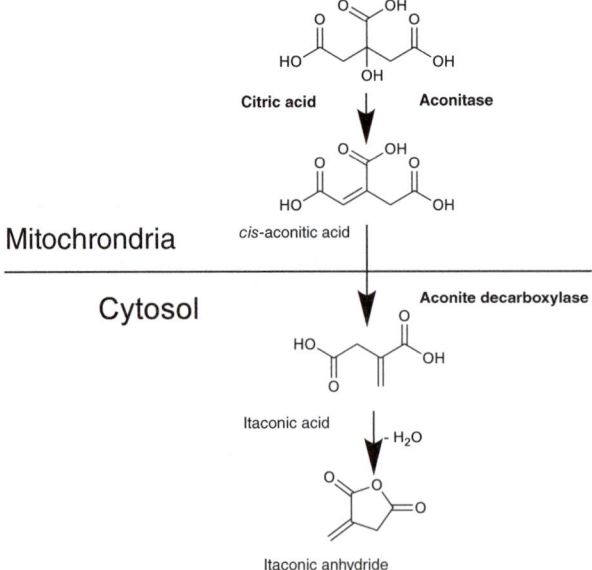

◘ **Fig. 8.9** Biosynthesis of itaconic acid

8.5 Oxidative Phosphorylation

The "silent combustion" of sugar in the body provides large amounts of energy with almost 32 mol ATP, which are not suddenly released all at once in living beings, but gradually in the form of energy cascades. Thus, even in the last step of energy metabolism, the reaction of hydrogen with oxygen to form water is broken down into individual steps. Oxidative phosphorylation is an electron transport chain that occurs in all aerobic organisms (◘ Fig. 8.10). In eukaryotes, this process takes place on the inner membrane of the mitochondria. The outer membrane contains the transport protein porin, which allows molecules up to a size of 5 kDa to pass through like a sieve. The inner membrane is impermeable to ions and most smaller molecules. However, the mitochondrial matrix has highly specific transport proteins, so the membrane can only be passed selectively. The electron transport takes place on the inner membrane side. This compartmentalization is important because the electron transport runs across the membrane and the driving force of phosphorylation is used as energy by an exothermic reaction to form ATP. The flow of electrons transports protons across the membrane, creating an electrochemical gradient (electron transport chain). The backflow of protons drives ATP synthase, which synthesizes ATP from ADP and phosphate (oxidative phosphorylation). Due to the ion gradient, the protons diffuse back again and again, otherwise too acidic an environment would develop in the mitochondria. This gradient is essential for the function of the electron transport chain. For each oxidized NADH, a proton is pumped into the intermembrane space, so the overall balance is 10 NADH, 10 protons, and 2 $FADH_2$.

ATP synthase as a coupled proton pump requires at least four protons for the synthesis of one ATP molecule. Thus, 2.5 ATP are produced from one NADH (◘ Fig. 8.11). As a result, 30 to 32 ATP are released per glucose molecule, although theoretically 38 ATP should be formed. The lower number is explained by self-consumption and regulation. In prokaryotes, the yield is slightly higher at about 34 ATP, as they do not have to resort to energy-consuming transport mechanisms due to the lack of compartmentalization. It should be noted that the quantities cannot be determined exactly, which may be due to the fact that the protein complexes do not function 100% ideally. Those who want to know more about the exact function and enzyme kinetics of ATP synthase are referred to the existing specialist literature, textbooks and - for historical reasons - also to the Mitchell hypothesis (chemiosmotic coupling).

Interestingly, ATP synthase can also be considered a molecular motor consisting of two units that resemble a rotor-stator complex. The F1 unit is an ATPase that binds ADP with Pi to ATP on the outside of the membrane. The F0 unit is directed inward and anchored in the membrane. It provides the driving force through the large proton gradient. But why are these enzymes referred to as motors? In the biosynthesis of 3 ATP, the complex must rotate once around its imaginary axis in three partial steps of 120° each (rotor principle). The stator in the middle is probably formed by membrane lipids. The catalytic subunits can be imagined as rotor blades, which rotate the protons by 120° over an asparagine residue. The trick is that the asparagine residue is in spatial proximity to an arginine, which stabilizes the negative charge of the asparagine (ASP-COO-). This interaction, which stabilizes the

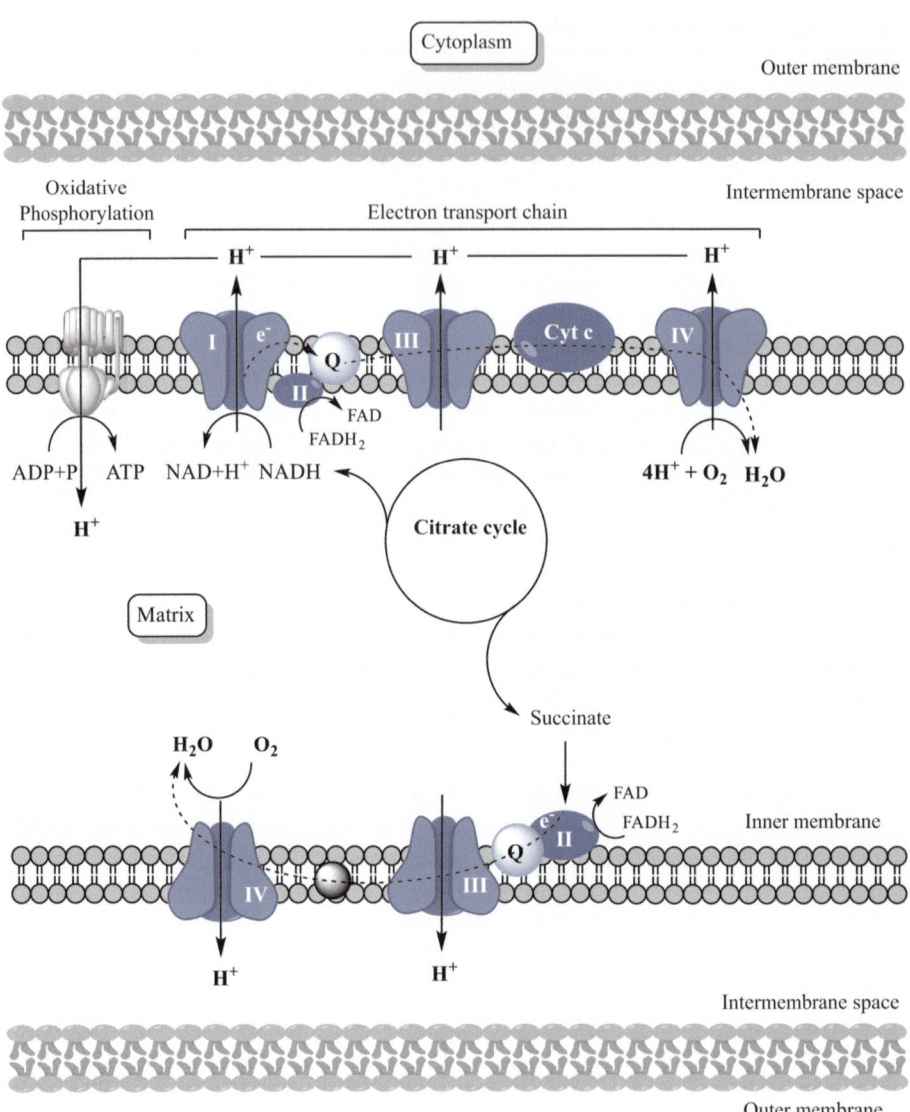

Fig. 8.10 Oxidative phosphorylation (respiratory chain) and electron transport chain, below connection to the citric acid cycle via succinate

deprotonated charge of a hairpin structure, resembles a spiral spring under mechanical tension. Protonation releases this tension through conformational change; the subunit rotates by 120° and releases the proton taken up from the outside into the cytosol. The rotation brings an ADP and a P anchored in the F1 unit into immediate proximity, so that the bond to ATP can be closed. The rotation speed of 6,000 rpm (100 r/s) is impressive and has motivated bioengineers to develop molecular motors in nanosystems.

8.5 · Oxidative Phosphorylation

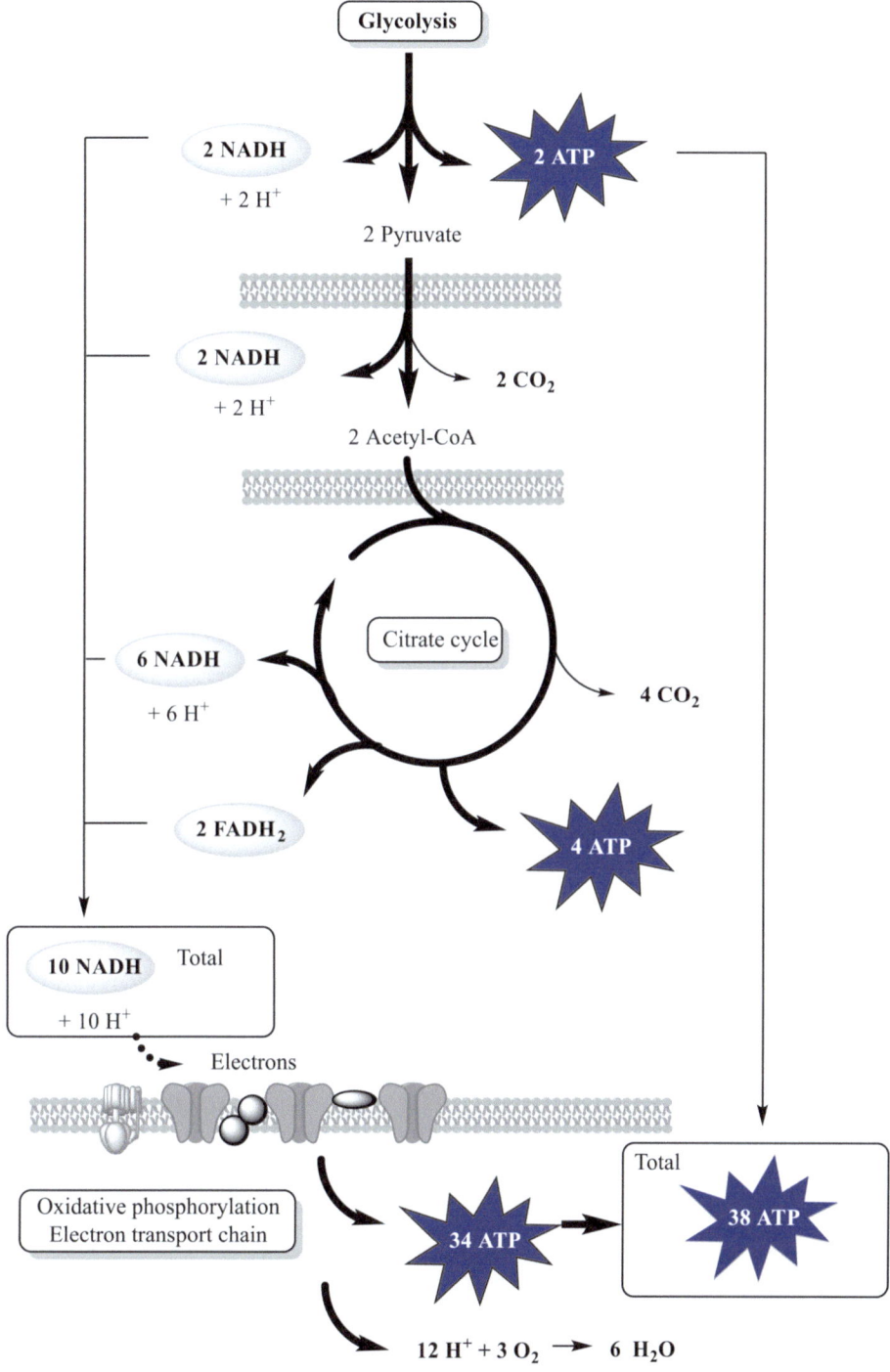

Fig. 8.11 Coupling of glycolysis, the citric acid cycle, and oxidative phosphorylation for the production of ATP and NADH

The respiratory chain cannot be directly attributed with any particular significance for technical biosynthesis. However, it is of great value in metabolic engineering to calculate and increase the provision of energy in the form of ATP and reduction equivalents. In other chapters dealing with the genetic optimization of metabolic pathways, the question of energy balance is often addressed. The introduction of a foreign (heterologous) metabolic pathway into a cell can lead to the cell being overwhelmed and only being able to execute the metabolic pathway suboptimally due to a lack of ATP or reduction equivalents.

8.6 Alcoholic Fermentation

8.6.1 Alcohol as Food

The extraction and enjoyment of alcohol (ethanol) as food is a cultural achievement that can be dated back to around 7,000 BC. In the course of the transformation of life from hunters and gatherers to that of farmers, ancient grains were cultivated in Mesopotamia and used as a food source. By chance or careless action with ancient grain or barley, fermentation with yeast occurred, converting sugar from fermentable yeast into ethanol. Recipes for brewing beer are well known from the high cultures of the Sumerians (around 3,000 BC) (Damerow 2011) and Egyptians. In addition to beer, the pressing of grapes to extract wine was an alternative that developed almost simultaneously with brewing. The origins of wine pressing can be found in Mesopotamia and on the northern edge of the Caucasus. In archaeological finds such as amphorae, dried remains of wine could be chemically analyzed. Wine as a food was very popular, but the cultivation of the grapevine Vitis vinifera only took place in cooler climates. This explains why wine was drunk in Egypt, but grown in present-day Anatolia. The Romans, for example, brought Cabernet Sauvignon from Bulgaria and Romania to Spain and France.

Regardless of the biological starting material (malt or grapes), the biochemical principle of ethanolic fermentation is identical (◘ Fig. 8.12). Under anaerobic conditions, glucose is not completely transferred by yeast as pyruvate (C3) into oxidative phosphorylation (respiratory chain). In a biochemical emergency program of metabolism, energy is obtained by splitting pyruvate into CO_2 and acetaldehyde by pyruvate decarboxylase, the latter metabolite being reduced to ethanol (C2) with the help of NADH. The net reaction (Reaction 8.1) of alcoholic fermentation is:

> **Reaction 8.1**
> Glucose + 2 Pi + 2 ADP + 2 H^+ → 2 Ethanol + 2 CO_2 + 2 ATP + 2 H_2O

Note that NAD/NADH are found as reduction equivalents, as these are energy-yielding and degrading reactions. In biochemistry, fermentation is understood to mean anaerobic fermentation without putrefactive fermentation. Putrefactive fermentation would be, for example, anaerobic methanogenesis, which is explained

8.6 · Alcoholic Fermentation

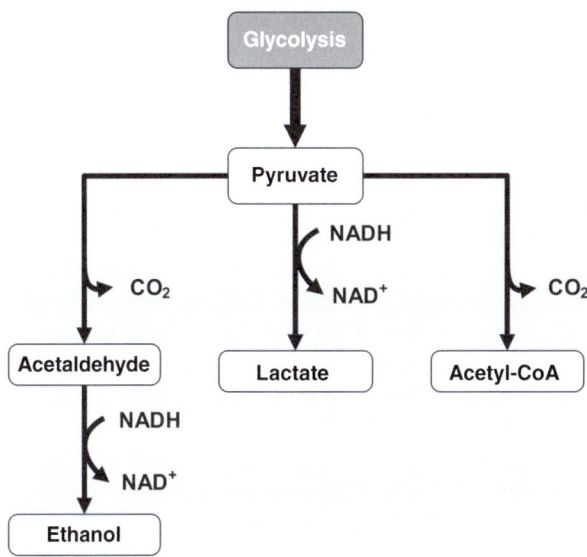

Fig. 8.12 Summary representation of the metabolic pathways of pyruvate

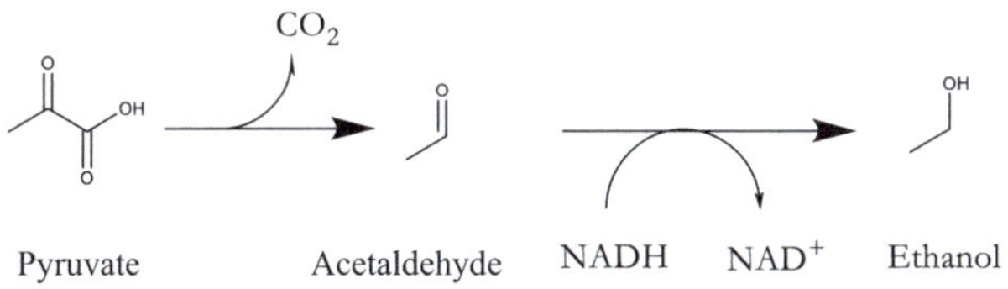

Fig. 8.13 Alcoholic fermentation, starting from pyruvate as the end product of glycolysis

in a later chapter. The two most important types of aerobic fermentation are the aforementioned alcoholic fermentation (Fig. 8.13) and lactic acid fermentation, which is discussed in ▶ Sect. 8.7. While ethanol and traces of methanol are produced in alcoholic fermentation, lactic acid fermentation is a biochemical degradation reaction in lactobacilli, with lactic acid being the end product. Lactic acid plays an important role in the production of sour milk products and bioplastics.

Alcoholic fermentation is a true metabolic pathway of anaerobic yeast. Baker's yeast shows a metabolic performance of up to 14% ethanol in wine, while the average beer has a low ethanol content of 4.5 to 5.5%. The vast majority of microbiologically obtained ethanol is used in the food industry and is subject to labeling requirements at concentrations above 0.5%.

8.6.2 Technical Alcohol

The classic alcoholic fermentation is found in food biotechnology and does not aim to deliver a high-proof alcoholic product, but to offer the customer a tastefully matured beer or a wine. During fermentation or pressing, great importance is placed on flavor and aroma substances that make up a delicacy.

In the extraction of technical alcohol or ethanol, the maximum yield is paramount. The aim is to obtain the ethanol as a pure substance and not as a mixture for consumption. This goal is of interest because ethanol is an interesting solvent for paints, disinfectants, fragrance carriers in perfumes, cleaning agents or antifreeze. However, the majority (80%) of the ethanol is burned as fuel in flexible-fuel engines of passenger cars. Additives to gasoline as E5 and E10 fuel in Germany, in the USA or in Japan, but also E25 in Brazil are known. The extraction of technical ethanol can be done via the classic alcoholic fermentation and will therefore not be further explained here (see above). A second biochemical process, which is preferably used in tropical countries like Brazil, will be addressed here.

In addition to yeasts, alternatives to glycolysis and classic fermentation are known in some bacteria and anaerobic (!) Archaea like *Zymomonas mobilis*. Nathan Entner and Michael Doudoroff (1911–1975) found in *Pseudomonas saccharophila* that glucose is metabolized via the pentose phosphate pathway to 2-keto-3-deoxy-6-phosphogluconate (KDPG). The metabolite KDPG is not known as a metabolite of the pentose phosphate pathway, but is formed from 6-phosphogluconic acid by oxidation and water loss. KDPG is an interesting metabolite consisting of six carbon atoms and is split into two C3 units. Pyruvate and glyceraldehyde-3-P are produced, both of which are provided for alcoholic fermentation via the energy-producing part of glycolysis.

The Entner-Doudoroff pathway begins with the phosphorylation of glucose to glucose-6-phosphate (◘ Fig. 8.14). In the next step, glucose-6-phosphate is oxidized by the Entner-Doudoroff dehydrogenase to 2-keto-3-deoxy-6-phosphogluconate (KDPG). KDPG is then cleaved by the KDPG lyase into pyruvate and glyceraldehyde-3-phosphate. Glyceraldehyde-3-phosphate is then further degraded in glycolysis. The Entner-Doudoroff pathway differs from glycolysis in two essential points. In the first step, glucose-6-phosphate is not oxidized to fructose-6-phosphate, but to KDPG. In the second step, KDPG is not cleaved into pyruvate and phosphoenolpyruvate, but into pyruvate and glyceraldehyde-3-phosphate. The advantages of the Entner-Doudoroff pathway over glycolysis are the lower NAD^+ consumption, as only one NAD^+-dependent reaction takes place, and the higher ATP production, as two ATP molecules are formed. The Entner-Doudoroff pathway is preferred by bacteria that live in an acidic environment. In this environment, glycolysis is inhibited, as the enzymes of glycolysis are inactivated by protons. The Entner-Doudoroff pathway, on the other hand, is less sensitive to this pH influence (◘ Table 8.1).

8.7 Lactic Acid Fermentation

The formation of lactic acid as an alternative reaction to alcoholic fermentation for the utilization of pyruvate under anaerobic conditions is a biotechnological process that has also gained great importance in the cultural history of mankind. The end product of lactic acid fermentation by lactobacilli does not lead to the release of carbon dioxide and the formation of ethanol as a C2 building block, but the C3 building block is retained (◘ Fig. 8.15). It is important that in lactic acid, in addi-

8.7 · Lactic Acid Fermentation

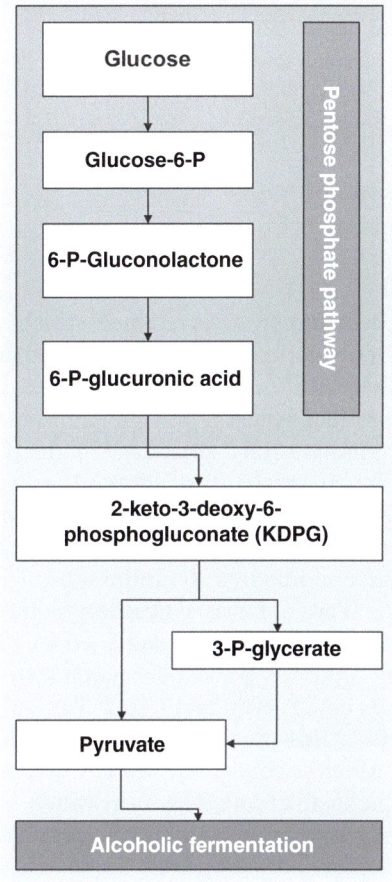

Fig. 8.14 KDPG pathway for the production of ethanol

Table 8.1 Comparison of S. cerevisiae and Z. mobilis as ethanol producers

Saccharomyces Oerevisiae	Zymomonas Mobilis
- Lower substrate requirements	- Faster and higher ethanol yield
- Stable pH value	- Unstable pH value, neutral pH must be maintained
- Limited substrate spectrum, no utilization of pentoses (xylan)	- Limited substrate spectrum, utilization of glucose
- Yield 2 ATP/glucose (M/M)	- Yield 1 ATP/glucose (M/M)
- Sites of cleavage of glucose into two ethanol molecules (blue)	- Sites of cleavage of glucose into two ethanol molecules (blue)

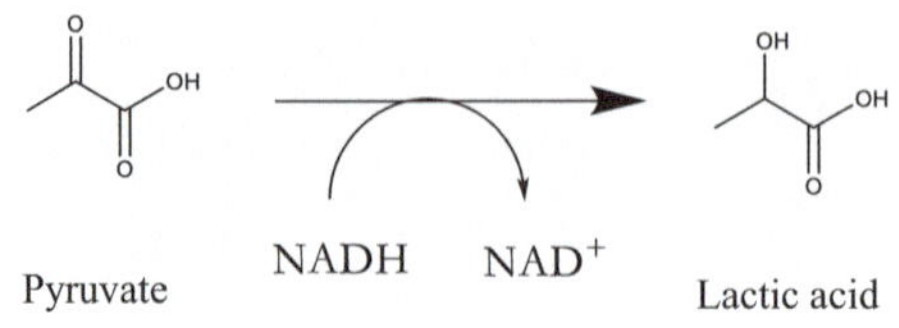

Fig. 8.15 Lactic acid fermentation

tion to the acid group, a hydroxyl group is retained, which allows an intermolecular reaction. Theoretically, a polymer of any length can be formed from this monomer, the polylactic acid or polylactate. However, of greater interest at present is the simple lactic acid as an end product, which is now indispensable in food production.

Through acidification, humans take advantage of the preservative properties of dairy products. From milk, coagulated products such as curd, yogurt, soured milk or cheese can be produced through controlled fermentation. The preservative effect is also used in meat products and plant products such as sauerkraut or silage for feed production. Today, several hundred thousand tons of lactic acid are produced. Starting from lactic acid, there are two synthetic polymerization routes to polylactides: In the first, lactic acid is directly condensed to PLA (Route 1), with one equivalent of water being produced per condensation (esterification). This is a disadvantage, as water causes chain termination and thus only low molecular masses can be produced. Therefore, this process is carried out step by step: First, oligomeric PLA is produced, which is condensed to PLA with a higher chain length after purification (azeotropic distillation). The most widely used method for producing PLA (Route 2) involves an intermediate stage, the cyclic diester lactide, which is produced by acid-catalyzed condensation from lactic acid or directly by fermentation. The lactide is then ring-opened by ionic polymerization on metal catalysts, often resulting in racemization of the lactic acid, which greatly influences the properties of the polymer. Through acidification, humans take advantage of the preservative properties of dairy products.

The properties of PLA, such as the degree of crystallinity and thus many other important physical properties, strongly depend on the optical activity of the polymers (they exist as D- or L-lactides, depending on whether they are derived from D-(+)-lactic acid or L-(−)-lactic acid). If the pure stereoisomers are used in the polymerization, crystalline polymers are obtained. In the polymer synthesis with racemic mixtures, only amorphous products are formed. Other property-determining parameters are the molecular mass and possibly the proportion of copolymers, as PLA is often used as a "blend" with other plastics (such as polyethylene and polypropylene). PLA, like PET, is biodegradable, but the conditions must be right. Under industrial composting conditions, both plastics are well degradable and biocompatible, but in nature, PLA is not degraded in most cases. Moreover, it is difficult to recycle PLA, so the often-praised sustainability of PLA is critically viewed due to these facts.

8.7 · Lactic Acid Fermentation

❓ Self-check Questions

1. Outline the basic steps of the citric acid cycle, including their cofactors.
2. Explain the purpose of the citric acid cycle in the energy balance of aerobic organisms.
3. In the industrial production of citric acid, *Aspergillus* is used. Name the growth phase in which the most citric acid is produced, and justify this.
4. Explain the preparatory phase and the pay-off phase in glycolysis.
5. Explain the naming of the pentose phosphate pathway. Into which two parts is it divided, and with which other metabolic pathway does it overlap?
6. Explain the physiological significance of coenzyme A.
7. Which metabolic pathways of primary metabolism does PEP participate in?
8. Explain the general principle of electron transport chains.
9. Provide the balance of oxidative phosphorylation.
10. Name other coenzymes in biochemical metabolism besides NADH/NADPH. In which foods can you find the mentioned coenzymes?
11. Assess the statement that vitamins are not coenzymes. Name two hydrophilic and two lipophilic vitamins.
12. Explain why up to 3 ATP are produced from one rotation of ATP synthase.
13. Give examples from biology where molecular motors are found to gain energy or enable movement.

Amino Acid Metabolism

Contents

9.1 Amino Acids as Building Blocks for Proteins – 78

9.2 Chemistry of Amino Acids – 79

9.3 Classification of Amino Acids – 81

9.4 Technical and Economic Importance – 82

9.5 Biosynthesis of Aliphatic Amino Acids – 82
9.5.1 Glutamate – 84
9.5.2 Lysine – 86
9.5.3 Threonine – 87
9.5.4 Methionine and Cysteine – 89

9.6 Biosynthesis of Aromatic Amino Acids – 93
9.6.1 Biochemistry of the Shikimate Pathway – 93
9.6.2 Phenylalanine and Tyrosine – 95

References – 97

© The Author(s), under exclusive license to Springer Fachmedien Wiesbaden GmbH, part of Springer Nature 2025
O. Kayser and N. J. H. Averesch, *Technical Biochemistry*, https://doi.org/10.1007/978-3-658-47121-7_9

> **Learning Objectives**
> - Essential Amino Acids
> - Chemistry of Amino Acids
> - Proteinogenic and Non-proteinogenic Amino Acids
> - Significance in Medicine and for Animal Feed
> - Biosynthesis of Lysine, Methionine, Threonine, Phenylalanine, and Tyrosine
> - Shikimate Pathway
> - Technical Processes with *C. glutamicum*

9.1 Amino Acids as Building Blocks for Proteins

Amino acids are the monomeric building blocks of proteins, also chemically called (poly-)peptides. Along with fats and carbohydrates, they are the third important biological group of building blocks in cell chemistry. As the name suggests, amino acids are chemically characterized by the presence of an acid function (-COOH) and an amino group (-NH$_2$), which are located in the alpha (α)-position in physiologically relevant amino acids. However, this is not necessarily the case in nature, as for example in paclitaxel (◘ Fig. 13.13) a β-amino acid also occurs. Both functional groups can be linked to form a peptide bond, resulting in oligomeric to polymeric chains that ultimately form proteins of almost unlimited chemical diversity, many conformations, and biological effects. The linkage of two amino acids to form an amide is shown in ◘ Fig. 9.1. The polymer formations to proteins, which are very important function carriers in the cell, are well known. It should be remembered that in technology and partly in our everyday products, polymeric amino acids form strong bonds and are used as polyamide and polyaramid fibers such as Nylon® or Perlon® (◘ Fig. 9.2) as plastics. In addition to these non-natural and synthetic fibers, highly elastic fibers also play an important role in our living environment. Interesting examples are silk or spider webs, which with their high elasticity and tear re-

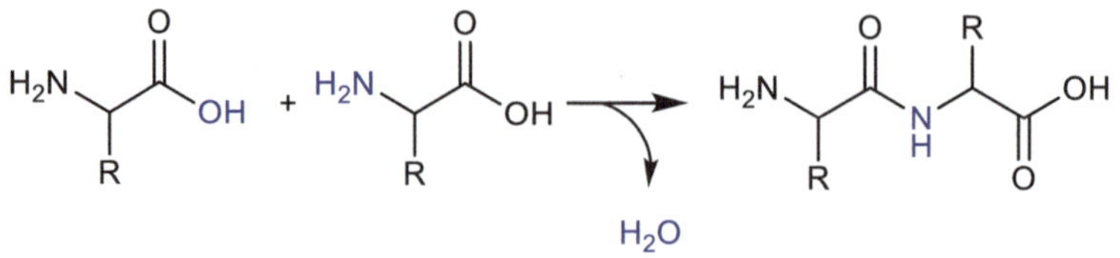

◘ **Fig. 9.1** Formation of Peptide Bonds

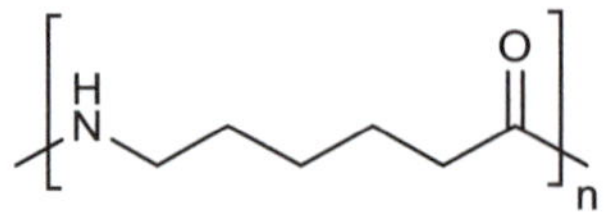

◘ **Fig. 9.2** Repeating Unit of the Polymer Nylon®

sistance underline the importance of polymeric amides. Amino acids are the monomeric building blocks of proteins. They are, along with fats and carbohydrates, the third important biological group of building blocks in cell chemistry. However, this is not necessarily the case in nature, as for example in paclitaxel a β-amino acid also occurs. Both functional groups can be linked to form a peptide bond, resulting in oligomeric to polymeric chains that ultimately form proteins of almost unlimited chemical diversity, many conformations, and biological effects. The linkage of two amino acids to form amides is shown in ◘ Fig. 9.1.

However, this chapter does not focus on the polymeric amino acids, but on their monomers, which are of particular importance as amino acids in medicine and pharmacy as well as in food and animal feed. Before biotechnology played a major role in the presentation of amino acids, most amino acids, which were very difficult to synthesize synthetically, were obtained by breaking down animal or plant waste. Today, these degradation pathways no longer play a significant role, as organic synthesis offers economic production routes for the vast majority of proteinogenic amino acids. What remains are a few amino acids such as glutamate, lysine, threonine, and phenylalanine/tyrosine, for which biotechnological production is still interesting. This chapter of the book aims to consider the industrially relevant biotechnological processes from their biochemical side, and it does not attempt to explain the possible biosynthesis pathways for all other amino acids. For those interested, the specialist literature is referred to at this point.

9.2 Chemistry of Amino Acids

Firstly, the division into proteinogenic and non-proteinogenic amino acids, of which there are about 250, should be mentioned. Amino acids that are building blocks of ribosomal protein biosynthesis are referred to as proteinogenic. The 20 proteinogenic "standard" amino acids, which are mediated in protein biosynthesis by the tRNA, are also shown with their structural formulas in ◘ Fig. 9.3. In the figure, the amino acids are again arranged according to their overlapping properties in the triplet code. A typical example of a non-proteinogenic amino acid is ornithine, which we will learn about in the chapter on the biosynthesis of tropane alkaloids. Another, rather simple division of amino acids is into essential and non-essential amino acids. The former are needed by the animal organism, which cannot synthesize them itself, but must take them in with food. In humans, these include, among others, the aromatic amino acids phenylalanine, tyrosine, and tryptophan, which are biosynthesized via the shikimate pathway.

Among the building blocks of proteins and many natural substances, amino acids occupy a prominent position. Formally, the molecules of this group are acids that have an amino group in L-configuration in C2 or α-position to the carboxyl group. These two formal chemical properties somewhat obscure the important role of L-amino acids in metabolism. In the context of this chapter, their importance as bio-building blocks for proteins will be less highlighted, but rather their function in the biosynthesis of natural substances, which is diverse (◘ Table 9.1). The multifaceted biosynthesis can be traced in standard textbooks. In this chapter, we focus

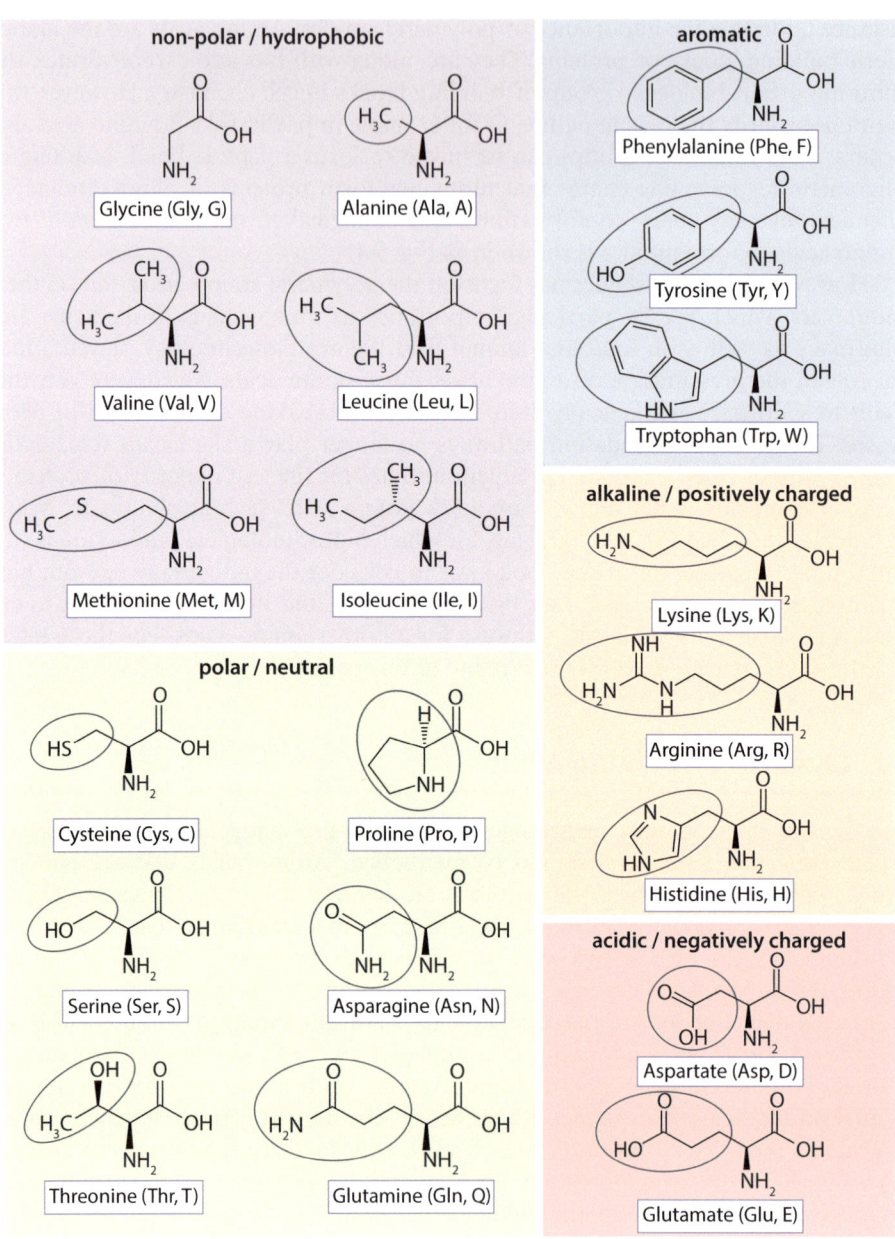

Fig. 9.3 The 20 proteinogenic standard amino acids and their classification according to structural chemical properties

on the technical importance and the biotechnical representation of the amino acids L-glutamate, L-lysine, L-threonine, L-cysteine, and L-phenylalanine.

A chemically oriented classification is based on the molecular structure and properties such as charge, polarity, and aromatic or aliphatic nature. Almost all amino acids are chiral (exception e.g. glycine). In nature, the L-form predominates,

9.3 · Classification of Amino Acids

Table 9.1 Important amino acids and the natural substances derived from them

Amino Acid (NP: non-proteinogenic)	Natural Substance
Arginine	Saxitoxin
Histidine	Pilocarpine
Lysine	Swainsonine, Castanospermine
Phenylalanine	Galantamine, Ephedra Alkaloids (Ephedrine)
Tyrosine	Dopamine, L-DOPA, Adrenaline, Mescaline, Morphine, Codeine, Tubocurarine, Berberine, Chelidonine, Colchicine
Tryptophan	Lysergic Acid, Psilocybin, Melatonin, Catharanthine, Yohimbine, Reserpine, Vinca Alkaloids (Vindoline, Vinblastine, Vincristine), Strychnine, Cinchona Alkaloids (Quinine, Quinidine), Camptothecin,
Ornithine (NP)	Cocaine, Scopolamine, Atropine, Epibatidine
Several different amino acids	Cyclopeptides, Cyclotides, linear oligopeptides Peptide hormones Peptide antibiotics (Daptomycin) Vancomycin Bleomycin Penicillins/Cephalosporins

so this book assumes that they are L-isomers. Of course, D-amino acids are also known, which often occur in microorganisms. As a solid and dissolved in water at pH 7, the amino group is protonated and the carboxyl group is deprotonated, the amino acid is a zwitterion (e.g. glycine as the simplest amino acid). Since the carboxyl group can react as a base, amino acids in basic solution exist as anions; due to the amino group, which can react as an acid, they exist in acidic solution as cations. In amino acids with several amino groups, the carbon atom whose amino group is closest to the carboxyl carbon atom determines whether it is an α-, β-, γ- etc. amino acid. Proteinogenic amino acids are always α-amino acids.

9.3 Classification of Amino Acids

Describing the biosynthetic pathways of all 20 proteinogenic amino acids or only the essential amino acids would far exceed the scope of this chapter and surpass the goal of providing engineers with necessary biochemical basic and detailed knowledge. Instead, some biosynthetic pathways that are useful for the engineering profession because they are implemented on an industrial scale in biotechnology or are significant in biotechnology will be discussed. The biosynthesis of amino acids can be divided into five groups based on the starting compound and the chemical similarities of the end products:

− The **Glutamate family:** Amino acids of this group such as glutamine, arginine, and proline are biosynthesized from 2-oxoglutarate (α-ketoglutarate).

- The **Aspartate family:** Amino acids of this group such as asparagine, lysine, methionine, threonine, cysteine, and isoleucine are biosynthesized from oxaloacetate.
- The **Alanine-Valine-Leucine group:** Amino acids of this group such as alanine, valine, and leucine are biosynthesized from pyruvate.
- The **Serine-Glycine group:** Amino acids of this group such as serine, glycine, and cysteine are biosynthesized from 3-phosphoglycerate.
- The **aromatic amino acids:** They are biosynthesized from phosphoenolpyruvate and erythrose-4-phosphate.

Often there are different biosynthetic pathways for the same amino acid. This depends on the respective organism and its specific metabolism and structure, which prefers or excludes one way or another.

9.4 Technical and Economic Importance

In addition to their incorporation into peptides and proteins, pure amino acids also have great importance in technical biochemistry, medicine, the food industry, or as feed additives (Mindt et al. 2020). The biotechnological production of amino acids has advantages, as unlike chemical synthesis, only L-enantiomers are produced. This is done from sugar-containing raw materials such as molasses, which allows for an environmentally friendly process. In ◘ Table 9.1, the important amino acids that are at the beginning as starting substrates in the biosynthesis of secondary natural substances are listed. The mentioned natural substances play an outstanding role in medicine.

Another look is at the economy and the economic importance of a few natural substances. In absolute quantities, L-lysine and L-glutamine are in the first place (◘ Table 9.2). When larger quantities are traded, these amino acids are mostly found in the feed industry (L-lysine, L-cysteine, L-methionine, L-threonine). An exception is L-glutamate in the food industry. All other amino acids have their application area in medicine or are used as synthesis precursors or as additives in dietary supplements (Ault 2004).

9.5 Biosynthesis of Aliphatic Amino Acids

An overview of the various biosynthesis pathways is shown in ◘ Fig. 9.4. For those who want to read detailed information about the individual synthesis pathways and steps, the "Biochemical Pathways" and the KEGG online database with the biochemical pathways are highly recommended. As can be seen in ◘ Fig. 9.4, there is not just one biosynthesis pathway for a particular amino acid, but alternative pathways depending on the organism. An example is the biosynthesis of lysine, which only occurs in plants and bacteria and is essential for vertebrates. In these organisms, lysine, like cysteine, methionine, and threonine, is synthesized from aspartate derived from oxaloacetate and 2-oxoglutarate (= α-ketoglutarate) from the citric

9.5 · Biosynthesis of Aliphatic Amino Acids

Table 9.2 Global production volumes per year and prices of amino acids

Amino Acid	Production Volume (2023)	Selling Price * (2023)	Applications
L-Lysine	1.5 million t	2,500 USD/Ton	Animal Feed Dietary Supplement
L-Glutamine	1.2 million t	1,000 USD/Ton	Flavoring
L-Arginine	800,000 t	1,500 USD/Ton	Pharmaceuticals Dietary Supplement
D-, L-Methionine	600,000 t	2,000 USD/Ton	Animal Feed Dietary Supplement
L-Valine	500,000 t	1,200 USD/Ton	Pharmaceuticals Dietary Supplement
L-Phenylalanine	400,000 t	1,500 USD/Ton	Sweetener Synthesis Material Dietary Supplement
L-Leucine	300,000 t	1,000 USD/Ton	Animal Feed Dietary Supplement
L-Isoleucine	200,000 t	1,200 USD/Ton	Pharmaceuticals Dietary Supplement
L-Tyrosine	100,000 t	1,500 USD/Ton	Dietary Supplement
Glycine	16,000 t	1,000 USD/Ton	Sweetener Synthesis Material
L-Aspartate	14,000 t	1,000 USD/Ton	Sweetener Synthesis Material

*International Amino Acid Institute. (2023). Amino Acid Market Report 2023–2027

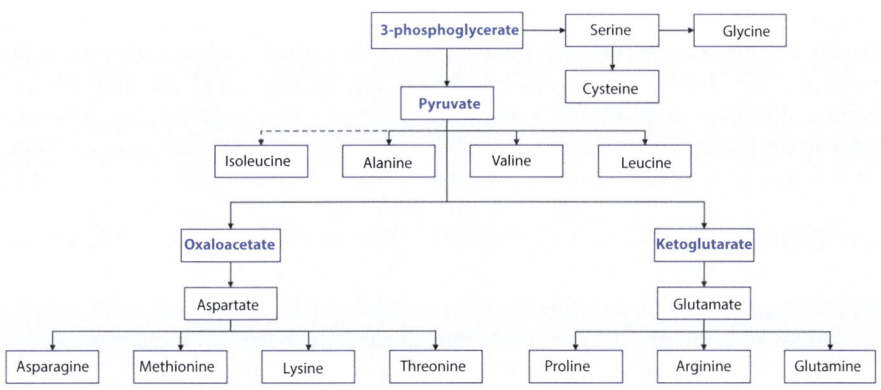

Fig. 9.4 Overview of the biosynthetic precursor metabolites of aliphatic amino acids

acid cycle, while in fungi the first step is the linkage of 2-oxoglutarate with acetyl-CoA.

Reduced sulfur is required for the biosynthesis of cysteine and methionine (◘ Figs. 9.8 and 9.9). Ferredoxin as an electron donor reduces the sulfur (taken up in the form of sulfate ions), so that it is incorporated as a thiol(SH)-group into cysteine. Valine, leucine, and isoleucine are also synthesized via the glycine, serine, and threonine metabolism, which are built up from threonine (◘ Fig. 9.7). Valine and leucine are formed independently of each other by chain extension of pyruvate, rearrangement, and subsequent transamination. Despite their structural similarity, the synthesis pathways of leucine and isoleucine are fundamentally different. Serine is formed in two steps from 3-phosphoglycerate. The first step is an oxidation, the second again a transamination.

The precursors of glycine and alanine originate from glycolysis. However, glycine can also be formed as a degradation product by decomposition of serine. This serves not only the production of glycine but also the conversion of tetrahydrofolate (coenzyme F) to N5-N10-methylenetetrahydrofolate. This is also necessary for the synthesis of thymine nucleotides, which are components of DNA. Conversely, glycine can serve the synthesis of serine by taking up a methyl group from N5-N10-methylenetetrahydrofolate, which is available for protein synthesis as the basic substance of choline or as pyruvate. Glycine is also often required for the synthesis of purines, which include the DNA bases adenine and guanine.

Glutamine and the closely related amino acid glutamate are also formed from 2-oxoglutarate. Subsequently, proline and arginine are derived from glutamine and glutamate. Glutamate is formed by the reaction of 2-oxoglutarate with an ammonium ion with the oxidation of an NADPH to $NADP^+$. In the subsequent reaction, ammonia is again linked as a protonated salt (NH_4^+) with the carboxy group of glutamate, using an ATP, to form glutamine. The binding of ammonia, which is produced by the degradation of peptides and proteins, serves detoxification, as it is a strong metabolic poison for many mammals. It damages the brain (CNS) at chronically elevated blood concentrations. Therefore, special attention should be paid to the enzyme group of transaminases, also known as liver enzymes: These are capable of transferring an amino group from glutamine to a 2-oxo or keto acid (like α-ketoglutarate), resulting in two moles of glutamate. This reaction is called transamination. Transaminases such as ASAT, ALAT, and γ-GT are diagnostically determined in liver diseases and poisonings. They also have significance in biotechnology in the production of amino acids.

9.5.1 Glutamate

This amino acid is familiar to us from everyday life, as glutamate is widely used as a seasoning and flavoring agent in kitchens around the world. It is indispensable as th umami flavor in Asian food culture. But it also determines the taste in European foods such as tomatoes and Parmesan and is associated with the "Maggi" flavor. Although it is an acid, it is converted into its sodium salt, which is why food chemists refer to sodium glutamate, 80% of which is biotechnologically produced in China, Japan, and Indonesia.

9.5 · Biosynthesis of Aliphatic Amino Acids

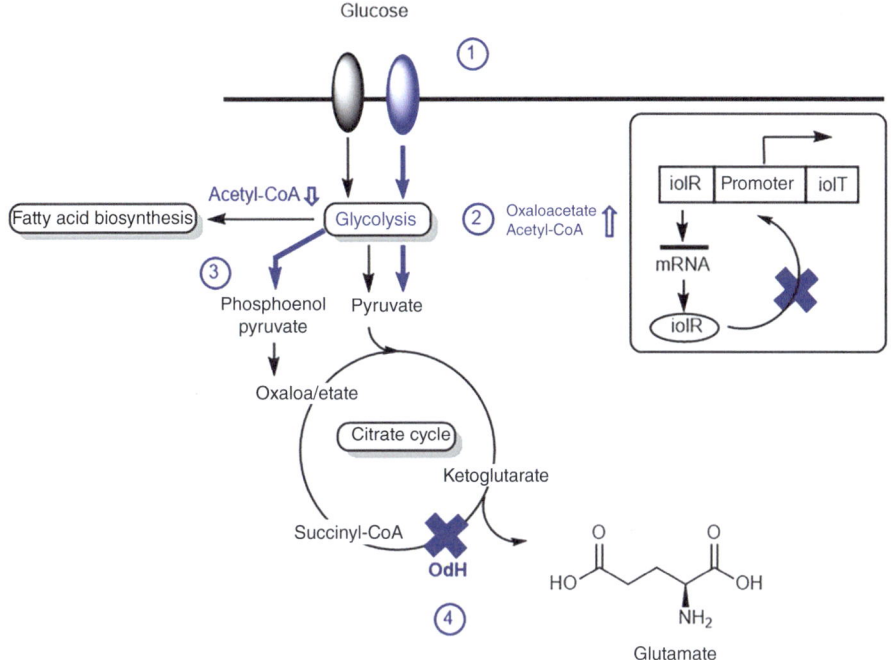

◘ **Fig. 9.5** Glutamate biosynthesis in *C. glutamicum*

Biosynthesis The synthesis in the cell is remarkably simple, as glutamate is directly available via α-ketoglutarate from the citric acid cycle. A transamination and reduction with NADPH yields glutamate (◘ Fig. 9.5). For subsequent biotechnological production, downstream metabolic pathways such as glycolysis and the pentose phosphate pathway must be considered or metabolically optimized (Sheng et al. 2021).

Biotechnology The biotechnical extraction has been well known since the beginning of the 20th century and was particularly advanced by Japanese scientists. Starting from the citric acid cycle, ketoglutarate is biosynthesized, which is converted to glutamate in a combined transaminase reaction with the enzymes glutamate dehydrogenase, glutamine synthetase, and glutamine-α-ketoglutarate aminotransferase. However, this biosynthesis does not provide enough glutamate in the industrial strains of the producer *Corynebacterium glutamicum*. Genetic modifications have improved the uptake of glucose into the bacterium ①, optimized the reduction of the forwarding of acetyl-CoA to fatty acid biosynthesis ②, the parallel increase in the provision of phosphoenolpyruvate for higher amounts of oxaloacetate ③, and the partial blockade of ketoglutarate dehydrogenase (ODH) ④. An increase in biosynthesis is conceivable not only through genetic changes but also through the design of fermentation media. The amount of the cofactor biotin leads to the excretion of glutamate at concentrations below 2 μg/L, as the biosynthesis of lipids is restricted. Concentrations above 5 μg/L do not lead to glutamate excretion, as sufficient biotin is now available as a cofactor of acetyl-CoA carboxylase. However, in the biochemistry of the technical production of glutamate, the inhibition of ketoglutarate dehydrogenase (ODH) ④ is significant. The protein OdHI is activated or controlled by phosphorylation via a kinase. The degree of phosphorylation depends on the state of the cell membrane. OdHI is associated with the cell membrane, and unphosphorylated OdHI is more prevalent in the case of cell wall damage. The unphosphorylated protein can interact with the enzyme OdH and lead to its inactivation. The result is that α-ketoglutarate is no longer sufficiently converted to succinyl-CoA, accumulates, and is available for the transaminase reaction to glutamate.

Corynebacterium glutamicum has mechanosensitive channels involved in regulating cell growth and division. They function like a kind of pressure relief valve and regulate cell homeostasis when the inflow of water and the parallel outflow of calcium ions need to be controlled. This causes water to flow into the cell, which could cause the cell to swell and burst. The mechanosensitive YggB channels in *C. glutamicum* prevent this by opening and allowing the outflow of ions from the cell. This balances the osmotic pressure, and the cell remains stable. Genetic modifications can alter the YggB channels to allow a higher permeability for glutamate.

9.5.2 Lysine

L-Lysine is used as a feed additive and to enhance lysine-deficient plant feed proteins. L-Lysine plays an important role especially in pig and poultry farming. In the feed industry, lysine is used as a supplement to protein-rich, but L-lysine-deficient feeds, to ensure sufficient supply of L-lysine to the animals. This is particularly important for animals that are fed with plant protein sources, as these often have a lower lysine content than animal protein sources. The main applications of lysine in the feed industry are:
— Improvement of muscle development in animals bred for meat production
— Improvement of bone health in animals bred for milk or egg production
— Improvement of fertility in animals

Biosynthesis The biosynthesis of lysine occurs via two main pathways: the diaminopimelic acid pathway (DAP pathway) and the α-aminoadipic acid pathway (AA pathway). Both pathways will be briefly introduced, even though in biotechnology L-lysine is produced on a large scale via the DAP pathway, as this is more efficient.

Diaminopimelic Acid Pathway The DAP pathway is the predominant pathway of lysine synthesis in bacteria and higher plants. It starts with the amino acid glutamine, which is converted by glutamine synthetase with ATP to glutamate-5-phosphate. Glutamate-5-phosphate is then reduced by aspartate semialdehyde dehydrogenase to aspartate semialdehyde. Aspartate semialdehyde is then converted by aspartate semialdehyde aminotransferase with glutamate to diaminopimelic acid (DAP). DAP is then decarboxylated by DAP decarboxylase to lysine.

α-Aminoadipic Acid Pathway The AA pathway is the predominant pathway of lysine synthesis in many fungi. It starts with α-aminoadipic acid, which is synthesized from pyruvate and aspartic acid. The α-aminoadipic acid is then converted by α-aminoadipic acid decarboxylase to acyl-D-glutamate. Acyl-D-glutamate is then converted by acyl-D-glutamate aminotransferase with glutamine to acyl-L-glutamate. Acyl-L-glutamate is then dehydrogenated by acyl-L-glutamate dehydrogenase to lysine.

Biotechnology For technical production, *Corynebacterium glutamicum* or *E. coli* is used. Here, biosynthesis takes place via the DAP pathway. *C. glutamicum* is a gram-positive bacterium with a high lysine synthesis rate, which has been optimized by genetic manipulation to increase lysine production. The enzymes of the DAP pathway are encoded on a chromosome in *Corynebacterium*. The regulation of lysine synthesis occurs via the allosteric inhibition of aspartate kinase, glutamine synthetase, and aspartate semialdehyde dehydrogenase (◘ Fig. 9.6), which are inhibited by lysine. Lysine synthesis is only activated when the lysine requirement of the bacterium is high. In addition to the biosynthesis of glutamate, which is undesirable but important as a nitrogen donor for N-succinyl-L-2-amino-6-heptanedione (not shown), threonine also plays a role, as this amino acid can also inhibit aspartate kinase. L-Lysine cannot passively pass through the cell wall, therefore an active exporter (LysE) is necessary (Buchholz et al. 2013).

9.5.3 Threonine

With the amino acid L-threonine, we come to the last interesting amino acid that can be biotechnologically produced in Corynebacteria. L-threonine is also important in animal feed mixtures. Many feeds such as grains and corn do not contain enough threonine to meet the needs of the animals. The addition of threonine to animal feeds can help improve the nutrient supply of the animals and boost their health and performance. The importance of threonine as a feed additive is particularly highlighted in the following areas, as threonine is

- ... an important amino acid for the formation of muscle protein. A deficiency of threonine can lead to a reduction in muscle mass and meat quality. The addition of threonine to livestock feeds can help increase meat production and improve meat quality.
- ... also important for the formation of milk protein. A deficiency of threonine can lead to a decrease in milk yield and milk quality. The addition of threonine to feeds for dairy cows can contribute to increasing milk production and improving milk quality.
- ... is important for the growth and development of poultry. A deficiency of threonine can lead to a decrease in weight gain and egg production. The addition of threonine to poultry feed can improve the growth and development of poultry.

Biosynthesis The biosynthesis of threonine in *Corynebacterium* occurs via the aspartic semialdehyde pathway (ASL pathway, ◘ Fig. 9.7). This pathway is predominant in bacteria and higher plants. The ASL pathway begins in *Corynebacterium* with the amino acid aspartic acid, which is reduced to aspartic semialdehyde by aspartic semialdehyde dehydrogenase. This short beginning is identical to the biosynthesis of glutamine or L-lysine. Aspartic semialdehyde is then converted to homoserine by homoserine dehydrogenase (thrA). Homoserine kinase (thrB) phosphoryl-

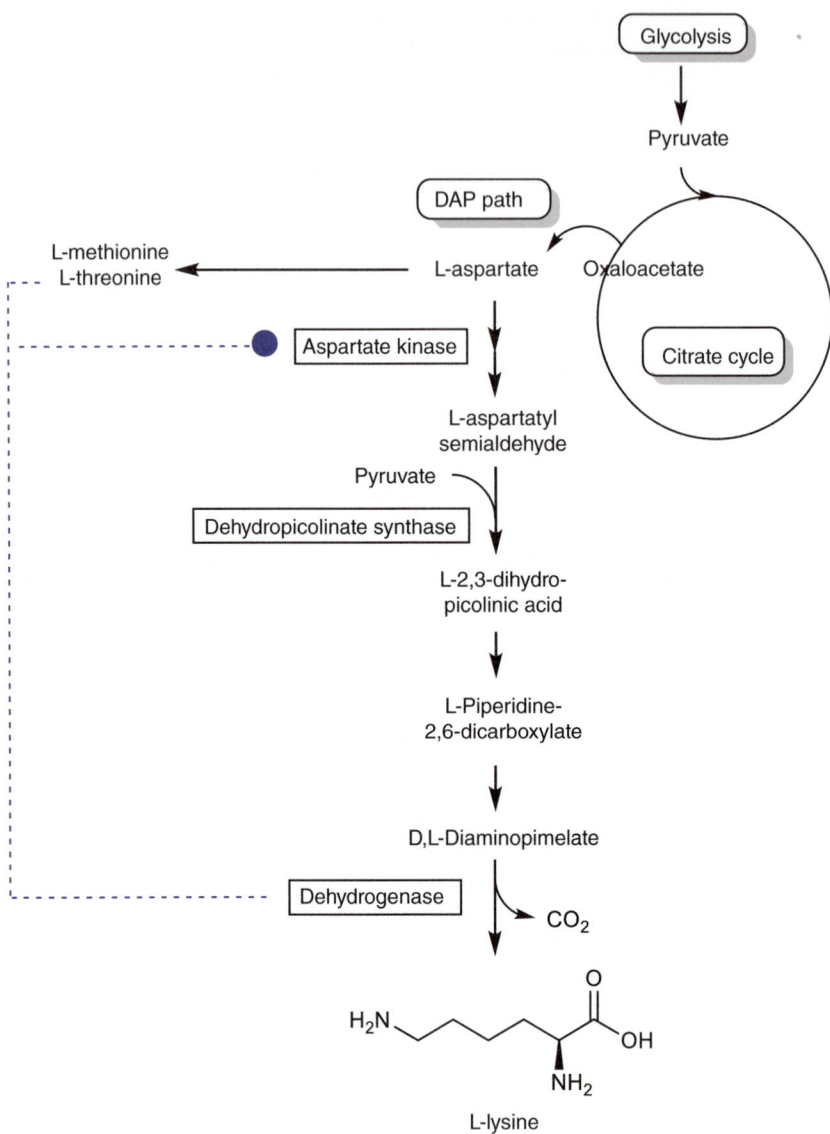

Fig. 9.6 Lysine biosynthesis in *C. glutamicum*

ates to homoserine phosphate, which is converted to L-threonine. The regulation of threonine synthesis is achieved by the downstream inhibition of the first two enzymes (thrA and thrB) of the pathway, which can be directly inhibited by threonine. Physiologically, threonine synthesis is only activated when the bacterium has a high demand for threonine and is therefore the starting point for industrial optimization of yields. If the biosynthesis of L-threonine is continued, the enzymatic conversion in further steps results in the amino acid L-isoleucine.

9.5 · Biosynthesis of Aliphatic Amino Acids

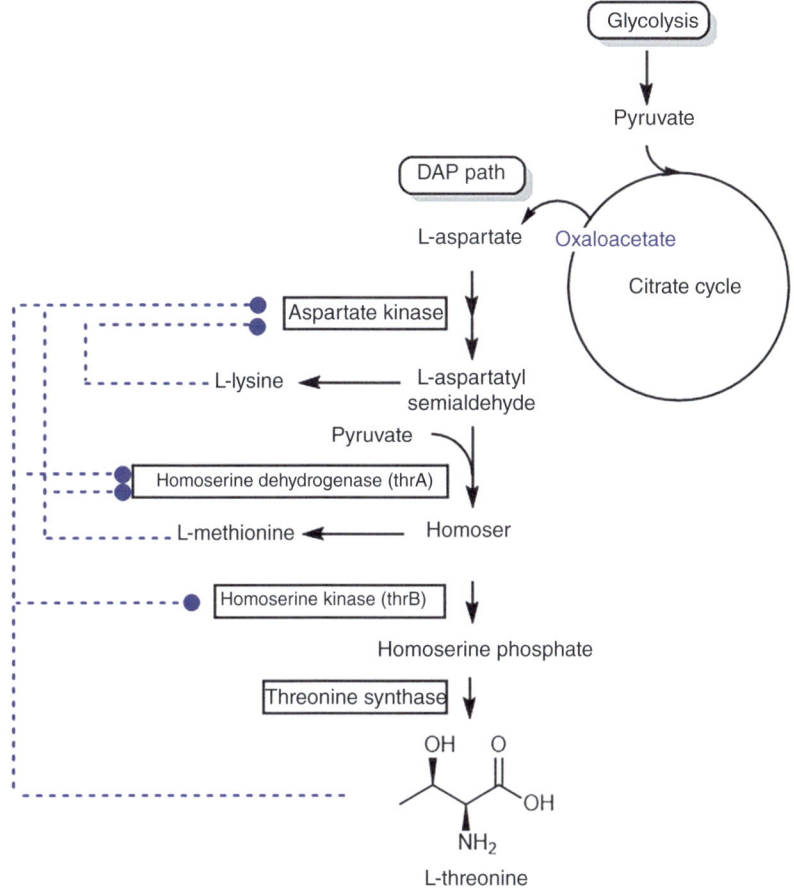

■ **Fig. 9.7** Threonine biosynthesis in *C. glutamicum*

9.5.4 Methionine and Cysteine

Both amino acids contain sulfur and are produced by chemical processes. Their importance also arises as an important component of animal feeds, especially for chickens. The biotechnological production is known as a technical prototype, but there is no established process of industrial significance. Let's first look at the attempt of biotechnological synthesis of L-cysteine.

L-Cysteine (■ Fig. 9.8) is widely used in the food, agro, and pharmaceutical industries. Due to the toxicity of L-cysteine and the complex regulation of its synthesis pathway, efficient microbial production of L-cysteine on an industrial scale has not yet been achieved. Most studies have focused on the production of L-cysteine from glucose by recombinant *E. coli* and *C. glutamicum*. However, *C. glutamicum* grows slowly, leading to a long production cycle. Compared to *C. glutamicum*, *E. coli* has a higher growth rate, and the genetic engineering methods for its modification are more advanced, making a technical process with this production organism

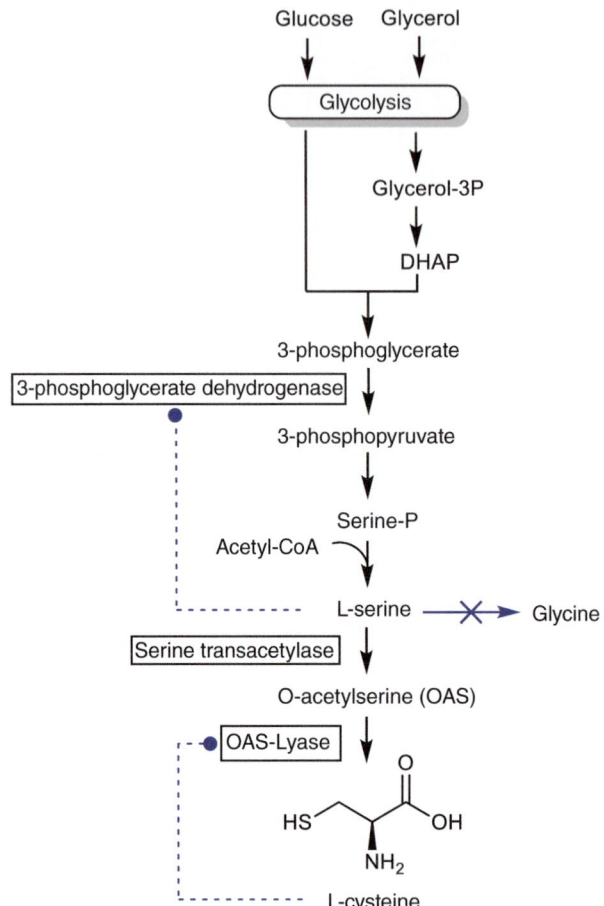

☐ **Fig. 9.8** Cysteine biosynthesis in *C. glutamicum*

more likely in the future. The precursor of L-cysteine in *E. coli* is L-serine. The biosynthesis of L-cysteine from L-serine in *E. coli* occurs via a two-step pathway catalyzed by L-serine acetyltransferase and L-cysteine synthase. The first reaction catalyzed by the acetyltransferase is a limiting step in the biosynthesis of L-cysteine in *E. coli*, which depends on the amount of L-serine formed. Due to its toxicity, L-cysteine undergoes degradation catalyzed by several L-cysteine desulfhydrases (not shown in the figure). Previous studies have shown that L-serine is an essential precursor for the biosynthesis of L-cysteine. Increasing L-serine synthesis is a necessary metabolic engineering strategy for L-cysteine accumulation (Zhang et al. 2022). So far, however, biosynthesis has not been transferred to an industrially relevant process (Cai et al. 2023a).

The global market for **L-methionine** is estimated at around 5 billion US dollars (as of 2019). The main consumer of L-methionine is the animal feed industry, as this amino acid is essential for the growth and development of animals. The worldwide industrial demand is so high because people's demand for animal protein is constantly increasing. Both isomers (D,L-methionine) can be metabolized for growth. The reason is that a conversion by diamino acid oxidase and a specific L-transaminase leads to L-methionine.

Organic synthesis only provides a DL-racemate, and the substrates used, such as acrolein, methanethiol, and hydrogen cyanide, are problematic for the environment and workers in the factories. Due to the rapid development of metabolic engineering, the biosynthesis in *E. coli* could be developed as an alternative to chemical synthesis (Cai et al. 2023b).

Chemistry L-Methionine is a nonpolar amino acid that occurs in nature in two mirror-image (chiral) forms, the D- and L-form. L-Methionine is the naturally occurring stereoform and part of most proteins. D-Methionine occurs in nature only in very small amounts. L-Methionine is found in a variety of foods, especially in animal protein, and is considered a good nutritional basis. In plant foods, L-methionine is usually present in smaller amounts. This is because plants cannot synthesize methionine themselves, but must absorb it from the environment (microorganisms). The biological functions of L-methionine are diverse. It is an important component of proteins, serves as a methyl group donor in various biosynthesis steps, and plays a role in detoxification.

Biosynthesis The biosynthesis of *C. glutamicum* shown in ◘ Figs. 9.8 and 9.9 shows that L-methionine, like threonine and L-lysine, belongs to the aspartate family (◘ Fig. 9.4). The biosynthesis can be divided into three steps. In the first step, oxaloacetate is provided as a C4 building block via glycolysis and the citric acid cycle. In the second step (not shown here), sulfur is assimilated and incorporated via homoserine. In the last and third step, the sulfur is methylated via SAM, completing the biosynthesis of L-methionine.

The goal of biosynthesis in the first step is to provide oxaloacetate for L-aspartate as an intermediate amino acid. Starting from this amino acid, homoserine is converted to homoserine-O-acetyl (OAHS) by homoserine-O-acetyltransferase with succinyl-CoA or acetyl-CoA. Acetylation is the preferred step for the formation of OAHS in *C. glutamicum*. Succinylation is crucial in *E. coli* and is used in serine biosynthesis with succinyl-O-homoserine as an alternative to sulfur assimilation. This seems not to be important, but it is for the biosynthesis of the enzymes aspartokinase (AK), aspartate semialdehyde dehydrogenase (ASD), and homoserine dehydrogenase (HD). These are crucial for the three different catalytic reactions to homoserine. But also in the producer itself, as shown in *E. coli* and *C. glutamicum*, the provision of sulfur is different. In *E. coli*, a transfer from L-cysteine via L-cystathion takes place. In *C. glutamicum*, L-serine is important for transsulfuration, in which acetylserine is formed, which is further metabolized to L-cysteine. The function of the sulfur donor is taken over in *C. glutamicum* by hydrogen sulfide (H_2S), which reacts directly with homocysteine and is involved in *E. coli* in the conversion of acetylcysteine to O-acetylhomoserine.

Biotechnology For industrial implementation, *E. coli* and *C. glutamicum* are suitable, as methods and techniques for genetic modification are available for both producers. However, challenges arise, as in addition to the basic metabolic pathways of

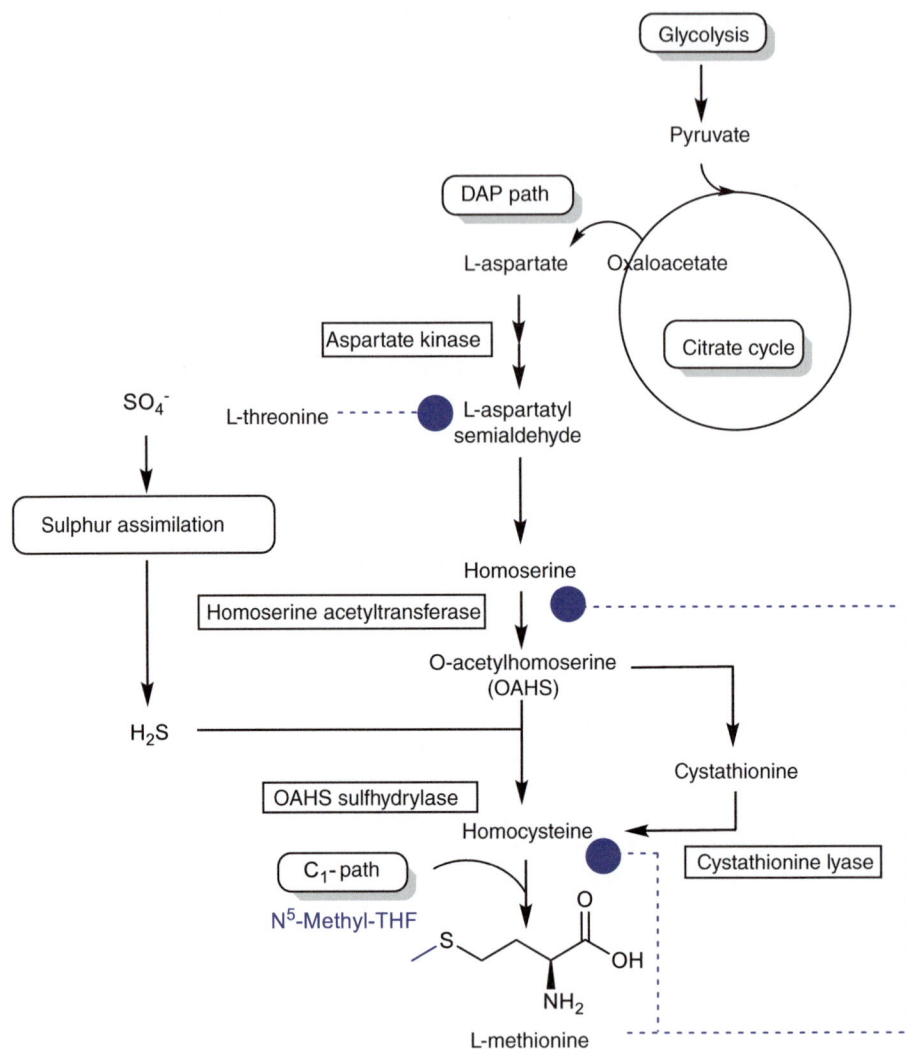

☐ **Fig. 9.9** Methionine biosynthesis in *C. glutamicum*

glycolysis and the citric acid cycle, sulfur assimilation from sulfate and the provision of C1-methyl must also be considered. Biosynthesis is strongly controlled and limited, therefore the removal of metabolic blockages in feedback inhibition is being investigated through promoter studies and exchange. By knocking out the promoter suppression, the biosynthesis of homocysteine can be significantly increased. The inhibition of the metabolic provision of homocysteine and homoserine was crucial to increase the production of a titer from 0.05 g/L to 2.97 g/L.

9.6 Biosynthesis of Aromatic Amino Acids

A prerequisite for understanding the biosynthesis of many aromatic natural substances and the amino acids phenylalanine and tyrosine to be explained here is knowledge of the shikimate pathway. This biosynthetic pathway is established in plants and many microorganisms, but is completely absent in higher mammals such as humans. Therefore, aromatic amino acids such as phenylalanine, tyrosine, and tryptophan are essential for humans and must be ingested with food. The shikimate biosynthetic pathway is the central biochemical pathway to aromatic amino acids and their phenolic derivatives. Starting from glycolysis and the pentose phosphate pathway, the three aromatic amino acids are provided via shikimic acid and anthranilic acid, which form the basis for the great variety of alkaloids, flavonoids, and cinnamic acid derivatives through structural modifications. But also important structural substances of technical importance, such as lignin and tannins, originate from shikimate biosynthesis.

9.6.1 Biochemistry of the Shikimate Pathway

In ◘ Fig. 9.10, the shikimate pathway is shown, starting from the precursors of the pentose phosphate pathway and glycolysis to the aromatic amino acids. From erythrose-4-phosphate, a metabolite of the pentose phosphate pathway, a C7 sugar is formed by linkage with phosphoenolpyruvate, which is converted to shikimic acid (C6C1) through various steps (cyclization, rearrangement, reduction, and dehydration). In the further course of the shikimate pathway, the aromatic amino acids phenylalanine and tyrosine are formed.

The simplified biosynthesis (◘ Fig. 9.10) will be briefly explained: The decisive building block for cyclization is provided by the linkage of the C3 building block phosphoenolpyruvate with the C4 sugar erythrose-4-phosphate to the C7 sugar sedoheptulose-7-phosphate, already mentioned above. Shikimic acid is formed in several steps by the elimination of phosphate and water and the consumption of the reduction equivalent NADPH. The shikimic acid-3-phosphate formed by further phosphorylation reacts with phosphoenolpyruvate to form chorismate, a central molecule of the shikimate pathway. Chorismate, which is converted to prephenate by a Claisen rearrangement, reacts further with the elimination of CO_2 and H_2O to form phenylpyruvate and *p*-hydroxyphenylpyruvate. The resulting compounds can be converted to phenylalanine and tyrosine by transaminases. Alternatively, prephenic acid is converted to arogenate by transamination, from which phenylalanine and tyrosine are directly formed.

In addition to the C6C3 compounds discussed so far, oxidative degradation reactions (oxidative decarboxylation) are also known, leading to C6C2 or C6C1 side chains. C6C1 acids, e.g., benzoic acid from cinnamic acid, are formed by β-oxidation, analogous to the process described above for fatty acids (C2). Most C6C1 compounds are formed directly from shikimic acid. The complete removal of the side chain by oxidative decarboxylation (C6) is also known (e.g., hydroquinone from *p*-hydroxybenzoic acid).

These three aromatic amino acids are structure-determining for many alkaloids and important precursors in further biosynthetic pathways to products such as

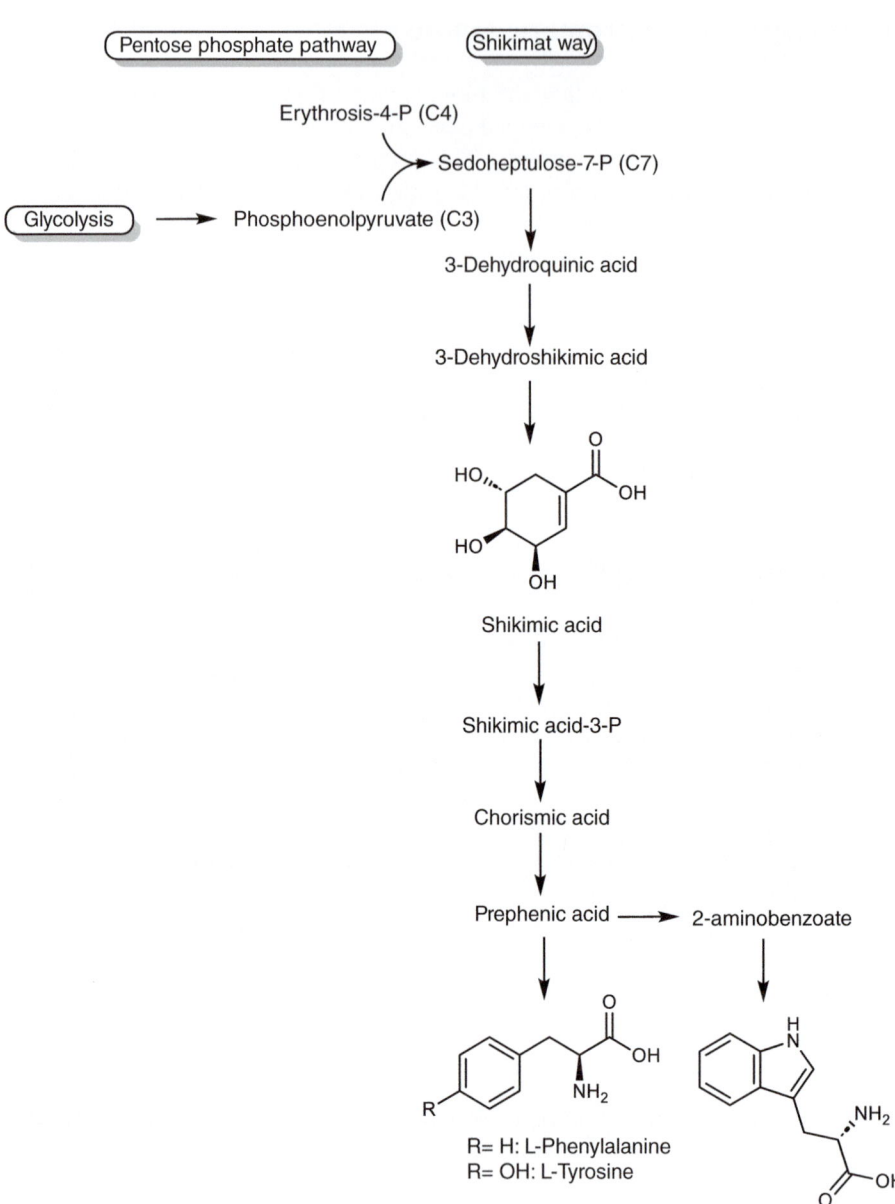

☐ **Fig. 9.10** Shikimate pathway and biosynthesis of phenylalanine, tyrosine, and tryptophan

vanillin, cinnamic acid derivatives, lignans, and lignins (see also the following chapters). This branch of the biosynthesis of aromatic amino acids also leads to neurotransmitters such as the catecholamines dopamine and adrenaline or to well-known representatives of pharmacologically active substances such as mescaline and morphine (for the biosynthesis of alkaloids from the shikimate pathway see ▶ Chap. 14).

Among other things, anthranilate, a precursor in the biosynthesis of tryptophan, is formed from chorismate and also leads to catechol (pyrocatechol). Catechol is the precursor for further natural and messenger substances, including psilocybin and serotonin, to name just two.

9.6.2 Phenylalanine and Tyrosine

The biosynthesis of phenylalanine is fundamentally preserved in biotechnological conversion. To achieve an industrially interesting titer (approx. 4 g/L phenylalanine), natural feedback inhibitors are switched off by genetic modification or metabolic engineering. In the biosynthesis pathway, the promoters (galP, glk) of the transport proteins for glucose and galactose are modified to achieve increased uptake and higher throughput. In the cell, phenylalanine biosynthesis is subject to feedback inhibition (◘ Fig. 9.11). Three important regulatory sites need to be considered and adjusted by genetic manipulation. Firstly, this is the provision of the starting substrate sedoheptulose (D-arabinoheptulonic acid-7-phosphate (DAHP)) from the pentose phosphate pathway, which is built up by the DAHP synthase. This DAHP synthase consists of the three isomers aroF, aroG, aroH, with aroG accounting for about 80% of the synthesis performance (Bang et al. 2021). Another important bifunctional enzyme at the end of biosynthesis is pheA, which has chorismate mutase and prephenate dehydratase activity. This enzyme can be altered by mutations so that the missing interaction due to lower specificity no longer allows feedback inhibition.

These two feedback-resistant enzymes aroG and pheA are of crucial importance for biotechnological production. However, since small amounts of tyrosine and tryptophan are also produced during production, these can inhibit the mentioned enzymes. For this reason, both enzymes are undesirable in phenylalanine biosynthesis. Although a large amount of glucose is desired for high throughput, an excess leads to the unwanted biosynthesis of acetic acid. Pyruvate is formed, which is broken down into acetate and carbon dioxide via other metabolic pathways. Intelligent bioprocess technology can control the process well and efficiently.

❓ Self-check Questions
1. Identify the chemical groups of an amino acid and explain their connection/linkage. Indicate the reaction.
2. Provide characteristics to characterize amino acids. Indicate at least six chemical properties.
3. Indicate how many amino acids there are. (There are several correct answers).
4. Sketch the classification of the biosynthesis pathways of the proteinogenic standard amino acids in central metabolism.
5. Describe the fermentative production of an amino acid of your choice.
6. Explain the advantages of producing an amino acid by transamination.

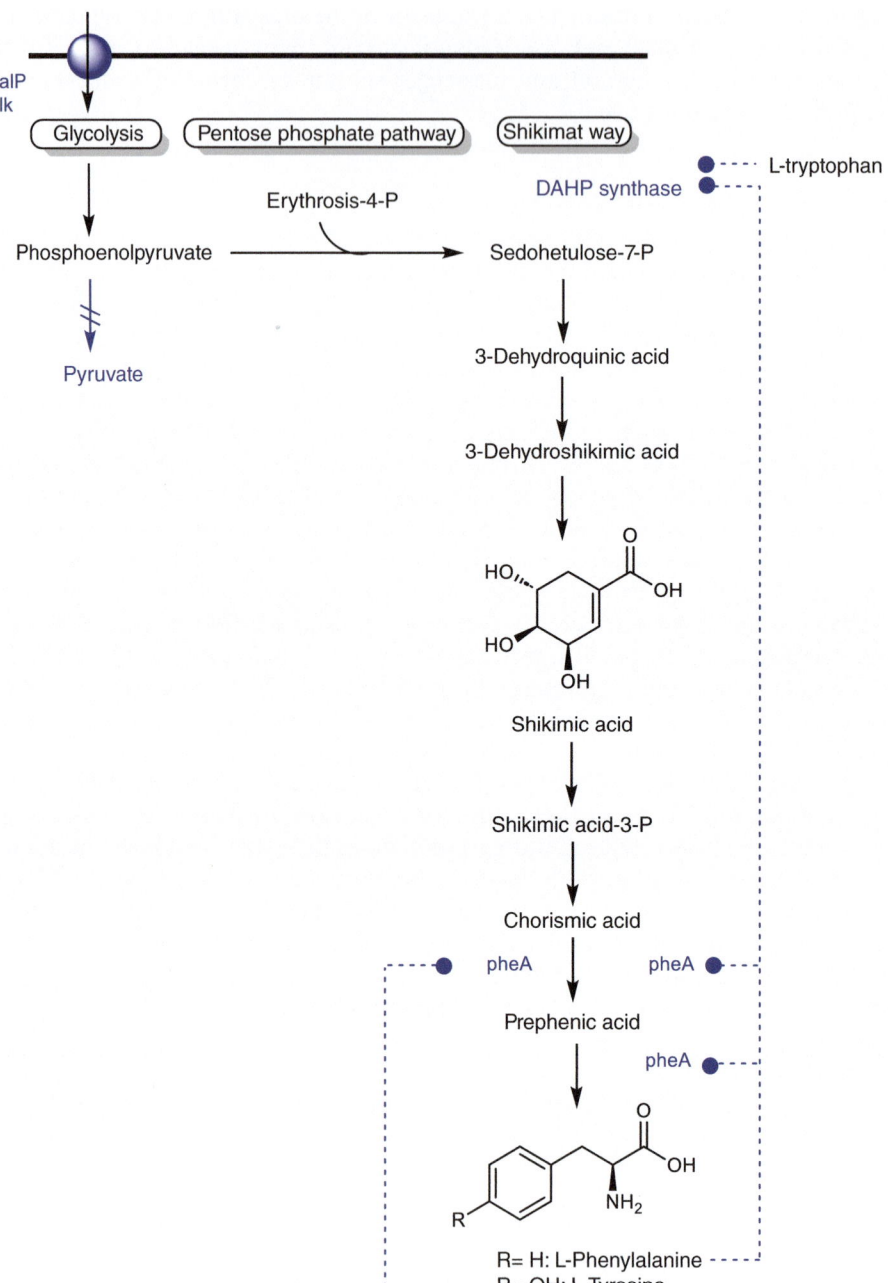

Fig. 9.11 Regulation of phenylalanine and tyrosine biosynthesis

Fatty Acid Biosynthesis and ABE Metabolism

Contents

10.1　Fats and Fatty Acid Biosynthesis – 100
10.1.1　Fatty Acid Biosynthesis – 100
10.1.2　Fat Biosynthesis – 102
10.1.3　Breakdown of Fats and β-Oxidation – 102

10.2　ABE Metabolism – 103
10.2.1　Isobutanol – 105

10.3　Technical Fats and Oils – 105
10.3.1　Biofuels – 105
10.3.2　Biodiesel – 107
10.3.3　Biokerosene – 108
10.3.4　Biogenic Lubricants – 108
10.3.5　Biobutanol – 108

References – 108

© The Author(s), under exclusive license to Springer Fachmedien Wiesbaden GmbH, part of Springer Nature 2025
O. Kayser and N. J. H. Averesch, *Technical Biochemistry*, https://doi.org/10.1007/978-3-658-47121-7_10

> **Learning Objectives**
> – Fatty acid biosynthesis
> – β-Oxidation
> – ABE metabolic pathway
> – Technical importance of
>   – Biodiesel
>   – Biofuel
>   – Isobutanol

Many natural substances consist of long carbon chains that are built from C2 units (acetyl-CoA) or C3 units (malonyl-CoA). This type of molecular linkage often reveals alternating (meta-positioned) keto groups, which can be traced back to acetate building blocks; they fall under the so-called acetate rule. Aromatic natural substances from this biosynthetic pathway are characterized by β-positioned hydroxyl groups, unless these have been eliminated in favor of a C-C linkage due to ring closure as a result of an intramolecular aldol reaction or Claisen condensation (see below). Although theoretically any number of linkages of the C2 units is conceivable, this potential is only used to a limited extent in nature. The molecular masses of important natural substances show that acetates are often built from 4 to 20 units. The most important representatives can be summarized in two groups: the fatty acids and the polyketides.

10.1 Fats and Fatty Acid Biosynthesis

Fats and fatty oils of natural origin consist of neutral esters of fatty acids and glycerin, the glycerides. Due to their natural origin, they often represent a mixture of substances, as small amounts of lipids, sterols, phospholipids, fat-soluble vitamins (vitamin E, D, K, A), colorants and flavorings can be included as impurities. Fats and fatty oils are differentiated according to their consistency at room temperature. Fats (e.g., butter, cocoa butter) are solid in various degrees of hardness, while fatty oils such as olive oil and peanut oil are liquid at room temperature. Physically, there is no difference between animal and plant fats. It is noticeable that plant fats qualitatively have more double bonds in the molecule and quantitatively more unsaturated fatty acids.

10.1.1 Fatty Acid Biosynthesis

Fatty acid biosynthesis is a biochemical reaction that occurs similarly in many organisms, catalyzed by a fatty acid synthase complex in a biotin-dependent carboxylation of acetyl-CoA to malonyl-CoA with ATP consumption (◘ Fig. 10.1). By repeatedly linking with malonyl-CoA with the release of CO_2 and reduction of the resulting unsaturated fatty acid, saturated carbon chains are formed, the length of which can vary depending on the fatty acid synthase complex. The degradation of fatty acids by β-oxidation is very similar to the reversal of fatty acid biosynthesis.

10.1 · Fats and Fatty Acid Biosynthesis

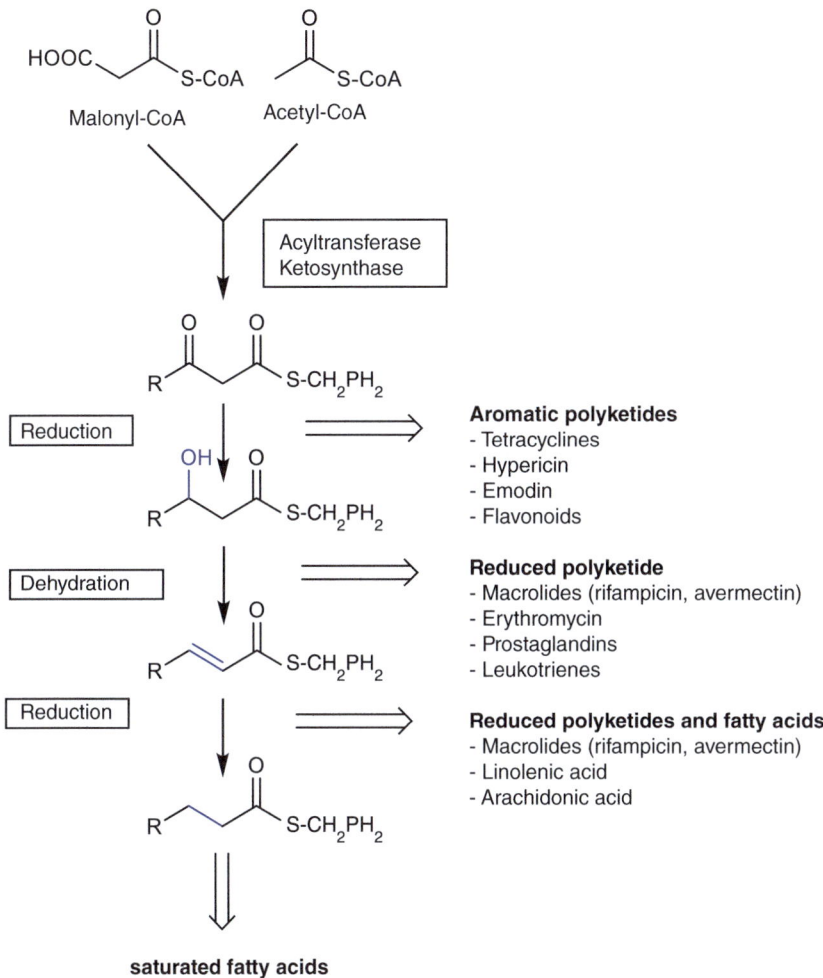

Fig. 10.1 Simplified representation of fatty acid biosynthesis and the natural substance groups derived from it

Fats, which are chemically esters of the trihydric alcohol glycerin and three fatty acids, are also referred to as triglycerides or more correctly triacylglycerols (short TAGs). This is a classic esterification, where the fatty acids can chemically differ by:
- their different chain length,
- an even or odd number of C-atoms in the chain (an even number is most common),
- branched or unbranched chains (some fatty acids also have a ring structure),
- the presence and number of hydroxyl groups,
- the number and stereochemistry of double bonds (saturated and unsaturated fatty acids).

10.1.2 Fat Biosynthesis

The glycerol component of fat biosynthesis originates as L-glycerol phosphate from glycolysis. In the further fat biosynthesis, which takes place in the microsomes of the cell, the linkage with this occurs via activated fatty acids (acyl-CoA). The principle of transferring a phosphate component to activate natural substances is also found here again. L-α-glycerol phosphate reacts with two molecules of acyl-CoA to form the diglyceride L-α-lysophosphoric acid. This compound is converted to D-α-β-diglyceride, which is now completed with a third molecule of acyl-CoA to form triglyceride.

10.1.3 Breakdown of Fats and β-Oxidation

In the "saponification" of fats, glycerides are split into glycerol and fatty acids; the latter are long-chain carboxylic acids with the general formula $CH_3C_nH_{2n}COOH$ (when saturated). Saponification refers to the splitting (hydrolysis) of fat into glycerol and fatty acids: chemically by bases (e.g., NaOH) or biologically by lipases. The enzymes involved in this hydrolysis, which chemically follows an addition-elimination mechanism, are also referred to as esterases. Lipases are very interesting enzymes that are located at the interface between water and fat and are capable of hydrolyzing fats in a hydrolytic reaction involving water in a stereospecific manner. Fats are gradually broken down from tri- to di- and monoglycerides. The resulting free glycerol is metabolized in glycolysis, the released fatty acids are broken down by β-oxidation back to C2 units and can then be fed into the citric acid cycle. A maximum of C18 units can be fed into β-oxidation. With each cycle of β-oxidation, a C2 unit is cleaved off and transferred to CoA; it stops at the latest when the body of the original fatty acid has been reduced to a minimum of C6. Since glycolysis in mammalian cells does not run reversibly to glucose, fats cannot be converted to glucose as a storage substance and are therefore suppliers of energy (ATP) and reduction equivalents (NADPH) in the citric acid cycle.

The properties of fats are mainly determined by the above-mentioned chemical structural features. The triglycerides of unsaturated fatty acids, i.e., the fatty acids with at least one double bond, have a lower melting point than those of saturated fatty acids without a double bond. An example is olive oil, which is liquid at room temperature, and palm fat, which is solid at the same temperature. Unsaturated fatty acids are more susceptible to oxidation by atmospheric oxygen, become rancid, polymerize and resinify, which is exploited by the paint industry. Unsaturated fats can technically be "hardened" so that they are no longer liquid but solid at the same temperature. Hardening technically refers to the hydrogenation of the double bonds with hydrogen at 160–200 °C using nickel catalysts. Hardened fats are used in the food industry as more durable lipids.

10.2 ABE Metabolism

The acetone-butanol-ethanol metabolism is an anaerobic fermentation in which acetone, butanol, and ethanol are produced from carbohydrates (preferably glucose) in an approximate ratio of 6:3:1 (◘ Fig. 10.2). The fermentation is preferably carried out in clostridia, with *Clostridium acetobutylicum* being noteworthy. However, the technical process is only of importance for the production of biobutanol and is currently the focus of research interest. Due to the special physiology of the clostridia, the ABE fermentation is divided into an acidogenic phase with pH values around 4–5 and a solventogenic phase with neutral pH value.

In the initial phase of acidogenesis, the cells grow exponentially and accumulate acetate, butyrate, and small amounts of CO_2 and H_2. A decrease in pH and other physiological factors lead to a metabolic shift to solventogenesis. Now, acetate and butyrate are converted into the corresponding alcohols. The formation of acetoin and lactate, both of which are not counted as part of the ABE metabolism, are by-products formed from one (lactate) or two (acetoin) molecules of pyruvate. The formation of the neutral products acetone, butanol, and ethanol is advantageous for clostridia, as they can break the steady pH decrease. As the acid concentration increases, the clostridia can no longer balance the intracellular pH, the cytosol becomes increasingly acidic, and metabolism comes to a halt as the proton gradient collapses. The cause is the transport of free fatty acids into the cytosol, where they dissociate and release protons. Toxicologically, the conversion of the acids into their neutral metabolites is only partially successful, as acetone and butanol are toxic to the cell.

As the butanol concentration increases, spore formation occurs, which is undesirable in a technical process. At low butanol concentrations, spore formation is low. This is due to the fact that *Clostridium acetobutylicum* has not yet entered the stationary phase of fermentation at low butanol concentration. In the stationary phase, the growth rate of the bacterium is lower, and more energy is expended for spore formation. In contrast, spore formation is strongly pronounced at high butanol concentrations. The reason for this is that butanol is a toxic metabolic

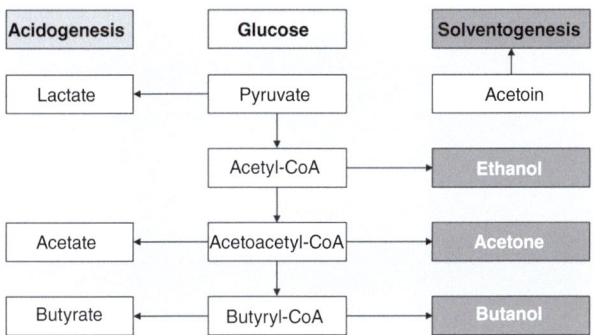

◘ **Fig. 10.2** Simplified representation of the ABE pathway with the derived technically important products

product and disrupts membrane fluidity. Corynebacteria now form spores to protect themselves from the toxic effects of butanol. Spore formation can be seen as a stress resistance mechanism. In summary, the relationship between butanol and spore formation can be explained as follows:

- **Low butanol concentration:** At low butanol concentration, the growth rate of *Clostridium acetobutylicum* is high. The bacterium requires energy for growth and reproduction. Spore formation is less important.
- **High butanol concentration:** At high butanol concentrations, the growth rate of C. *acetobutylicum* is low. The bacterium uses more energy for spore formation. Butanol is also a toxic metabolic product that can damage cells. The formation of spores is a mechanism to protect against the toxic effects of butanol and ensure survival.

Technical significance: When discussing the technical use of biobutanol today, it is a rediscovery of a solvent from the research time of Louis Pasteur, who produced butanol in 1861 using ABE fermentation. Industrial use began in 1916, in the middle of World War I, by Chaim Weizmann, who isolated C. *acetobutylicum* and patented the production process. With the advent of fossil fuels, the bioprocess became unprofitable after World War II and was replaced by petrochemical plants. The need for sustainable solvents and the strong development of genetic engineering have prompted Celtic Renewables in Scotland to build a new bio-plant by 2022, which, unlike traditional methods, uses waste materials such as molasses as a starting material. Molasses is a thick, dark brown sugar syrup that is a by-product of sugar production from sugarcane, sugar beets, and also sorghum. In addition to about 60% sugar (sucrose or raffinose), molasses contains organic acids, betaine, vitamins, and about 3% inorganic salts.

In the last 10 years, it has been possible to express the genes of the ABE metabolism from clostridia in *E. coli* (◘ Fig. 10.3). Starting from the early fatty acid biosynthesis to acetoacetyl-CoA in *E. coli*, the subsequent genes are cloned heterologously from *C. acetobutylicum* (Abdelaal and Yazdani 2022).

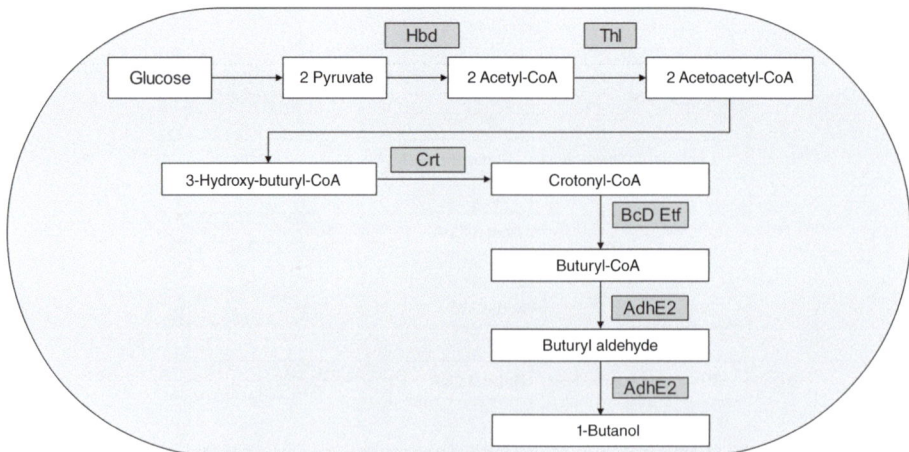

◘ **Fig. 10.3** Metabolic implementation of ABE fermentation in *E. coli* (Thl:, Hbd: , Crt:, BcD etf:, AdhE2: Alcohol dehydrogenase E2)

10.2.1 Isobutanol

This alcohol has a terminal branching in the side chain (◘ Fig. 10.4). The biosynthesis does not follow the ABE metabolism, but is an interesting continuation of the biosynthesis of 2-ketoisovalerate, the precursor of valine. Until recently, ethanol and 1-butanol were considered the only two industrially producible biofuels. Although isobutanol and other higher alcohols have long been identified as by-products in various fermentation broths, the quantities were too small to be of interest for biofuel applications. With the help of modern metabolic engineering techniques, specific enzymes of the metabolic pathway could be overexpressed to increase titer, yield, and productivity to an industrially relevant level (Smith and Liao 2011). In particular, the production of isobutanol was achieved by diverting 2-keto-isovalerate, an intermediate of valine biosynthesis, into alcohol synthesis. Starting from glycolysis, 2-acetolactate is formed via pyruvate, which is converted to 2,3-dihydroxy-isovalerate by reduction. The conversion of ketoisovalerate to 2-aceto-2-hydroxy-butyrate (AHB) by the enzyme ketose decarboxylase is carried out by a reductase (Adh/Adh2), which forms isobutyraldehyde (◘ Fig. 10.4).

With this synthetic biology approach, the biosynthesis in *Clostridium* spp. could be improved. Of great interest is also the transfer to *E. coli* with a titer of 22 g/L and a theoretical yield of more than 80%, which is comparable to the biosynthesis performance in the original host for the production of 1-butanol (Atsumi et al. 2009). In the genetically modified *E. coli* microorganisms, an acetolactate synthase from *B. subtilis* is cloned, which replaces the first enzyme in valine biosynthesis and redirects the flow via the 2,3-dihydroxyvalerate to 2-ketovalerate. The subsequent ketoisovalerate decarboxylase (KivD) from *Lactococcus lactis* converts the 2-ketoisovalerate to isobutyraldehyde. The final reduction by an alcohol dehydrogenase can occur directly (left, Adh) or via isobutyryl-CoA (right, Adh2). The right pathway is active when host-specific enzymes cause an activation of the short fatty acid with coenzyme A.

Biotechnologically, the production in *E. coli* certainly has the advantage that no spore formation occurs as in clostridia, however, the toxicity of 6-8/L remains unless working in solvent-resistant strains (up to 15 g/l).

10.3 Technical Fats and Oils

10.3.1 Biofuels

Modern fuels will no longer be of fossil origin. In a world that wants to consistently become climate-neutral, most fuels and lubricants will be obtained from sustainable biomass in the future. These are usually liquid and are used in mobile combustion engines of cars and trucks. The source of the biomass is differentiated between first-generation raw materials such as cereals, sugar beets, and oil plants, which were originally widely used as food, and second-generation raw materials such as Jatropha oil, algae, or manure for biomethane (◘ Table 10.1). Biobutanol, which was already used as aircraft fuel by the Royal Air Force during the Second World War,

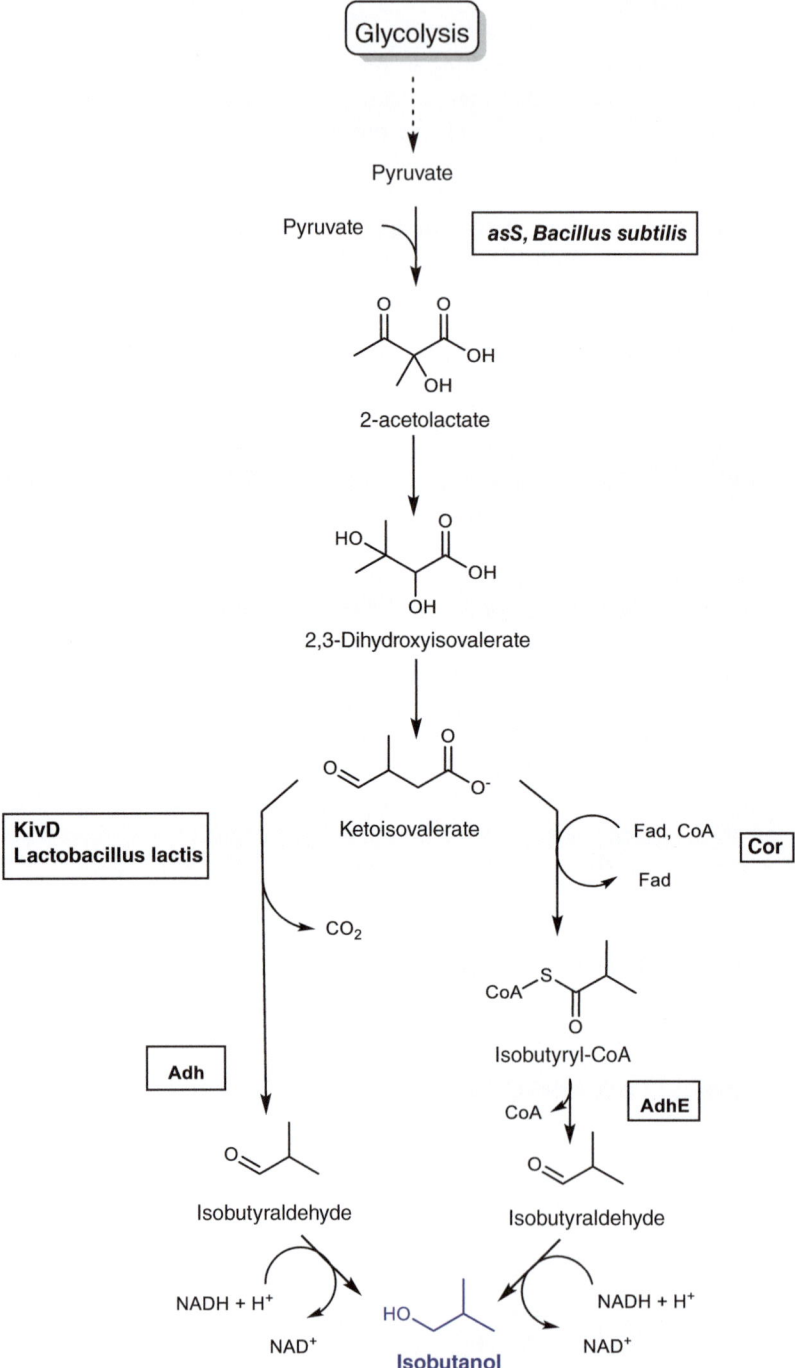

Fig. 10.4 Heterologous biosynthesis of isobutanol in *E. coli*. Representation of the 2-keto acid pathway, left reductive via Adh, right via activation with coenzyme A

was already mentioned in ▶ Sect. 10.2. However, the importance of biofuels is seen as a replacement for gasoline, diesel, or kerosene.

10.3.2 Biodiesel

This fuel is also referred to as agrodiesel and is chemically a fatty acid methyl ester, which can be used as diesel fuel due to their physical calorific values. The production is not a biological process, as the fats are chemically transesterified. In this process, the bond of the fatty acid to the glycerin is broken and converted into a methyl or ethyl ester. In Europe, rapeseed oil is mainly used for biodiesel production, in the USA soybean oil, and in Asia palm oil. Technically, for chemical extraction, the vegetable oil is mixed with methanol and the catalyst and kept at 30 to 70°C for several hours. The biodiesel phase is obtained by distillation, washed, cleaned, and dried.

Alternative and renewable energies are not based on fossil petrochemical sources, but on renewable raw materials. A plant that can be considered a second-generation energy plant (not a food plant) is *Jatropha curcas* from the Euphorbiaceae family. This plant, which is known in ethnomedicine for its purgative effect and is therefore also referred to as purging nut, is known more for its toxic than its healing effect. Originally, the wild plant comes from Mexico and Latin America; today it is cultivated in South America, but also in Africa and Asia. About 80% of today's global cultivation area of approx. 1 million hectares is in Asia. It plays an important role in Indonesia as a potential source of truck diesel, and its cultivation is considered a task of national interest. In Asia and Europe, airlines are experimenting with the plant oil of *J. curcas* as a replacement for kerosene, as it has a similar calorific value to the kerosene used so far. The yield is good with an average of 2 to 3 t per hectare. The plant can be grown on so-called degraded areas and is therefore not in competition with cultivation areas for food. Of particular interest is the seed, which can contain up to 40% oil in its embryonic tissue, but also proteins (curcin) and toxic phorbol esters. The oil is obtained by pressing and could be used directly as fuel. However, saponification to fatty acids often takes place, which are then used as fuel to optimize engine performance. The oil of *J. curcas* would actually be unproblematic and a welcome fuel if it did not contain the potentially carcinogenic phorbol esters. These natural substances, which chemically belong to the diterpenes,

Table 10.1 First and second generation biofuels

First Generation	Plant oil fuels: Rapeseed oil,
	Biodiesel: Fatty acid methyl ester
	Bioethanol: Sugar, wheat
Second Generation	Biomethane: Manure
	Biomass to Liquid (BtL) fuels
	Cellulose Ethanol
	Biokerosene: *Jatropha* oil, algae fuel

are of concern because they can cause skin cancer in the long term. Inactivation would be conceivable by heating the oil, but this is not economically viable. Therefore, intensive work is being done on process engineering detoxification methods.

10.3.3 Biokerosene

A fuel for aviation is hydrogenated vegetable oil from rapeseed, palm or jatropha oil, which is intended to replace fossil kerosene. Fossil kerosene is a light petroleum oil that is obtained by distillation from crude oil.

10.3.4 Biogenic Lubricants

Lubricants serve the function of solid, usually moving parts, to reduce wear through friction, cool them, or prevent corrosion. In contrast to petroleum-based lubricants, biogenic lubricants are made from vegetable fats or animal fats (beef tallow). In the case of vegetable fats, rapeseed oil is predominantly used in Germany (approx. 46,000 t). Due to the requirements, vegetable oils cannot completely replace mineral lubricants, but are added to them.

10.3.5 Biobutanol

The solvent biobutanol is obtained through the ABE metabolism with *Clostridium acetobutylicum* or *Clostridium beijerinckii*. Fermentation is an anaerobic process that takes place in the absence of oxygen. Butanol is obtained from the fermentation approach by distillation. The production of biobutanol is an approach to the production of a sustainable and climate-friendly fuel. Biobutanol has a higher energy density than ethanol and can therefore serve as a replacement for gasoline. In addition to its use as a fuel, butanol is an important basic chemical as a solvent for paints, varnishes, adhesives, and plastics.

Secondary Metabolism and Biochemical Pathways of Significance

Contents

11.1 Polysaccharides – 109
11.1.1 Homogeneous Polysaccharides – 110

11.2 Heterogeneous Polysaccharides – 116
11.2.1 Xanthan – 116

11.3 Aminopolysaccharides – 117
11.3.1 Chitin – 119
11.3.2 Heparin – 120
11.3.3 Hyaluronic Acid – 122

© The Author(s), under exclusive license to Springer Fachmedien Wiesbaden GmbH, part of Springer Nature 2025
O. Kayser and N. J. H. Averesch, *Technical Biochemistry*, https://doi.org/10.1007/978-3-658-47121-7_11

Chapter 11 · Secondary Metabolism and Biochemical Pathways of Significance

> **Learning Objectives and Key Topics**
> - Chemical Diversity of Secondary Metabolites
> - Systemic Coordination of Biosynthesis Pathways
> - Biological and Physiological Properties
>
> **Technical Applications**
> - Synthetic Biotechnological Production of Natural Substances
> - Systems Biology in Natural Substance Biotechnology
>
> **Biological Activities**
> - Polymeric Sugars
> - Phenols
> - Terpenes
> - Alkaloids
> - Antibiotics

All biologically active organisms have biochemical metabolic pathways to convert chemical energy according to the laws of thermodynamics and use it in various ways for basic needs (energy), movement (kinetic energy), and cellular communication. The previous chapters described the role and function of primary metabolism; it was shown how sugars, fats, proteins, and nucleic acids are built up, converted, or broken down almost ubiquitously in most organisms.

In contrast to primary metabolism (= biochemistry), secondary metabolism (= biochemistry of natural substances) fulfills other species-specific tasks. It is always linked to primary metabolism, but does not serve energy production and the maintenance of life functions. Its secondary metabolites are produced via specialized metabolic pathways depending on temperature, pH value, tissue, age of the organism, and external environmental factors (predators), to name just a few. The secondary metabolism is the scene of chemically and pharmacologically very diverse structures, which we humans use in various ways as colorants, flavorings, and active substances. Of course, not a few of them are toxic, but the vast majority are important for our nutrition and well-being. However, the two metabolic pathways cannot be sharply separated from each other. There are always overlaps, such as steroids and vitamins, where a clear assignment is not possible (◘ Fig. 3.1).

> **Important Characteristics of Secondary Metabolites for Organisms**
> Secondary metabolites are…
> - indispensable for the existence and further development of a species
> - not generally spread in all organisms
> - chemically very diverse, but they can often be traced back to biosynthetic building blocks
> - can have strong pharmacological effects or be toxic
> - species-specific molecules that take on specialized functions in the organism

11.1 Polysaccharides

Learning Objectives
- Differences between homogeneous and heterogeneous polysaccharides
- Structure of linear polysaccharides
- Technical importance of
  - Starch
  - Cellulose
  - Dextrans
  - Dextrins, Cyclodextrins
- Biosynthesis of Hyaluronic Acid
- Biosynthesis of Xanthan
- Biosynthesis of Chitin

Polysaccharides are complex high-molecular compounds that can be split by hydrolysis. If polysaccharides are composed of the same monomers, they are referred to as homogeneous polysaccharides (e.g., cellulose from cellobiose/glucose). When a combination of different monomer units is used, they are referred to as heterogeneous polysaccharides (e.g., marshmallow slime as a cough remedy, xanthan as a thickener). Polysaccharides are found in all living tissues and play an important role as structural substances (e.g., cellulose) or storage substances (e.g., starch, glycogen) and have a protective effect like gums (botanical term for rubber-like compounds). Simple polysaccharides are linear like cellulose or inulin, but they also form three-dimensional, helical chains like the amylose in starch. In nature, a distinction is made between **storage polysaccharides** and **structural polysaccharides.** Glucose belongs to the first group, which serve as energy storage and reservoir in the form of glycogen and starch and are provided by degradation of the polymer in case of deficiency. Structural polysaccharides give the cell or tissue stability and enable the function of the unit. These polymers, such as murein, can be present in the cell wall or connected with the cell wall inside or outside. The chemistry and structures of some polysaccharides are briefly explained (◘ Table 11.1).

Table 11.1 Important polysaccharides of various organism groups (storage polysaccharide (RP), structural polysaccharide (SP))

Higher plants	Algae	Fungi/Yeasts	Animals
Cellulose (SP)	Cellulose (SP)	Cellulose (SP)	α-Glucans
Arabinoxylanes	Alginates	α/β-Glucans	(Glycogen, Starch
Pectins (SP)	Galactanes	Mannans	(RP))
	β-Glucans		Heparin
	L-Fucans		
	Mannans		
	Xylanes		

11.1.1 Homogeneous Polysaccharides

11.1.1.1 Cellulose

Chemistry Plant cell walls are predominantly composed of cellulose fibers, which are similar to starch in that they are made up of glucose. The main difference is that the smallest building block of cellulose (◘ Fig. 11.1), the disaccharide cellobiose (β-D-glucopyranosyl-β-D-glucopyranose), is linked with a β-bond of the two glucose molecules. Shorter chains of cellobiose are referred to as hemicellulose, which forms the framework for the incorporation of lignin. However, lignin is not a sugar, but a phenylpropan polymer from the shikimate pathway, which is responsible for the lignification of the plant and gives it more strength and stability.

Biosynthesis The provision of activated glucose is essential for biosynthesis. This can be activated by transferring sucrose to uridine triphosphate. The C1 position of the sugar is bound to the diphosphate of the UTP, and with a transfer potential of $\Delta G^{o'} = -33.5$ kJ/mol, glucose has become a very good leaving group. By multiple chaining, a cellulose chain consisting of thousands of glucose molecules is formed. Several of these cellulose chains aggregate into cellulose fibrils, which give the plant the necessary tensile strength (◘ Fig. 11.2). A look at the microstructure shows that the cellulose fibrils are subject to a linear order, which can be explained by the directed movement of the cellulase complex along the microtubules near the cell surface. Cellulose is insoluble in water, but can swell and assume a gel-like structure. This is important to understand that in technical extraction processes, organic solvents such as methanol or ethanol and a water content of 20 to 30% are often used. The water content is important to allow the dried cell walls to swell, significantly improving the diffusion of the natural substances to be extracted.

The extraction of pulp plays an important role in the paper industry. Cellulose, as a fibrous substance, forms the basis for paper production and is obtained from wood, wood residues, pulp, or waste paper. It is important for paper production that the lignin, as a disturbing component of wood, is removed.

Biological Effect Humans cannot break down cellulose into glucose in the gastrointestinal tract, as they do not possess cellulases. Cellulose belongs to the indigestible dietary fibers, which play an important role in stool formation and the microflora of our digestive system. Foods containing cellulose are also an important part of the dietary nutrition of people with obesity. In ◘ Table 11.2, the chemical and physical differences between both glucose polymers are summarized.

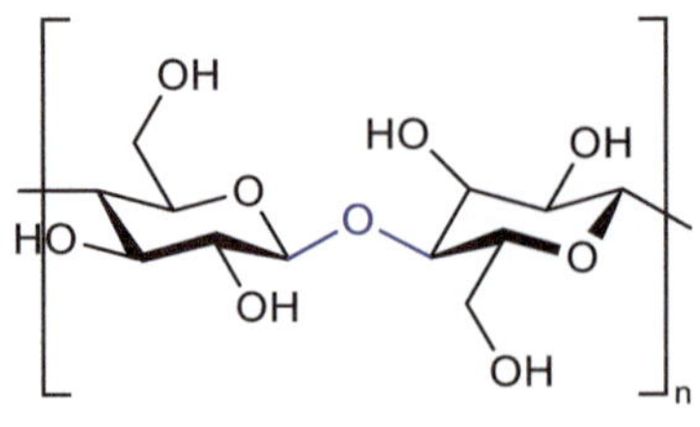

◘ **Fig. 11.1** Cellulose, with the β-glycosidic bond in blue

11.1 · Polysaccharides

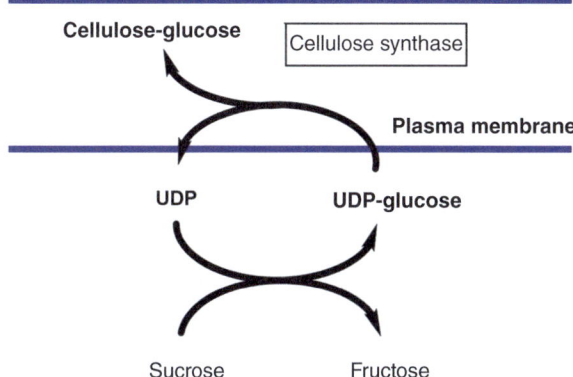

Fig. 11.2 Biosynthesis of cellulose by membrane-bound cellulose synthase

Table 11.2 Differences between Starch and Cellulose

Characteristic	Starch	Cellulose
Chemical Structure	α-D-Glucose monomers linked by α-1,4-glycosidic bonds	β-D-Glucose monomers linked by β-1,4-glycosidic bonds
Structure	Mixture of Amylose and Amylopectin	Unbranched chain
Molecular Mass	100,000 to 1,000,000 g/mol	162,000 g/mol
Solubility	Insoluble in water	Insoluble in water
Digestibility	Digestible by starch-splitting enzymes	Indigestible for humans
Occurrence	Plants, grains, potatoes	Plants, wood, pulp
Usage	Food, adhesives, paper, fertilizer	Paper, textiles, packaging, building materials

11.1.1.2 Starch

Starch is a typical homogeneous sugar polymer, consisting of 25% amylose and 75% amylopectin. Although both show chemically similar basic building blocks, they differ in their physical and structural properties. Amylose is more water-soluble and turns dark violet with iodine (Table 11.2). Amylose is long-chain and linear, forming a characteristic helical structure. It typically consists of 300 to 600 glucose units. In contrast, amylopectin is highly branched with side chains on the C6 hydroxyl group. This offers the advantage of a large number of terminal maltose units, which can be split more quickly by amylases and glucosidases and made available as energy carriers. With 0.62 to 4.3 million repeating units, amylopectin is the largest naturally occurring polymer. Starch is found in the typical storage organs of plants such as roots, seeds, and fruits. Humans also store glucose in the form of glycogen (approx. 500 g) in the liver and muscles. Like amylopectin, glycogen is characterized by a high degree of branching. Starch is broken down by the α-amylase, which is found in plants, but also in human saliva and the pancreas. The α-amylase randomly splits the α-1,4-linkages into maltose and further fragments of 6 to 7 oligomers. The enzymatic hydrolysis by another β-amylase leads to glucose and "limit" dextrins (polysaccharides with α-1,6-linkages), which occur in small amounts. The R-enzyme only splits α-1,6-linkages, and each resulting glucose molecule is activated for further biochemical conversion by phosphorylation in the C6 position.

The raw material starch plays an important role in pharmacy and technology. As an excipient, starch ensures the cohesion of tablets and granules, in technology it is used as glue and adhesive. Amylopectin hydrolysates can also exist as ether derivatives and are then referred to as hydroxyethyl starch. On average, there are 7 2-hydroxyethyl residues for every 10 glucose units. The average molecular weight of the commercial preparation is 450 kDa.

11.1.1.3 Dextrans

Chemistry Dextrans are polymeric, water-soluble sugar compounds with a molecular weight of MW 10,000 to 10 million. Dextrans are highly branched polysaccharides with glycosidic bonds to neighboring glucose molecules through 1→6 (main chain), 1→4 (branching) or 1→3 (branching), less commonly 1→2 (branching) bonds. The difference between dextrans and dextrins lies in the glycosidic bond. Dextrins have a predominant α1→4 bond and dextrans a predominant (95%) α1→6 bond.

Biosynthesis Dextrans are polymeric sugars of microbiological origin. *Leuconostoc williamsii* and *L. mesenteroides* produce α-1,6-linked polymers with a broad size distribution. These microorganisms do not synthesize dextrans as a storage substance, but as an exopolysaccharide, which provides the microorganisms with protection against desiccation and adhesion to surfaces. Here too, biosynthesis starts from sucrose, which provides activated UDP-glucose. Using glucosyltransferases, which show homology to cyclodextrin glucosyltransferases, the activated glucose units are α-1,6-linked to build up the polymer backbone. The side chains have an α-1,3 and α-1,4 linkage (◘ Fig. 11.3).

Medical Application In medicine, dextran 10–60 kDa (Dextran 60 = Macrodex®) is used as a plasma expander in case of blood loss (Blood plasma replacement: 6% solution). The size specification corresponds to the average molecular weight and varies between approximately 25 and 110 kDa. In case of non-critical blood loss, solutions can be used instead of bodys own plasma substitutes (Whole blood transfusionn, plasma protein solutions) that are not physiologically identical to blood, but osmotically and colloid-osmotically correspond to the oncotic pressure of the plasma. Dextrans are technically obtained by cultivating the above-mentioned strains in sucrose solutions. Subsequent fractionation allows classification by polymer weight.

Technical Applications Dextrans are used as thickeners in the food industry (ice cream, reduced-fat cheese), in cosmetics, paints and as aids in the paper and textile industry. Under the trade name Sephadex, cross-linked dextrans are sold for gel chromatography, which are 2→2 cross-linked with glycerin.

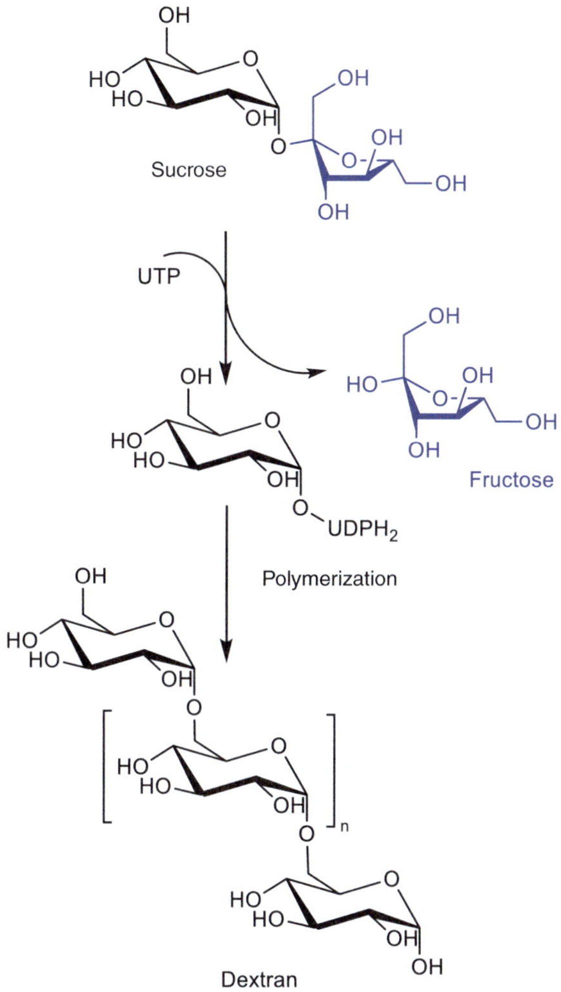

11.1 · Polysaccharides

Fig. 11.3 Dextran biosynthesis

11.1.1.4 Cyclodextrins

A technically interesting group of oligosaccharides are the cyclic sugars, such as cyclodextrins, which find diverse applications. Cyclic dextrins were first discovered in 1891 by A. Villiers (Fig. 11.4). They are dextrins as sugar compounds, usually consisting of 6(α), 7(β) or 8(γ)-sugar units, which are hydrophilic on the outside and lipophilic towards the pore on the inside. Schematically, the structure can be understood as a cone. With increasing number, the diameter increases and the physical properties change. These cyclic oligosaccharides are α-1,4-glycosidically linked, with β-cyclodextrins finding the greatest technical application. Biosynthetically, they are obtained by enzymatic cleavage and internal ring closure of starch by bacterial fermentation of *Paenibacillus macerans*, which possess cyclodextrin glucosyltransferases (CGTases). The starting material is corn starch. The fact that the ring

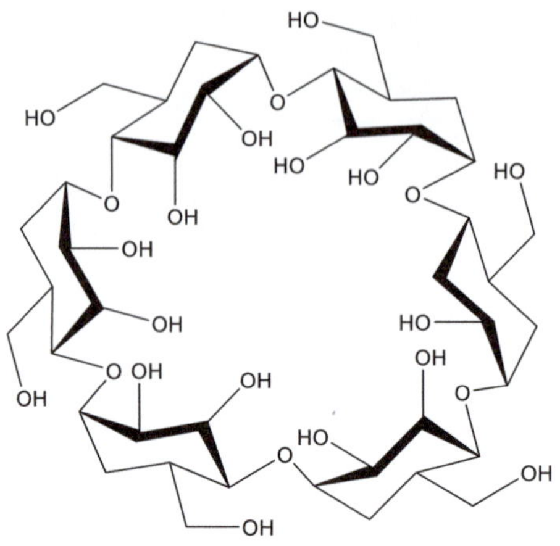

Fig. 11.4 α-Cyclodextrin

structure preferably consists of six to eight glucose molecules is explained by the helical structure of amylose, which is characterized by the same number of glucose units per turn. Lipophilic ligands or chemical substances can be complexly stored in the hydrophobic cavities. The enclosed complexes have different physical properties; in particular, solubility is increased, which is why they are used in pharmacy as excipients to improve solubility. Examples are prostaglandins in α-cyclodextrins, as in Prostavasin® for improving the circulation of smoker's legs, or the conversion of garlic extracts (Tegra®) into β-cyclodextrins for taste masking.

Applications In addition to their medical applications, cyclodextrins are an indispensable substance in the food and cosmetics industry. They serve as oxidation protection, for binding essential oils, for binding unpleasant odors (e.g., cat litter), as indigestible sugar polymers in dietetics and as retardants in the sense of a controlled release of drugs in the human body. As food additives, they are used in mayonnaise, baked goods, instant juices, and potato chips. Their use in household chemicals is also interesting. The odor-eliminating effect of the room spray Febreze® is based on β-cyclodextrins, according to the manufacturer.

11.1.1.5 Pectin

Pectins are sugar polymers, preferably obtained from citrus fruits and apples. These sugars are partially methylated and are isolated as methoxypolygalacturonic acids with an average molecular weight of 100–250 kDa. Pectin consists of α-1,4-glycosidically linked galacturonic acid units, which form a heterogeneous polymer via Ca^{2+}- and Mg^{2+}-ions. It is of great importance as a thickening agent in food technology and as a dietary supplement. As a thickening agent, it is used, for example, in ketchup, ice cream and fruit spreads. Depending on the degree of methylation, pectins can also be gelled with sugar and are therefore found in jams.

11.1 · Polysaccharides

11.1.1.6 Carrageenan

Carrageenan (also carrageen) is isolated from Irish moss, the dried red algae *(Chondrus crispus)*. It occurs as κ-, ι- and λ-forms, which are characterized by different gelling properties. In addition to being used as a thickening agent, it is used in food technology and increasingly in special applications, e.g., in the immobilization of biocatalysts such as enzymes.

11.1.1.7 Alginate

Alginate (◘ Fig. 11.5) is produced by brown algae and some bacteria (e.g., *Acetobacter*). In the algae, it is the structural element of the cell walls. The intercellular gel matrix gives the algae both flexibility and strength. Alginate acts as a thickener, stabilizer, and gelling agent and is often found as a component of biofilms. Alginate is a polysaccharide consisting of a mixture of the two uronic acids α-L-guluronic acid (GulUA) and β-D-mannuronic acid (ManUA), which are 1,4-glycosidically linked in varying proportions to linear chains. It forms homopolymeric regions in which mannuronic acid or guluronic acid is present as a block (hence it is also referred to as a block copolymer), which are referred to as GG or MM blocks. Gelation occurs by the incorporation of calcium ions into the GG blocks, which form a regular zigzag structure. The average molecular weight is between 48 and 86 kDa. It is mainly used as a thickening or gelling agent in the food industry, but also in various medical applications, e.g., as a component of wound patch, due to its ability to exchange ions and store large amounts of water. Another application is the immobilization of enzymes or cells in biocatalysis.

Technical Application Alginates are used as a material for the microencapsulation of adhesives. The classic technique of microencapsulation involves dropping the adhesive dissolved or suspended in an alginate solution with $CaCl_2$ solution, so that spherical aggregates of defined size and size distribution are formed. Under the influence of divalent metal ions, the acid function of the alginate forms an insoluble complex; gelation occurs. Typical diameters of these immobilizates are between 1 and 2 mm; however, significantly smaller particles (0.1 mm) or hollow spheres can also be produced to improve substrate diffusion. The technique of hollow sphere encapsulation is interesting because the enzyme "floats" in a solution in the middle of the sphere and the substrate must overcome the sphere wall by diffusion. Alginate micropastilles are also used in technology to secure screws against self-loosening by gluing the screw and nut. The thread of the nut is coated with a film that contains adhesive and hardener separately. When screwing in, the microcapsules are destroyed by shear forces, so that adhesive and hardener can leak out, mix and harden.

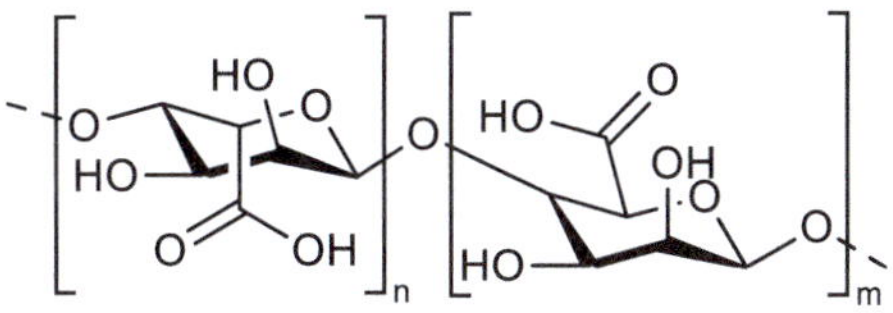

◘ **Fig. 11.5** Section of the alginate polymeric chain

11.2 Heterogeneous Polysaccharides

11.2.1 Xanthan

Chemistry Xanthans are water-soluble, high-molecular, acidic polysaccharides with a complex structure. The building blocks are glucose (β-1,4-D-glucan) for the main chain and D-mannose acetyl ester, D-glucuronic acid in the side chain (◘ Fig. 11.6). Mannose pyruvate ketal always closes the side chain. Glucose, mannose, and glucuronic acid are in a molar ratio of 2:2:1 to each other and form a repeating unit as a pentasaccharide. Xanthans are highly effective binders even at very low concentrations below 1%. The thickening property is very stable in a wide pH range from 2 to 11 and is not restricted by heating. Xanthan solutions are shear-thinning (thixotropic), which explains their use in the food industry.

Biosynthesis The biosynthesis of xanthan is a complex, multi-step process (◘ Fig. 11.7). It begins with the uptake of sucrose or glucose from the environment by bacteria of the genus *Xanthomonas*. The first two steps ① and ② involve the binding of activated glucose to the lipid carrier, which is oriented towards the cytosol on the inside of the membrane. In the third step ③, mannose is condensed with the second glucose molecule. Acetylation takes place, which does not need to be mentioned separately in biosynthesis. The extension with glucosamine ④ presupposes the amination of glucose, which is carried out in advance. There is no transamination in the gene cluster that includes the enzymes Gum B, C, D, F, G, H, I, K, L, M. The biosynthesis of the pentasaccharide is formally completed with the last linkage of mannose ⑤. The terminal mannose is acetylated and reacted with phosphoenolpyruvate ⑥. Polymerization is carried out in the cell wall and the elongating xanthan polysaccharide is released into the extracellular space ⑦ or anchored in the outer cell wall. In biotechnology, work is mainly done with Xanthomonas camp-

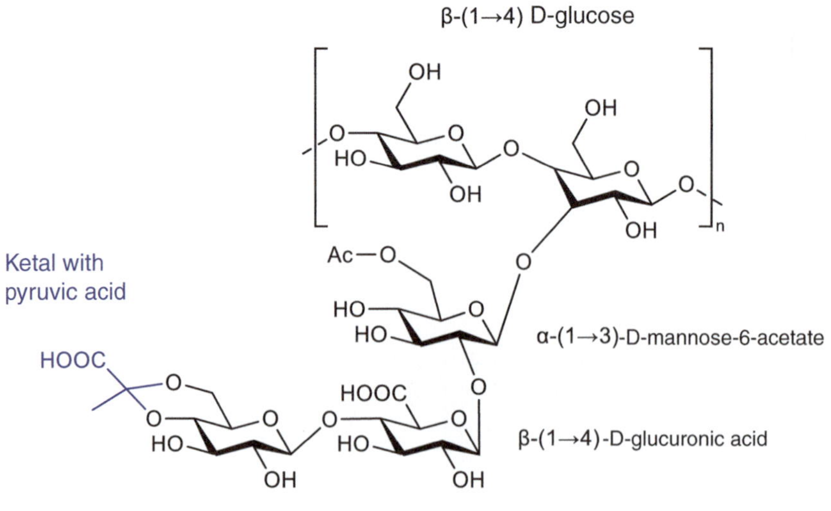

◘ **Fig. 11.6** Section of the xanthan polymeric chain

11.3 · Aminopolysaccharides

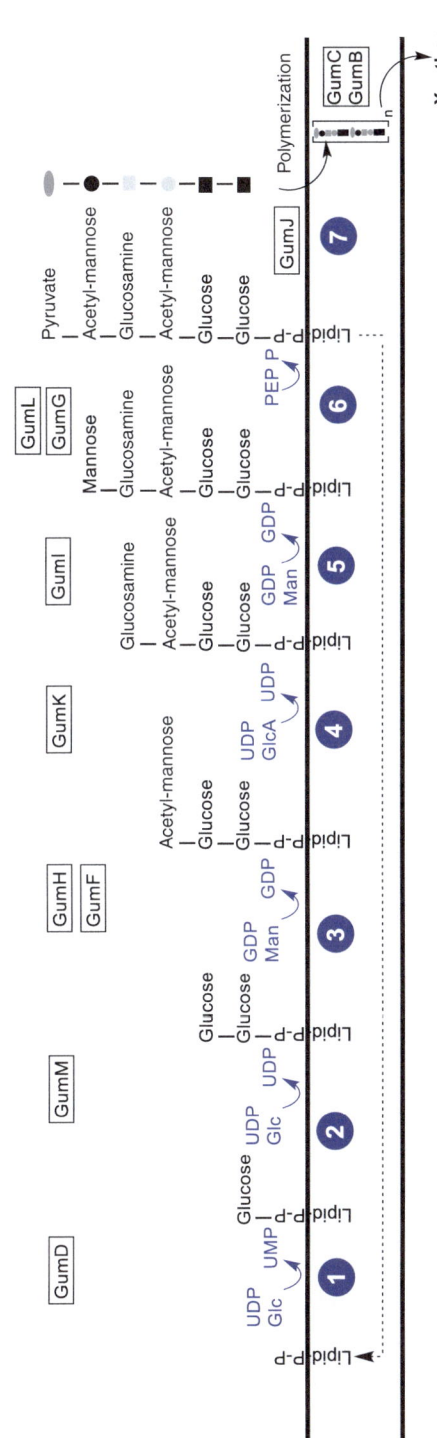

Fig. 11.7 Xanthan biosynthesis in *Xanthomonas campestris* (Genes B-I, K, L, M are in a cluster.)

estris. Sugars are converted into sugar nucleotides in the cell, which serve as building blocks for the synthesis of xanthan. The actual xanthan synthesis takes place in the cell membrane. This requires several enzymes that work together in a specific order and are iteratively coupled. This is also a "conveyor belt system", which will be discussed later as an example for the biosynthesis of polyketides (chapter antibiotics). The first step is the formation of a UDP-glucose molecule from glucose and uridine diphosphate (UDP). This molecule is converted by the enzyme GumM into a UDP-mannopyranosyl molecule. In the next step, the UDP-mannopyranosyl molecule is bound to the backbone of the xanthan polymer by the enzyme GumH. The backbone consists of β-(1→4)-linked glucose units. The side chains of the xanthan polymer consist of β-D-mannopyranosyl-(1→4)-β-D-glucuronopyranosyl-(1→2)-6-O-acetyl-α-D-mannopyranosyl units. The side chains are attached to the backbone of the xanthan polymer by the enzymes GumH, K, I. The length of the side chains is variable and can be between 1 and 15 units. *Xanthomonas* produces the pentasaccharide as a biosynthesis building block, which is repeatedly strung together. After the synthesis is completed, the xanthan polymer (GumC,B) is exported from the cell membrane into the surrounding medium. The exact mechanism of export is not fully understood (Becker et al. 1998).

Biotechnology   This is an aerobic fermentation with glucose or sucrose as the substrate (EU, USA) or starch (Asia). Biotechnology and production have been well known since the 1950s. The fermentation is classified as solid-state fermentation, which is not without problems due to rheology. An important point is the oxygen supply, which is why the fermentation approach is kept in motion mechanically and less by aeration. The formation of xanthan is not uniform and can be influenced by the fermentation. Thus, the average molecular weight is variable due to the fermentation time and the substitution with acetate and pyruvate. After a fed-batch production of 48-72 h, a yield of up to 25 g/L can be expected. The xanthan is precipitated in a complex downstream process with isopropanol and spray-dried. The annual production is 30,000 tons.

Applications   Xanthans are used in the food industry as thickeners (E415) for sauces, dressings, and desserts. In diabetics, xanthans can trigger allergies. The alternative would be pectin. Also in the oil industry, its property as a viscosity regulator in boreholes is appreciated.

11.3 Aminopolysaccharides

In addition to the classic polysaccharides, in aminopolysaccharides a hydroxyl group can be replaced by an amino group, which can also be acetylated or sulfated. The introduction of the amino group or the sulfate group changes the charge, so that in living nature both positively charged (aminopolysaccharides) and negatively charged (sulfation) polysaccharides occur. The most important properties of aminopolysaccharides are:
- **Chemical Structure:** Aminopolysaccharides consist of a linear or branched chain of sugar monomers, which are connected by glycosidic bonds. The sugar mon-

11.3 · Aminopolysaccharides

omers can be glucose, galactose, mannose, xylose, fucose or other sugars. The amino acid residues can be connected to the sugar monomers through various bonding patterns such as ester, amide or peptide bonds.
- **Molecular Weight:** The molecular weight of aminopolysaccharides can vary. It can range from a few thousand to several million Daltons.
- **Solubility:** Aminopolysaccharides have varying solubility in water. Some aminopolysaccharides are very soluble in water, while others are moderately or poorly soluble.
- **Physical Properties:** Aminopolysaccharides have a number of physical properties that depend on their chemical structure and molecular weight. These include viscosity, elasticity, strength, and water-binding capacity.

11.3.1 Chitin

The polysaccharide most widely distributed in nature after cellulose is chitin. It is the animal analogue to cellulose in plants and serves as a component of the exoskeleton in insects and marine organisms such as crustaceans for structure formation. The composition of chitin varies depending on the organism, season, sex, age, habitat, and other environmental conditions. Based on X-ray diffraction studies, chitin microfibrils appear to be assignable to three crystalline allomorphic forms (α-, β-, and γ-chitin). Chemically and physically, the microfibrils also differ in orientation, number of chains, degree of hydration, and size of the units. The α-chitin crystal structure is the most common form in the exoskeletons of arthropods of krill, lobster, crab as well as in the cuticle of insects.

Chemistry Chitin, very similar to cellulose, consists of N-acetyl-D-glucosamine, which is β-1,4-glucosidically linked. The essential difference to cellulose is therefore the presence of an acetamide group. Chitin is also found in fungi, arthropods, and mollusks. In arthropods, it is the main component of the exoskeleton. Chitin is the starting material for the technical production of chitosan and glucosamine. Chitin is not a uniform polymer, but a mixture of statistical copolymers of D-glucosamine (GlcN) and N-acetyl-D-glucosamine (GlcNAc). This means that not all amino groups are acetylated. Chitosan contains significantly fewer, ideally no acetate groups. If the degree of acetylation is more than 50%, it is referred to as chitin.

Biosynthesis The chitin biosynthesis pathway is conserved in all organisms, from algae to crustaceans and from fungi to insects. The biosynthesis can be summarized in five steps (◘ Fig. 11.8): In the first step, the synthesis of N-acetylglucosamine-6-phosphate from sugars such as glucose, glycogen or trehalose via the hexosamine pathway ①, in the second step follows the substitution of the C2 hydroxyl group and the build-up of the amino sugar ②. The glucosamine is acetylated and activated with the help of UTP to uridine diphosphate N-acetylglucosamine (UDP-N-acetylglucosamine) ③. In the fourth step, the formation of the disaccharide by the chitin synthase and the polymerization ④. Formally, the last step is not attributed to biosynthesis, as the final step consists of the deposition along the cell membrane and the assembly into Nanofibrils ⑤.

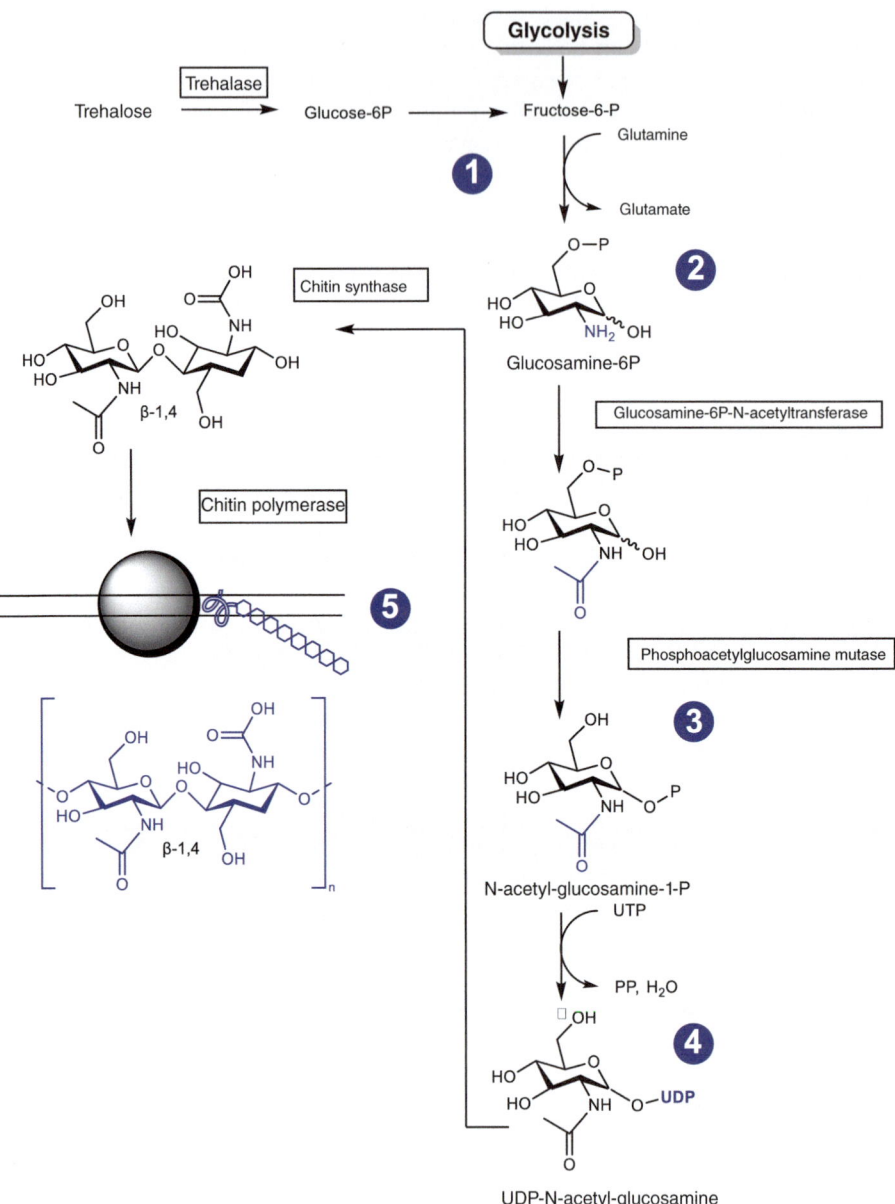

◘ **Fig. 11.8** Chitin biosynthesis

11.3.2 **Heparin**

Chemistry With the intake of heparin (◘ Fig. 11.9), we cross a boundary of the biological realms, as in addition to plant substances, an animal natural substance is now being discussed. Heparin is a polysaccharide, which is composed of a variable number of amino sugars. The molecular weight of heparin lies between 4,000

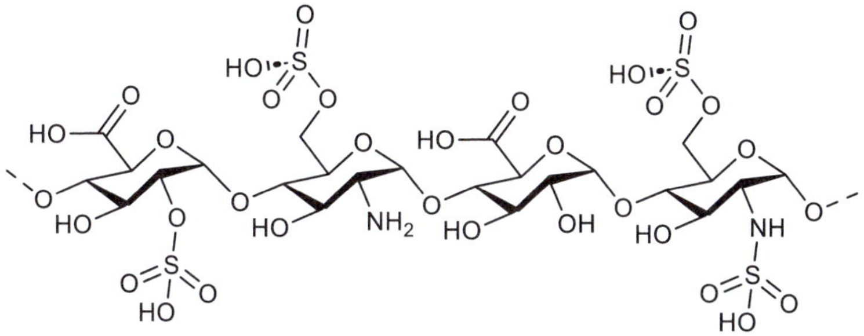

11.3 · Aminopolysaccharides

Fig. 11.9 Heparin

and 40,000 Dalton, with the most common molecular weight being about 15,000 Dalton. The linear structure of heparin consists of alternating sequences of D-glucosamine and an uronic acid (D-glucuronic acid or L-iduronic acid), which are O-glycosidically linked as amino sugars. The uronic acids can be linear or cyclic. The linear uronic acids are usually D-glucuronic acids, while the cyclic uronic acids are usually L-iduronic acids. The chain length of heparin is important for its biological effect. With a chain length of five monosaccharides (three D-glucosamines and two uronic acids), heparins are already anticoagulant. Heparins are formed in the liver and in the skin. They also occur in other tissues such as lungs, kidneys, and vascular walls.

Biosynthesis The biosynthesis of heparin is a complex biochemical process that takes place in the Golgi apparatus of mast cells and basophilic granulocytes. The basic structure of N-acetylglucosamine (GlcNAc) and glucuronic acid (GlcA) is built up and then postbiosynthetically modified by various enzymes. The biosynthesis begins with the transfer of GlcNAc to a lipid intermediate. GlcA is then added and the chain is extended by alternating transfer of GlcNAc and GlcA. During chain extension, the chain is also modified by various enzymes. The most important bioorganic reactions are:
- **Epimerization:** N-acetylglucosamine can be epimerized to iduronic acid (IdoA).
- **Sulfation:** Sulfate groups can be attached at various positions of the GlcNAc and IdoA residues. Sulfation plays a particularly important role in the formation of heparin. Hep*arin* is much more sulfated than Hep*aran*sulfate, which occurs in all mammalian cells. The sulfation gives heparin its characteristic properties such as its anticoagulant activity.
- **N-Deacetylation:** The N-acetyl group on the GlcNAc residues can be removed.

The biosynthesis of heparin (Fig. 11.10) is a non-template process, i.e., there is no DNA template which determines the structure of heparin. Rather, the structure is determined by the activity of the various enzymes involved in the biosynthesis. The finished heparin is stored in vesicles and released into the extracellular space as needed. There it can perform its various functions, e.g., inhibit the blood clotting or regulate cell growth.

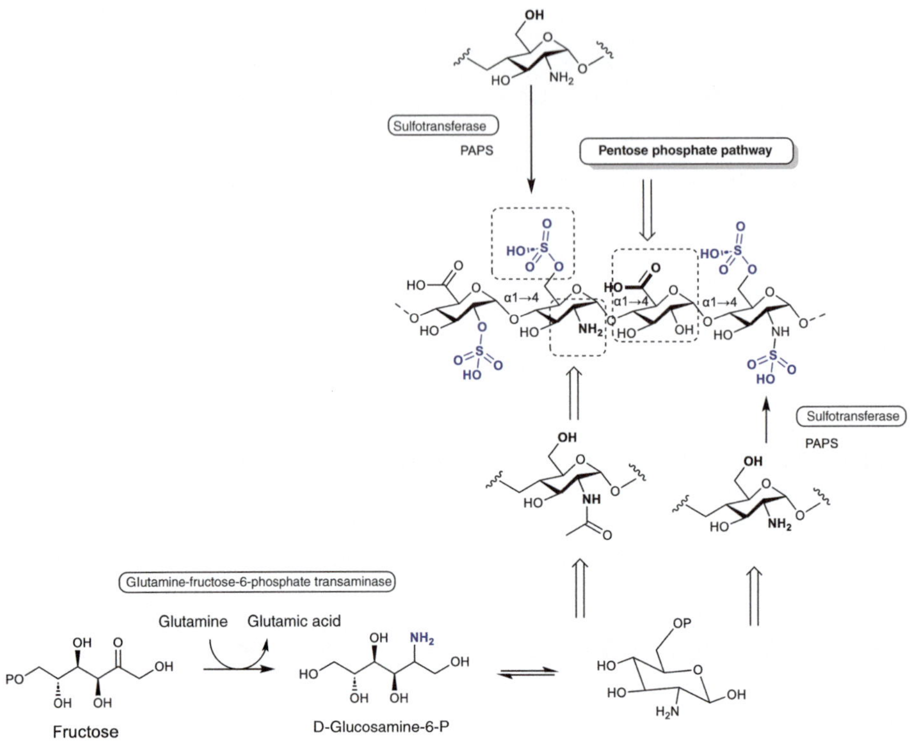

◘ Fig. 11.10 Heparin biosynthesis

Biological Effect Heparin is an important component of the blood clotting system. It inhibits blood clotting by binding to the clotting factor thrombin and blocking its activity. Heparin is therefore used as a medication for the prevention and treatment of thrombosis and embolisms. However, heparin has other functions in the body. For example, it plays a role in cell communication, wound healing, and the development of blood vessels. Heparin biosynthesis is vital for humans. Disturbances in heparin biosynthesis can lead to severe diseases such as thrombosis and bleeding.

11.3.3 Hyaluronic Acid

Chemistry Hyaluronic acid (HS, ◘ Fig. 11.11) is a linear biopolymer from the group of glycosaminoglycans, which is not sulfated. The molecule consists of a repeating disaccharide unit of N-acetyl-D-glucosamine and D-glucuronic acid, which are β-1,3-glycosidically linked. The polysaccharide has a chain length of 2,000 to 25,000 repeating units, in which the monomers alternate. The chain formation, in turn, is based on a β-1,4-glycosidic linkage of the repeating units. Since hyaluronic acid has a lower acid strength than heparin, the protein binding is correspondingly lower. The molar masses of the unbranched chains vary between 10^4 and 10^7 Da, resulting in molecule sizes of up to 5 µm. The properties of HS vary depending on the molecule size due to molecular interactions, such as Van der

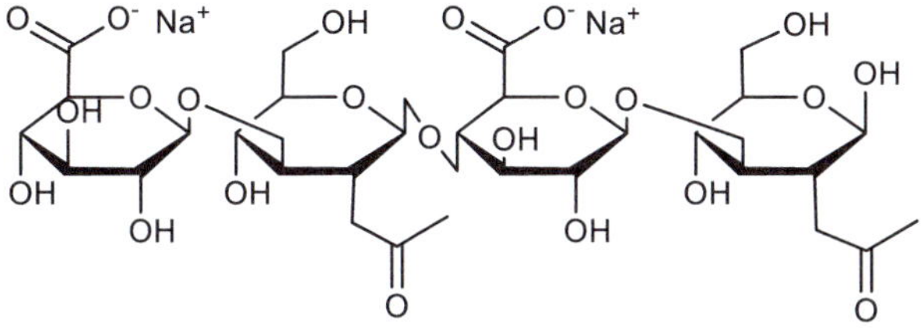

11.3 · Aminopolysaccharides

☐ Fig. 11.11 Section of the hyaluron chain

Waals forces or hydrogen bonds. Hyaluronic acid can exist in both solid and liquid states. The solid form of the biopolymer is referred to as sodium hyaluronate. This is the salt of HS, in which sodium ions form a salt with the carboxyl group of the polymer (Dicker et al. 2014).

Biosynthesis The biosynthesis of hyaluronic acid begins with the provision of the building blocks N-acetyl-D-glucosamine and D-glucuronic acid. These building blocks are provided from glucose via glycolysis in the cell (☐ Fig. 11.12). The two building blocks are activated with the help of UTP and fed to the hyaluronic acid synthase located in the cell wall, which has the task of linking the two building blocks, extending them, and providing them as an exopolysaccharide for excretion. The biosynthesis of hyaluronic acid is a continuous process. As soon as a disaccharide building block is attached to the chain, the next disaccharide building block is added. The hyaluronic acid synthase is a very efficient enzyme. It can attach up to 1,000 disaccharide building blocks per minute to a chain. The biosynthesis of hyaluronic acid is regulated by a series of genes. These genes encode for enzymes involved in biosynthesis, as well as proteins involved in the regulation of biosynthesis.

The technical biosynthesis is carried out with short chain lengths in Streptococcus zooepidemicus, which will be explained further below. In the mammalian organism, hyaluronic acid is found in fibroblasts, chondrocytes, and keratinocytes. Hyaluronic acid is an important structural component of the extracellular matrix. It contributes to the strength, elasticity, and fluid binding of the extracellular matrix.

Biotechnology The industrial extraction or production of hyaluronic acid is based on two methods: on the one hand, extraction from animal cells, and on the other hand, microbial fermentation (Brown and Pummill 2008; Sugahara et al. 1979). In the first method, which is no longer of great industrial interest due to the possible presence of prions, animal cells from rooster comb, bovine synovial fluid (joint fluid), or human umbilical cord were used. The contamination of the product with prions meant that the bovine spongiform encephalopathy (BSE, a disease that leads to the regression of brain substance in cattle) posed a risk to humans.

The advantage of this old process was that high molecular weight hyaluronic acid (from 1.2 MDa in rooster comb to 14 MDa in bovine synovial fluid) was produced in animal cells, which would have been very interesting for use in eye drops. However, a disadvantage of the method was that the hyaluronic acid had to be extracted

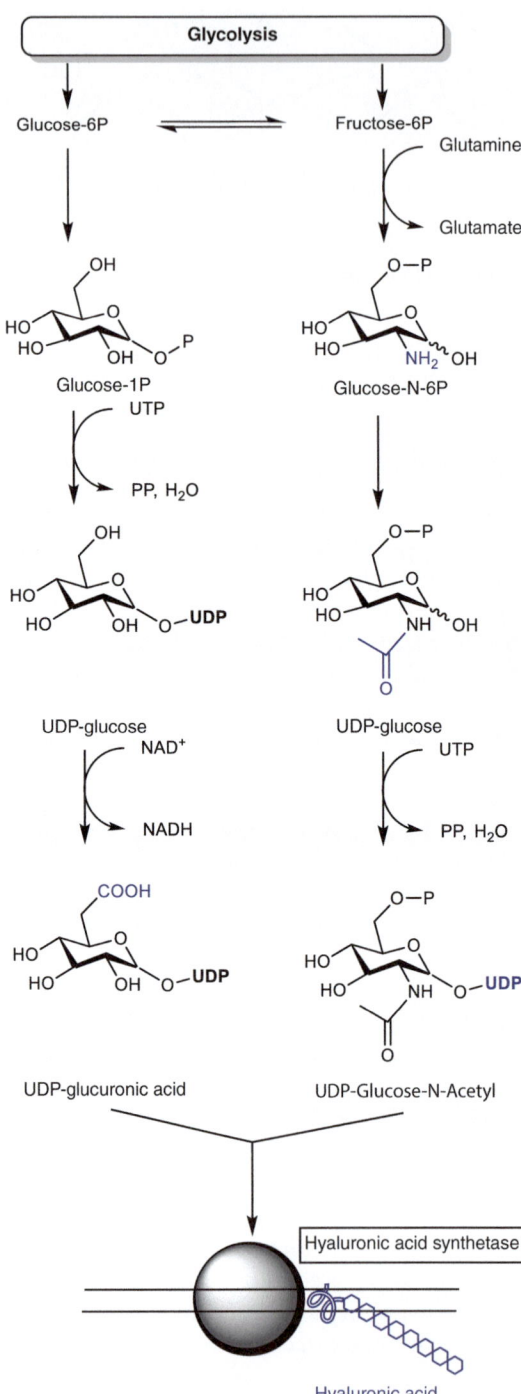

Fig. 11.12 Biosynthesis of hyaluronic acid

in a complex process due to its high viscosity. The protocols include steps such as the use of proteolytic enzymes (e.g., papain), ion pair precipitation (e.g., with cetylpyridinium chloride), or precipitation with organic solvents, which are problematic to remove. To avoid these disadvantages, hyaluronic acid is now produced by microbial fermentation. Organisms used include bacteria of the genus *Streptococcus* spp. or *Bacillus* spp., which, however, only biosynthesize shorter chains (1.6–2.5 MDa). Through fermentation, it is possible to produce high molecular weight hyaluronic acid in the order of 1–4 MDa very efficiently, depending on the species and substrate. Since the fermentation broth of *Streptococcus zooepidemicus* with an HS concentration of 4–5 g/L has a high viscosity (400–500 mPas), the production of high molecular weight hyaluronic acids is limited by the restricted oxygen transport. However, the oxygen supply is essential for bacterial metabolism. Another disadvantage when using streptococci is the presence of pathogenic endotoxins (peptidoglycans of the cell wall), which pose a pharmaceutical challenge.

Application   In medicine, hyaluronic acid is used as a storage for drugs and proteins, as a 3D scaffold for cell cultures, for wound healing, for bone and cartilage regeneration, and in cancer therapy. Due to its biocompatibility, its water-binding properties, and its permeability to nutrients, oxygen, and other water-soluble metabolic products, hyaluronic acid is suitable for treatments in many areas. Particularly in medicine, the viscosity enhancer is used in eye drops and in aesthetic medicine for injection into wrinkles, for skin deformities, and for lip augmentation (Saranraj and Naidu 2013).

❓ Self-check Questions

1. Explain the name "carbohydrates". What is their role in the organism?
2. In what form is glucose available in metabolism? What is the reason for this?
   a. Sketch the involved/related substances and metabolic pathways.
3. Name carbohydrates that are monosaccharides. Which C-bodies are considered true sugars under the narrower term?
4. Name the most common isomer of carbohydrates in nature.
5. Sketch the structure of a ring-shaped D-hexose.
6. What is the maximum yield in the fermentative production of ethanol? What is the energy yield from one mole of glucose in alcoholic fermentation compared to aerobic energy gain?
7. Define oligosaccharides and polysaccharides.
8. Name three important polysaccharides and their function in the organism. Which polysaccharide is present in all living beings?
9. Explain why "roughage" is not a source of energy for humans. (Why is the human not a side dish eater?) Name three animal species that can use cellulose as an energy source. What physiological role does muesli play as a basic food for humans?
10. What property is common to almost all polymers built from saccharides and makes them biocompatible?
11. Explain the chemical structure of heparins. Which functional group makes them biologically active?
12. Explain the structure of chitin and elucidate the chemical difference to chitosan.

Phenolic Natural Products

Contents

12.1 Simple Phenols and Phenylpropanes – 128
12.1.1 Vanillin – 130

12.2 Biosynthetic Classification – 132

12.3 Lignans – 133

12.4 Lignins – 134

12.5 Coumarins – 136

12.6 Flavonoids – 137

12.7 Styrylpyrones and Stilbenes – 141
12.7.1 Styrylpyrones – 142
12.7.2 Stilbenes – 142

12.8 Tannins (Tannins or Polyphenols) – 143
12.8.1 Hydrolyzable Tannins – 144
12.8.2 Condensed Tannins – 145

© The Author(s), under exclusive license to Springer Fachmedien Wiesbaden GmbH, part of Springer Nature 2025
O. Kayser and N. J. H. Averesch, *Technical Biochemistry*, https://doi.org/10.1007/978-3-658-47121-7_12

Chapter 12 · Phenolic Natural Products

> **Learning Objectives and Key Topics**
> - Nomenclature of Phenols
> - Shikimate Biosynthesis Pathway
> - Radical Polymerization
> - Chemical Diversity of Phenols
>
> **Technical Applications**
> - Synthetic biotechnological production of vanillin, flavonoids, and resveratrol
>
> **Biological Activities**
> - Lignans
> - Tannins

The biosynthesis of shikimic acid is the central biochemical pathway to the aromatic amino acids and their phenolic derivatives. Starting from glycolysis and the pentose phosphate pathway, the three aromatic amino acids phenylalanine, tyrosine, and tryptophan are provided via shikimic acid and anthranilic acid, which form the basis for the great variety of alkaloids, flavonoids, and cinnamic acid derivatives through structural modifications. But also technically important framework substances such as lignin and tannins can be traced back to shikimic acid biosynthesis. In ◘ Fig. 9.10 the shikimate pathway starting from its precursors from the PP pathway and glycolysis to the above-mentioned aromatic amino acids has already been explained. From erythrose-4-phosphate, a metabolite of the pentose phosphate pathway, a C7 sugar is formed by linking with phosphoenolpyruvate, which is converted to shikimic acid through various steps (cyclization, rearrangement, reduction, and dehydration). In the further course of the shikimate synthesis pathway, the aromatic amino acids phenylalanine, tyrosine, and tryptophan are formed.

In the biosynthesis of phenolic natural substances, the amino group of phenylalanine can be replaced by a keto group either by a hydroxyl group or completely omitted in favor of a double bond through ammonium lyases. If a double bond is present, these biosynthetic products are referred to as cinnamic acids or phenylpropanes (C6C3). Many phenols and their structural variants such as lignans or flavonoids are built from cinnamic acid derivatives. In addition to the previously discussed C6C3 compounds, oxidative degradation reactions (oxidative decarboxylation) are also known, which lead to C6C2 or C6C1 side chains. Through β-oxidation, as described above for fatty acids, cinnamic acid (C6C3) forms C6C1 acids, e.g., benzoic acid (C6C1).

12.1 Simple Phenols and Phenylpropanes

Phenols are aromatic natural substances in which at least one hydrogen of the aromatic compound is replaced by a hydroxyl group (◘ Table 12.1). Depending on the number of hydroxyl groups, we distinguish between mono-, di-, tri- and other

12.1 · Simple Phenols and Phenylpropanes

Table 12.1 Benzoic acid derivatives (top) Hydroxybenzoic acid derivatives (bottom)

Substituents	Name
R = H	Benzoic acid
R_4 = OH	Salicylic acid
R_2 = OH	p-Hydroxybenzoic acid
$R_2 = R_3$ = OH	Protocatechuic acid
$R_1 = R_2 = R_3$ = OH	Gallus acid

Substituents	Name
R_3 = OH	*para*-Coumaric acid
R_1 = OH	*ortho*-Hydroxybenzoic acid
$R_2 = R_3$ = OH	Caffeic acid
$R_2 = OCH_3$, R_3 = OH	Ferulic acid

polyvalent phenols. In addition to free hydroxyl groups, these can also exist as O-prenyl or methoxy groups. Phenolic aromatics often occur as aldehydes (e.g., vanillin) or acids (e.g., benzoic acid). Structural variations can greatly influence the physicochemical properties, which shape the further biosynthesis of these precursors. Most aromatics are biosynthesized via the shikimate pathway. Another variant is the polyketide pathway, and a third pathway leads via terpene biosynthesis, where only gossypol, thymol, and estrogen play a minor role and will not be further discussed. Phenolic natural substances shape the plant kingdom and occur either freely or bound to sugar (phenol glycosides). Free phenols also occur in petrochemical products, essential oils, and resins.

This group of natural substances is characterized by an aromatic ring (C6) and a propane side chain (C3). Due to the substitution of protons by hydroxyl groups, phenylpropanes are also referred to as phenols. Hydroxyl groups are reactive substituents that can influence the pH value in their free form and can exist in a deprotonated state (R-O⁻) in basic conditions. Further functionalizations of the hydroxyl group through etherification or glycosylation are ubiquitous, which is why phenylpropanes often exist in nature in a structurally modified form. Various derivatives are formed by reducing the terminal carboxyl group (cinnamic acid) and transaminating the original amino acid to an aldehyde compound (cinnamaldehyde) or an alcohol (cinnamyl alcohol), which undergo further secondary biosyn-

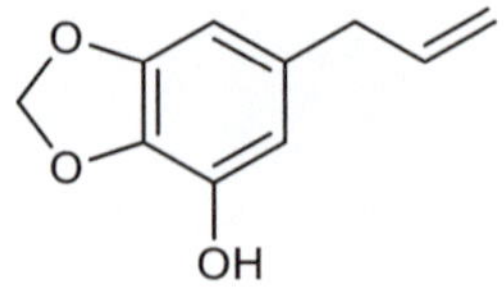

Fig. 12.1 Myristicin

thetic modifications (Table 12.1). The pharmacological effect of the mentioned natural substances is moderate. Phenols can be toxicologically problematic if they are very lipophilic, are not rapidly detoxified in the liver, have a high redox potential like quinones (oxidized phenols), or enter the bloodstream in too high doses.

Simple phenylpropanes are valued in the food and cosmetics industry as flavor and fragrance substances. Estragole (methylchavicol) is a flavoring substance found in basil (Ocimum basilicum) and is typical for the taste and smell of pesto. Myristicin is a natural substance with an interesting structure: In the nutmeg (Myristica fragrans), myristicin (Fig. 12.1) is the flavor determinant. Upon closer inspection, it is revealed that its structure is very similar to the psychotropic 3,4-methylenedioxy-N-methylamphetamine (MDMA). The intoxicating effect of myristicin, which is also an MAO inhibitor, is based on its metabolization to 3-methoxy-4,5-methylenedioxyamphetamine (MMDA). However, the usual amounts are too small for a euphoric effect. An effect is only to be expected from an intake of about 500 g of nutmeg. From this dose, however, severe side effects such as vomiting can be expected, so excessive consumption is strongly discouraged.

12.1.1 Vanillin

Vanilla and vanillin (Fig. 12.2) are the most important and sought-after flavor and fragrance substances with an annual production of approximately 18,000 tons (equivalent to approximately 600 million US$). Only a small part of this is obtained naturally. Interestingly, the great popularity of the vanilla flavor is associated with an imprinting in early childhood. In the 1960s, vanillin was added to infant food as a flavoring and thus shaped the taste of babies. Natural vanillin is a mixture of various components of the vanilla pods of the vanilla orchid (Vanilla planifolia). The plant originally comes from Mexico; today, *V. planifolia* is mainly cultivated in Madagascar and Indonesia. Natural vanilla is obtained by fermenting the pods, which turn black in the process. Vanilla is present as a glycoside, which is converted into the odor-intensive aglycone by fermentation. For mass applications, the main flavoring substance vanillin is synthetically produced due to its limited availability from natural sources and the associated costs. A large part of the synthesis precursors comes from sulfite waste from paper production.

Biotechnologically, vanillin can be produced from ferulic acid by *Amycolatopsis* or *Streptomyces* strains, as well as from eugenol by *Pseudomonas* strains in the fed-batch process. Eugenol is a natural substance derived from clove oil, which is available in sufficient quantities. Unlike fully synthetic substances, which must be referred

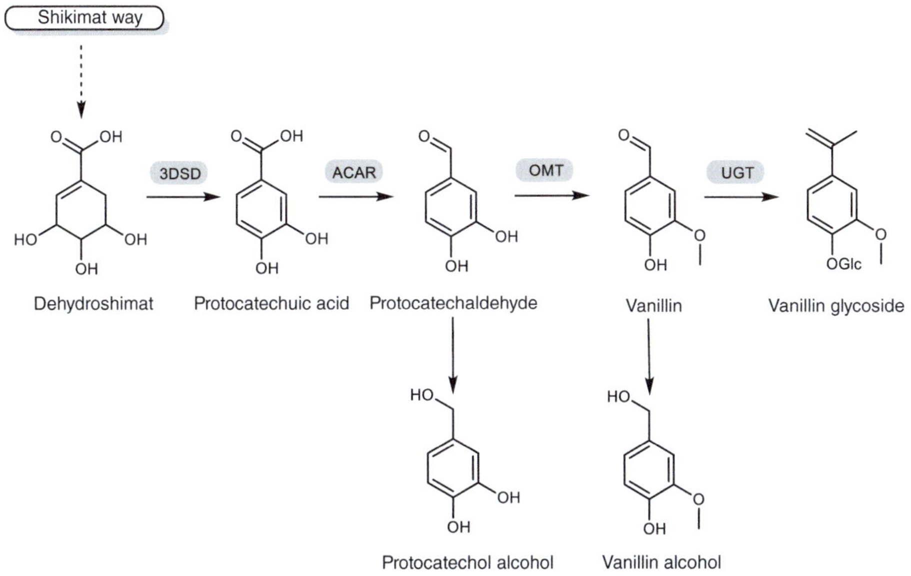

12.1 · Simple Phenols and Phenylpropanes

Fig. 12.2 Recombinant vanillin biosynthesis (Evolva process) DSD: 3-Dehydroshikimate dehydratase, OMT: O-methyltransferase, ACAR: Arylcarboxy reductase, UGT: UDP-glucosyltransferase

to as "nature-identical", biotechnologically produced products can be declared as "natural". Biotechnologically produced vanillin is cheaper than the extraction and extraction of real vanilla, but still about 60 times more expensive than fully synthetic vanillin. The Swiss company Evolva AG has developed a new biotechnological process in which both vanillin and other vanilla flavors are produced via the shikimate pathway in yeast. The genetically modified organism uses a biosynthesis pathway in which the end product is completely obtained from the central metabolism. The synthesis does not start from a precursor of vanillin, and the developed biosynthesis pathway is not known in nature (◘ Fig. 12.2).

The core of the technology is the construction of artificial yeast chromosomes (YACs), into which a large number of genes (50–150) can be integrated. These are treated by the yeast like natural genes. By the directed expression of the encoded genes, entire biosynthesis pathways can be built or supplemented. The constructed strains are tested on high-throughput systems with up to 1 million screenings per day for the biosynthesis performance of the desired products. By combining metabolic engineering, molecular biology, and flux analysis (fluxomics), the metabolic pathway can be identified and optimized through gene recombination. In ◘ Fig. 12.2, the synthesis route to vanillin based on a modified shikimate pathway is shown. Vanillin is obtained from dehydroshikimate through several biochemical transformations that are artificially built into yeast. This biosynthesis does not occur in nature. A scale-up is planned for 2025, with commercial production on an industrial scale according to the company's information for 2028.

12.2 Biosynthetic Classification

The biosynthesis of aromatic phenols from simple precursors such as sugars only occurs in microorganisms and plants. Since the shikimate pathway does not occur in animal organisms, phenolic structures such as the three aromatic amino acids are absent. In principle, the biosynthesis of aromatic compounds, which can also include phenols, is possible in three ways:

- *Shikimate biosynthesis* with sedoheptulose from the pentose phosphate pathway as a precursor
- *Acetate-Malonate Pathway*, also known as the polyketide pathway, which can build aromatic rings from C2 units via polyketides through Claisen condensations
- *Acetate-Mevalonate Pathway*, which is rather unusual as it builds isoprenoids that can be dehydrated to aromatics like gossypol or thymol.

A biosynthetically interesting property of phenols is their ability to undergo oxidative coupling (Fig. 12.3) to direct carbon compounds (C-C). In condensable catechins and lignans, phenols are oxidized to aryl radicals. The unpaired radical electron migrates stabilized from the oxygen to the aromatic carbon and allows two

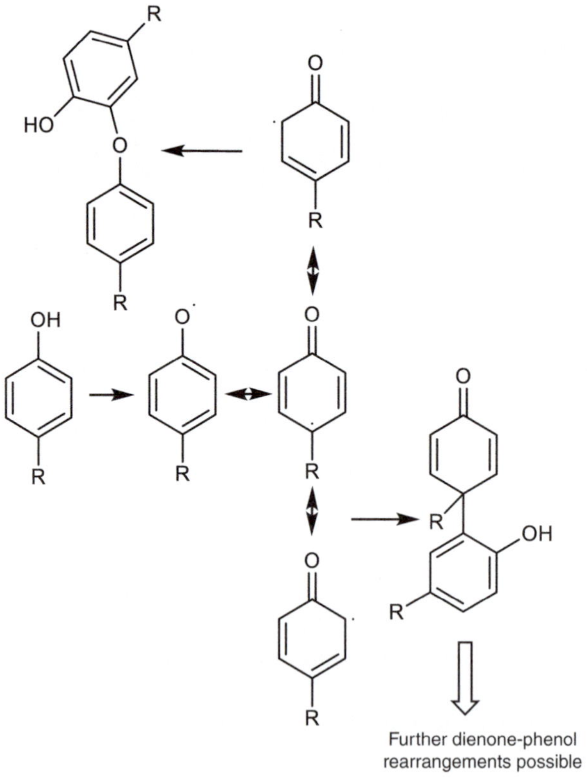

Fig. 12.3 Radical oxidative coupling

aromatics to react and couple. Of course, coupling to aryl ethers (aromatic ethers) is also possible. This chemical reaction is to be distinguished from the enzymatic browning reactions in which monophenol monooxygenases catalyze browning reactions in the presence of oxygen, as we know them from cut apples and bananas. The preferred substrate of these enzymes are *ortho*-diphenols, which are initially oxidized to *ortho*-quinones, thus enabling the polymerization of phenols, very often catechins. A biosynthetically interesting property of phenols is their ability to undergo oxidative coupling (◘ Fig. 12.3) to direct carbon compounds (C-C). In condensable catechins and lignans, phenols are oxidized to aryl radicals. The unpaired radical electron migrates stabilized from the oxygen to the aromatic carbon and allows two aromatics to react and couple.

12.3 Lignans

Through oxidative coupling of two C6C3 units of two phenylpropanes via the beta-C atom of the side chain, lignans are formed (◘ Fig. 12.4) such as podophyllotoxin (◘ Fig. 12.5). Lignans should be distinguished from the polymeric lignins, which play an important role in the lignification of plants and will be discussed separately below. Lignans are often a component of plant resins, but they are fundamentally present in all plant organs both freely and glycosidically bound. Lignans have strong cytotoxic effects and are used as drugs in cancer therapy (e.g., matairesinol or podophyllotoxin). They also show antiviral effects, e.g., against the HIV virus. Podophyllotoxin and related lignans with a *trans*-positioned lactone ring inhibit mitosis by inhibiting the assembly of the spindle apparatus (cf. colchicine). Podo-

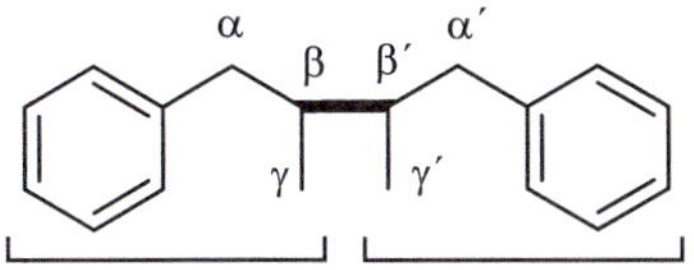

◘ **Fig. 12.4** Simple structure of a lignan from two C6C3 units

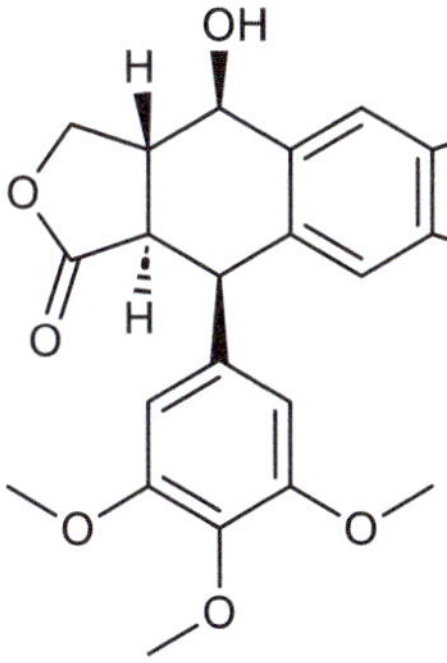

◘ **Fig. 12.5** Podophyllotoxin

phyllotoxin is applied externally, incorporated into ointments against genital warts, and used as a precursor for the semisynthesis of etoposide and teniposide, which are used as topoisomerase II inhibitors in approved cancer drugs, among other things, against small cell bronchial carcinoma, lymphomas, and Ewing's sarcoma (Pang et al. 2018; Shah et al. 2021).

Due to increasing demand, the provision of podophyllotoxin is becoming increasingly difficult. To obtain an alternative biological source, the weed *Anthriscus sylvestris*, also known as cow parsley, was genetically modified. The aim of the research is to create a transgenic line that contains the human cytochrome P450 3A4 to metabolize the naturally occurring substance desoxypodophyllotoxin directly to podophyllotoxin in the plant. The trick is that the existing biosynthesis pathway is extended in the last step with the help of human cytochrome, and thus the desired substance is produced. In this way, a "weed" can become a medically valuable plant.

Biotechnology The biotechnological construction in heterologous microorganisms seems simple, as formally two phenylpropanes need to be connected by oxidative coupling. The biosynthesis is well known, but has not yet been successfully transferred to microorganisms. Biotechnological studies are trying to produce lignans in plant cell cultures such as forsythia. The genetically modified callus cultures show basic problems in stabilizing the phenylpropane radicals and the correct spatial arrangement. So far, no technically relevant process has been realized.

12.4 Lignins

Lignins are polymeric phenylpropanes of the monomer coniferyl alcohol (◘ Fig. 12.6). Like lignans, which are dimers, polymerization in lignins also occurs via a radical metabolism, so that a C-C bond is formed via the C'-C atom or the biopolymer (molecular mass 5,000 to 10,000) is built up via hydroxyl groups (O-C). Since lignins grow in all spatial directions, it can be assumed that in a plant or a tree there is probably only a single molecule present, which can weigh several tons. When speaking technically of wood (a composite of cellulose and lignin), biologically it is the lignin that is meant, which causes the woodiness and strength of the plant as an inclusion in the cell wall (the lignin content of most terrestrial plants is about 1%). The hardness and strength of wood depend on the degree of woodiness and differ significantly between tree species such as softwood (27–32%) and birch wood (19–20%). The lignin content of wood waste products (wheat straw, bagasse) is low and is between 15 and 20%.

The extraction of lignin from wood is of interest to the paper industry. The applied technical processes are the sulfate process (kraft process) and the sulfite process in pulp production, which deliver lignin-free papers. These should technically be correctly referred to as lignin-free. The further technical use of the lignins obtained from wood is limited. Through ammonolysis, the lignins are further modified, with the end products being used as humic acid in fertilizers. Technically impure lignin mixtures are often used as adhesives, binders (particle boards, animal feed, wood pellets) or dispersants (paints, varnishes). Originally, lignins were problematic residues in the utilization of plant biomass for biogas, as they could

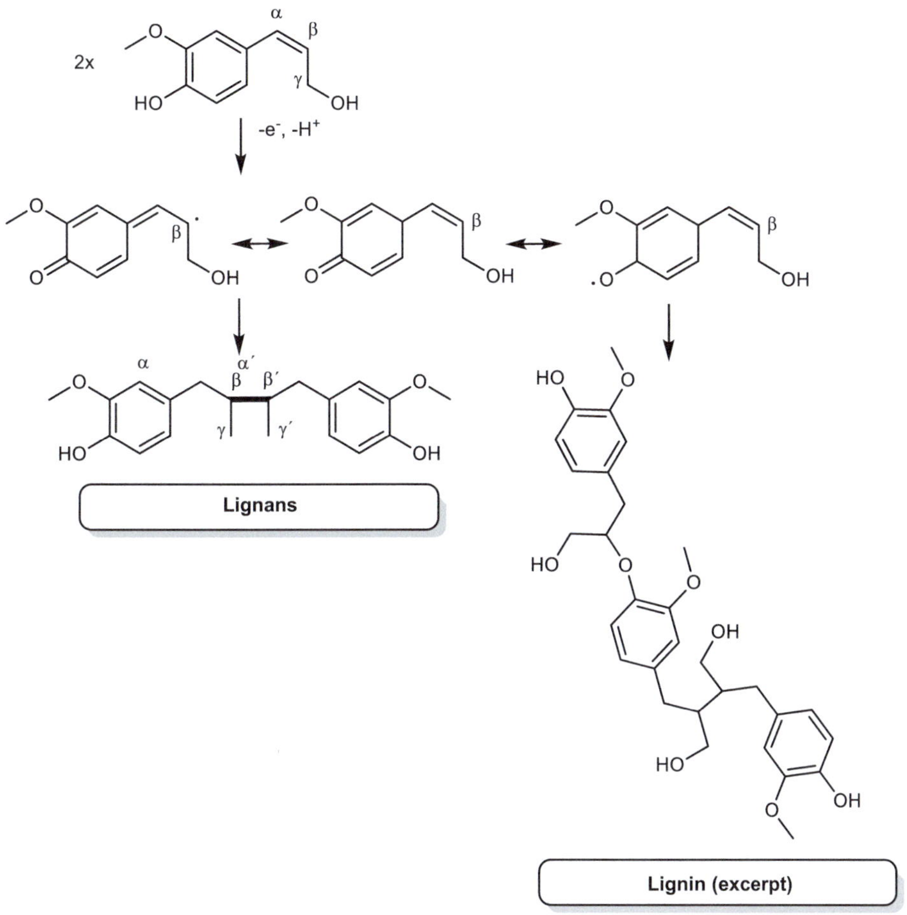

12.4 · Lignins

◻ Fig. 12.6 Lignin biosynthesis through oxidative coupling

not be degraded. Today, lignins are interesting natural substances for the production of biofuels such as wood pellets.

Biosynthesis The structure is formed by high degree polymerization, in which not only C-C-, but also C-O-couplings take place (◻ Fig. 12.6). The composition is determined by the type and number of phenylpropane monomers (monolignols) that originate from phenylalanine and tyrosine from the shikimate pathway. The most important precursors are *para*-coumaryl alcohol, coniferyl alcohol, and sinapyl alcohol.

Biotechnology The degradation by microorganisms is known and is now used for biotechnological utilization, especially for the extraction of fats. The so-called bio-funneling (English funnel = funnel) starts from lignin, which is broken down into the three cinnamyl alcohols mentioned above. With the help of bacteria such as *Rhodococcus* species, protocatechuic acid and catechol are obtained, which are transferred by other species such as Acinetobacter calcoaceticus or Bacillus alcalophilus into the β-ketoadipic acid metabolism. It should be noted that this is not a purely biological process, as the lignin must be chemically and thermally pretreated. In industrial practice, co-cultures of microorganisms are preferably used.

12.5 Coumarins

The basic structure of coumarin is 1,2-benzopyrone (◘ Fig. 12.7), which was first isolated in 1822 from the tonka bean Coumarouna odorata. As a lactone, this compound is strongly pH-dependent. At acidic pH, the coumarins (◘ Table 12.2) are ring-closed with a lactone as an internal ester.

At a basic pH value, the lactone ring opens and the coumarin is present as o-hydroxy-cinnamic acid. Often these hydroxycinnamic acids are glycosylated and stored in the cell vacuole. Only when the cell is damaged and the vacuole is broken open, these glycosides are split by glucosidases into the coumarin aglycone, which then forms the ring-shaped coumarin.

Often this degrading step occurs "post mortem" through wilting or mechanical cutting. This explains the typical hay or woodruff smell, which is caused by the conversion of the glycoside to a pleasantly smelling coumarin. Coumarins play no technical role. They are found as fragrances in woodruff products such as jelly, May punch or in the Polish vodka Żubrówka, which is mixed with coumarins. A characteristic feature is the stalk of bison grass (*Hierochloe odorata*) in the bottle, which is more decorative in nature. The biosynthesis of coumarins is derived from the shikimate pathway. From phenylalanine, a *trans*-cinnamic acid is formed, which is rearranged under the influence of UV light into a *cis*-cinnamic acid and forms a lactone, an internal ester. However, this first and not further hydroxylated coumarin is rare. Further derivatives are formed by cytochrome P450 reactions, in which ar-

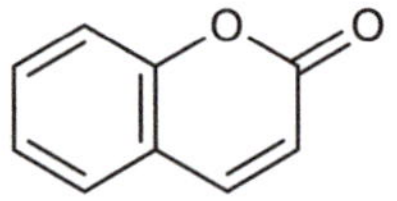

◘ **Fig. 12.7** 1,2-Benzopyrone (Coumarin)

◘ **Table 12.2** Coumarin derivatives

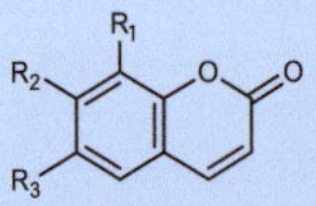	
Substituents	**Name**
R_1 = OH	Umbelliferone
R_2 = R_1 = OH	Esculetin
R_2 = OCH_3, R_1 = OH	Scopoletin

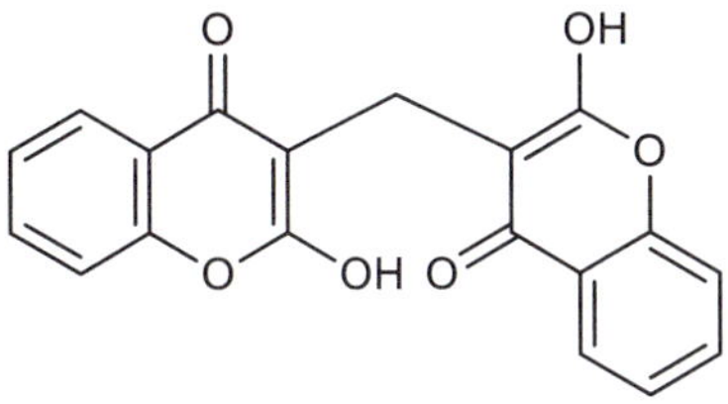

12.6 · Flavonoids

Fig. 12.8 Dicoumarol

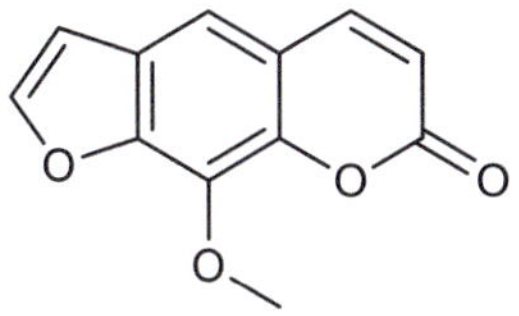

Fig. 12.9 Psoralen

omatic hydroxylations are preferred over the 1,2-double bond. Umbelliferone as a 4-hydroxycoumarin is formed directly from tyrosine; further biotransformations follow as described above.

In addition to monomeric coumarins, dimeric coumarins are known, which often arise through microbiological biotransformations. Dimeric coumarins such as Dicoumarol Dicumarol (Fig. 12.8) show effects on blood clotting, which is why they are used as anticoagulants in humans, but also as rat poison. Prenylated coumarins play an important role as furanocoumarins in medicine. They have photosensitizing properties, are activated by UV-A light, and can intercalate into DNA. These photochemical properties make furanocoumarins interesting for the treatment of psoriasis and Skin cancer. The most important furanocoumarin is 8-methoxypsoralen (Fig. 12.9) from the fruits of Ammi majus, Apiaceae, which is used in the so-called PUVA therapy.

12.6 Flavonoids

Flavonoids (2-Phenylbenzopyrones) are synonyms for a large group of plant phenols, consisting of two aromatic rings and a pyrone ring (Fig. 12.10). The correct designation for this class of natural substances is flavones, which are responsible for the yellow color in plants (lat. flavus = yellow). If there is a keto group at the C4 position, it is a Flavanone; if there is a hydroxyl group, it is a flavanol. In the biosynthesis (Fig. 12.11), structural formulas of the respective basic structures are given, which clarify the nomenclature. Flavonoids are formed by a polyketide synthase(PKS) complex. In plants, this synthase is of type III. In bacteria, all three types (I, II, III) occur (see erythromycin biosynthesis), in fungi type I. Biosynthetically, flavonoids have a C15 skeleton, which can also be understood as diphenylpropane C6C3-C6. Regardless of the color and other physiological properties, all flavonoids with this skeleton are counted as part of the flavonoid group. The great

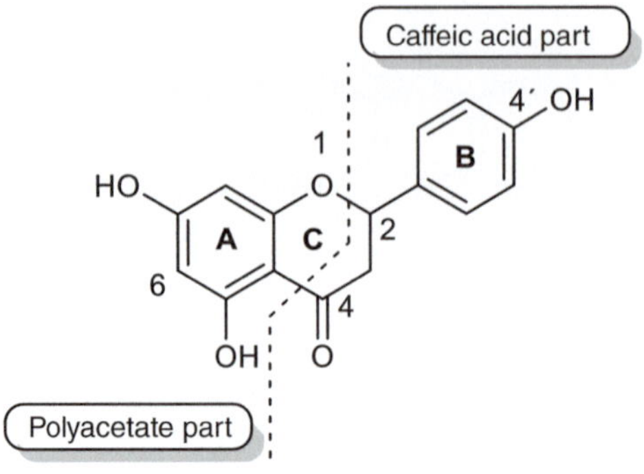

◘ Fig. 12.10 Basic structure and construction of flavonoids

structural diversity can be explained by the type, number, and localization of the hydroxyl and methoxy groups on the aromatic rings A and B (5, 7 and 3', 4', 5') as well as by the type, number, and linkage of the sugars as glycosides. Flavonoids occur in all higher plants; their biosynthesis is not known in animal organisms. The most common flavonoid is quercetin. Although the structure of the monomeric flavonoids varies greatly, other interesting structures such as biflavonoids and lignan flavonoids are known. The latter have great importance in the therapy of fungal poisonings (e.g., Death cap) to prevent liver failure. In the plant cell, flavonoids mainly occur as glycosides in the vacuoles and in the cytosol.

The biosynthesis of flavonoids (C6C3–C6) takes place via two metabolic pathways (◘ Fig. 12.11). The acetate pathway provides the C2 building blocks for the A-ring (C6), the shikimate pathway forms the B-ring with the C6C3 building block. After linking the three malonyl-CoA units with the cinnamic acid-CoA unit, a polyketide is formed, which is biotransformed to chalcone in a Claisen condensation by naringin. A chalcone is a natural substance that contains two aromatic rings connected by a propane chain. Through nucleophilic attack of the hydroxyl group at C9 on C1 of the double bond, the third ring, a pyran ring or pyran-3-one, is formed in the presence of the keto group. The resulting naringenin is the first flavonoid, which can now undergo various structure-determining biotransformations. Two biotransformations of great importance will be briefly mentioned: firstly, the biotransformation to the proanthocyanidins, the monomeric precursors of the tannins, and secondly, the biosynthesis of the isoflavonoids, which have great importance in medicine. In the case of the isoflavonoids, the B ring is not at position C2, but at position C3. This shift occurs after the completion of flavonoid biosynthesis through a cytochrome P450 reaction with O_2 as a cofactor. The cytochrome deprotonates at C3, and allows the rearrangement of the aromatic compound into the so-called *iso*-position.

Catechin, a flavan-3-ol or simply flavanol, is a hydrogenated flavone. Another important derivative is epicatechin with its stereoisomers. Catechins are found in the leaves of many types of tea, including green tea (Camellia sinensis). They are attributed with the health-promoting effect of tea as antioxidants. However, they are also responsible for the bitter taste. Structurally, they belong to the proanthocyanidins, which are also attributed with other positive effects such as tumor prevention.

12.6 · Flavonoids

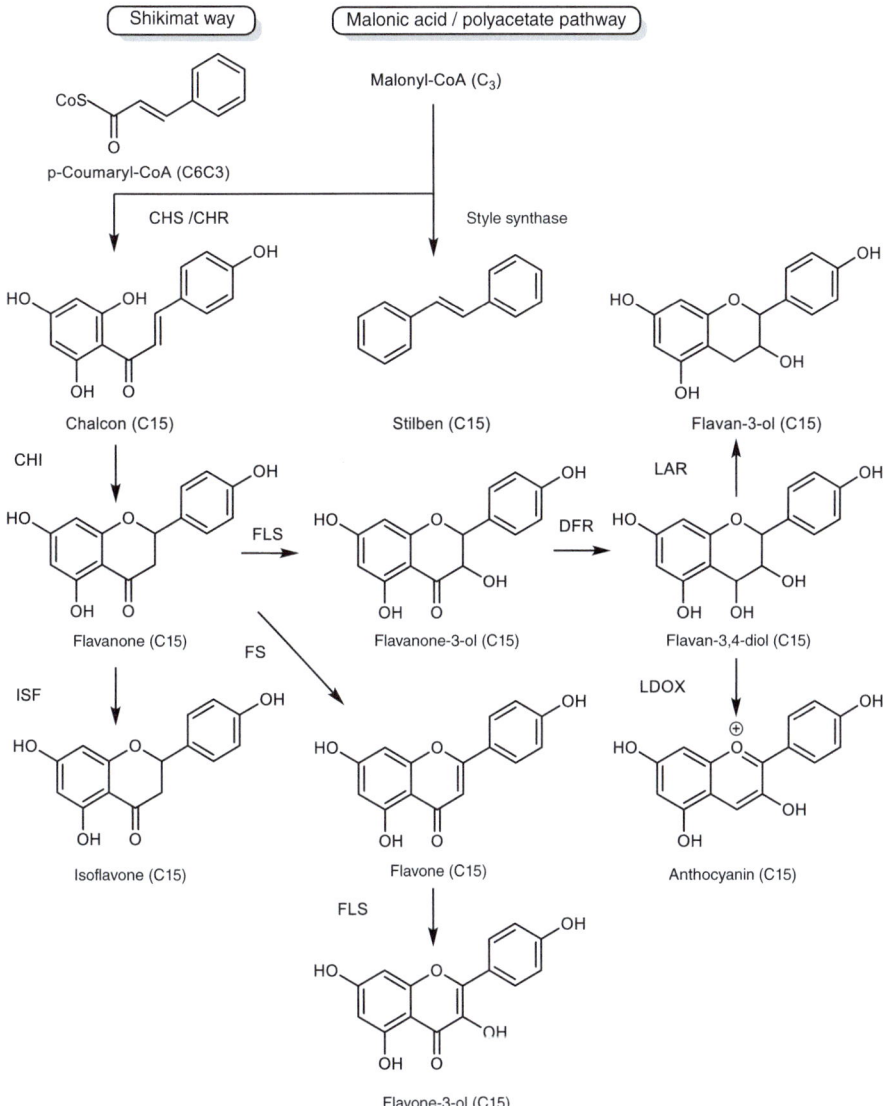

Fig. 12.11 Overview of the individual flavonoid subclasses. Abbreviations of the enzymes: CHS (Chalcone synthase), CHI (Chalcone flavanone isomerase), F3H (Flavanone 3-hydroxylase), FS (Flavone synthase), ISF (Isoflavone synthase), FLS (Flavonol synthase), DFR (Dihydroflavonol reductase), LAR (Leucoanthocyanidin reductase), LDOX (Leucoanthocyanidin dioxygenase)

Starting from flavanon-3-ols, a reduction of the keto group in C4 position to flavanol-3-ols occurs, where the hydroxyl group at C4 is removed by a reductase. The remaining flavanol-3-ol is highly reactive and tends to polymerize (usually about 20 to 100 units are linked together). The resulting polymers are called tannins or tannic acids. They also contribute to the flavor formation in black tea or cocoa. They are also referred to as "condensed tannins". Tannins are biologically intended inclusion compounds in the wood or bark of the plant. Their antibacterial effect serves as protection against fungal infections, and their astringent effect keeps herbivores away. That's why oak woods with a high tannin content were often used in shipbuilding. Other technical properties include:

- Woods with a high tannin content (e.g., oak) are used in wooden shipbuilding because they rot slower and have antifouling properties.
- In the tanning of animal hides in the leather industry, tannins have been and are again being used. The metal salts traditionally used in technical tanning (aluminum salts) can cause allergies, which is why sheepskins for children are now being tanned naturally again.
- Manufacture of barrels for the storage and maturation of wine and whiskey. The age and origin of the wood (usually French oak) have an influence on the taste.
- However, tannins can also be problematic, e.g., in biogas plants, as microorganisms cannot or only hardly degrade them. Because of their antibiotic properties, they can kill important microorganisms that are essential for the operation of the plant.

The pharmacology and application of flavonoids are not always clear and uniform, as there are very large structural and also biological differences. Many so-called "anti" effects such as antibiotic, antineoplastic or anti-inflammatory (anti-inflammatory) have been described, even if they are not always supported by rational studies. However, one of the many effects can be considered proven with great certainty, namely the activity as radical scavengers (Eng. free radical scavengers), which capture or prevent the spread of so-called free radicals as harmful cell poisons in the body. Organisms with a high flavonoid content are typically yellow-colored plants such as chamomile (Matricaria recutita), goldenrod (Solidago virgaurea) and linden (Tilia cordata), but also wine (Vitis vinifera). Flavonoids have so far only been detected in plants, not in microorganisms.

Biotechnology   Recently, the biosynthetic pathways to these flavonoids in *E. coli* were cloned and apigenin and kaempferol were heterologously produced (Zha et al. 2019). The artificial biosynthesis (◘ Fig. 12.12) shows a combination of enzymes from very different domains. Starting from the two known biosynthetic pathways, the shikimate and the polylactate pathway, the construction of coumaric acid-CoA is carried out by the 4-coumarate-CoA ligase from Streptomyces coelicor, a microorganism. The construction to chalcone is catalyzed by the plant chalcone synthase (CHS) from Glycyrrhiza echinata to form apigenin in the final step, which is formed chemically and not enzymatically.

Two flavonoids are of biotechnological interest: moringenin (◘ Fig. 12.13) is used as a dye and naringin (◘ Fig. 12.14) as a sweetener in the food industry. The flavonoid is a glycoside in grapefruit (*Citrus paradisi*), which also occurs in oranges. Medically, it can be a problem in food-drug interactions, as it is an inhibitor of the liver's CYP450 enzymes (CYP3A4, CYP1A2, and CYP2A6).

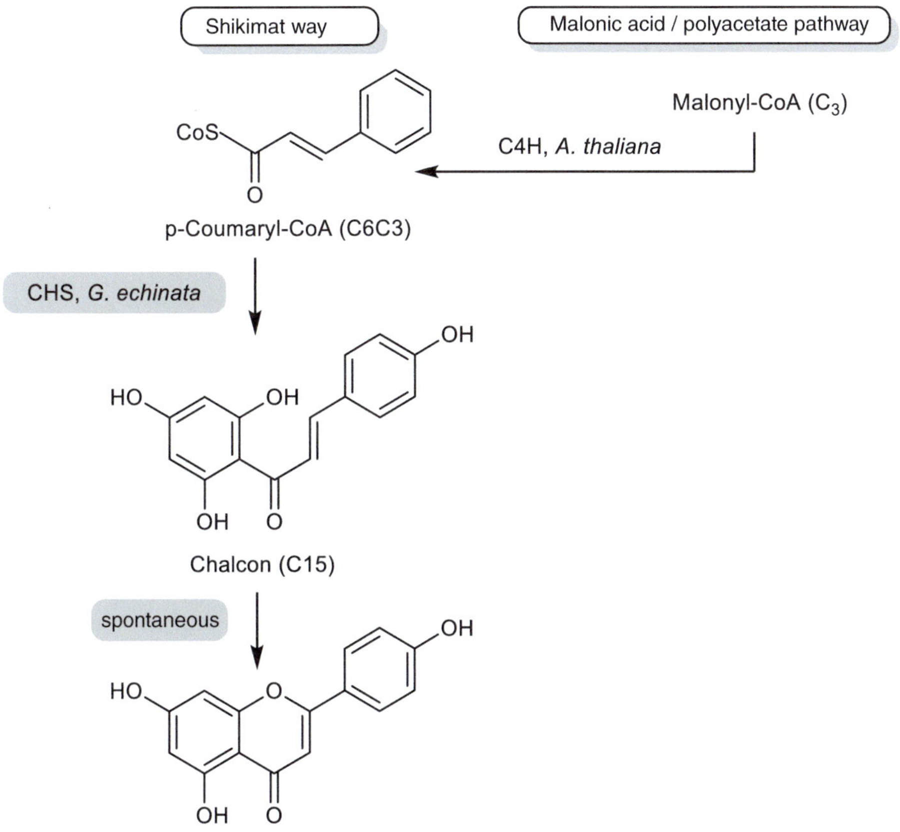

12.7 · Styrylpyrones and Stilbenes

Fig. 12.12 Recombinant flavonoid synthesis in *E. coli*

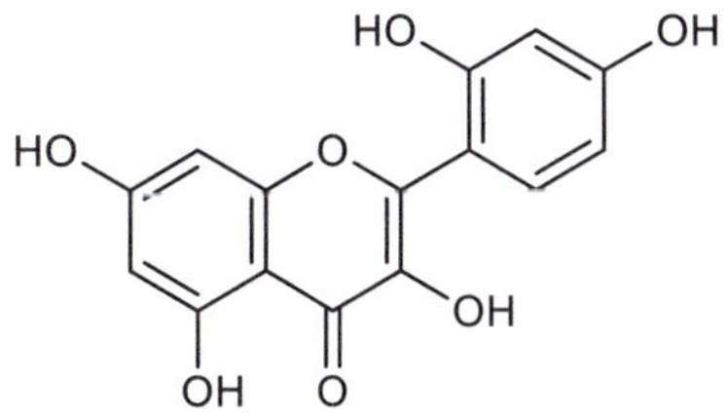

Fig. 12.13 Moringenin

12.7 Styrylpyrones and Stilbenes

In addition to flavonoids, other natural substance groups are also built up via the shikimate (C6C3) and acetate (C2) pathways. Although new compounds are constantly being discovered, only two large groups will be discussed here.

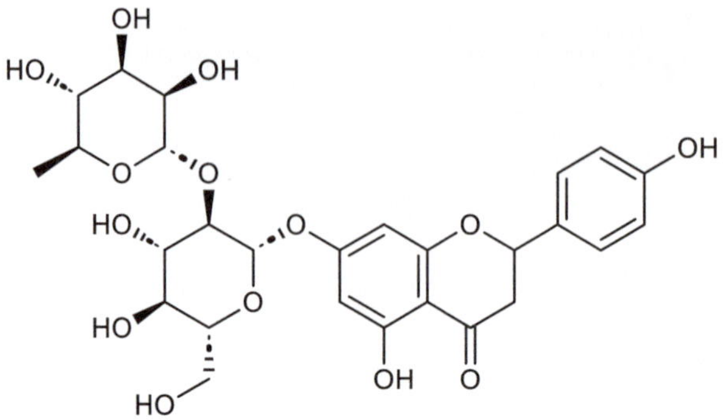

◘ **Fig. 12.14** Naringin

12.7.1 Styrylpyrones

This rare group of natural substances is found in the family of pepper plants (Piperaceae), particularly the genus *Piper* should be mentioned here. The most well-known source for styrylpyrones is the plant *Piper methysticum*, which is consumed in Polynesia and Melanesia as kava-kava due to its anxiolytic and antidepressant effects. Due to the plant name, pyrones such as kavain (◘ Fig. 12.15) are also referred to as kavapyrones. Pharmacologically, they are characterized as active ingredients that act in the brain similarly to the neurotransmitters glutamine, GABA, dopamine, and serotonin. Because of their intoxicating effect, they are also called "intoxicating pepper" and pose a societal problem as a drug in combination with alcohol in Australia and New Zealand. In Germany, kavain-containing medicines were withdrawn from the market in 2015 due to potential liver damage.

12.7.2 Stilbenes

Stilbenes at first glance look similar to styrylpyrones, but the biosynthesis proceeds differently. Typical representatives are resveratrol (◘ Fig. 12.16) in wine and the an-

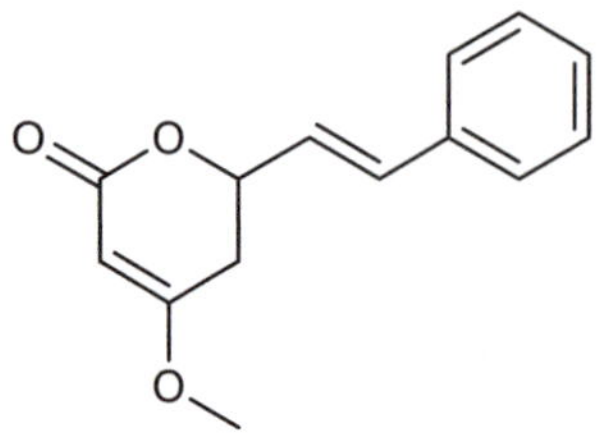

◘ **Fig. 12.15** Kavain

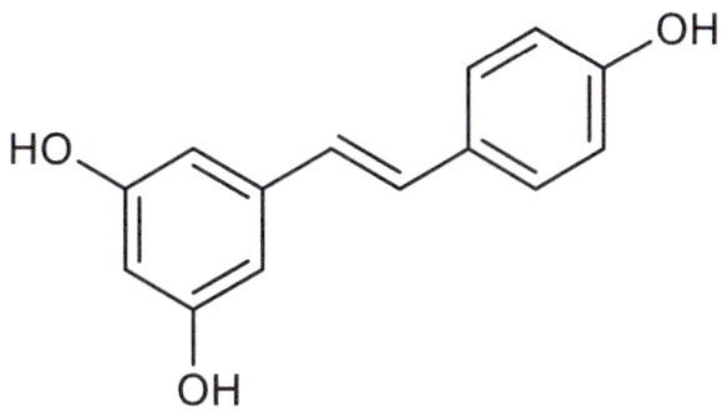

Fig. 12.16 Resveratrol

tineoplastic combretastatin. Today, over 300 stilbene structures are known, many of which are found in mosses and ferns. Stilbenes are often found in bark tissue, suggesting a possible protective function (feeding deterrent). Similar to flavonoids, 4-hydroxycinnamic acid (p-coumaric acid) and three malonyl-CoA building blocks are structural components of a polyketide synthase complex (PKS). After decarboxylation, the first and simplest stilbene, resveratrol, remains, which occurs as a natural substance in red grapes and accumulates in red wine. In the past, resveratrol was considered an important natural substance for explaining the so-called French Paradox. This paradox refers to the unexplained correlation between red wine consumption and the lower risk of heart attacks and coronary heart disease among the French, who simultaneously consume above-average amounts of butter and meat products. Today, it is known that in addition to resveratrol, the tannins (polyphenols) in red wine and a balanced diet with lots of fruits and vegetables are also important. However, this knowledge should not lead to excessive consumption of red wine, as too much alcohol can negate this positive effect. A "daily dose" of one glass of red wine is therefore sufficient, and for those who want to avoid alcohol, red grape juice is recommended, which contains equal parts tannins and resveratrol.

Resveratrol can also be obtained biotechnologically today. The biosynthesis follows the pathway already discussed in Fig. 12.12. In initial studies, a heterologous biosynthesis pathway was established in *E. coli*. The yield is well below 1% (titer: 5.8 mg/kg) and is therefore considered very low compared to the natural substance content in wine. Ongoing work is concerned with protein design, increasing the pool, and providing malonyl-CoA, the amount of which probably limits metabolic capacity (Ibrahim et al. 2021).

12.8 Tannins (Tannins or Polyphenols)

The natural substance name tannins is derived from their technical application, as they were used for tanning animal hides in the leather industry. Tanning refers to the preservation of the skin (dermis). Today, predominantly chromium salts (chromates) are used, but due to the risk of allergies, there is also a return to the "gentle" tanning with plant tannins, especially for "baby fur" for toddlers and shoes, where the chromates dissolved by sweat come into direct contact with the skin. Tannins are widespread in the plant kingdom: Oak roots (*Quercus* spp.), Ratanhia roots, and rose plants (Rosaceae, such as blueberries and raspberries) are particularly rich in tannins. Tannins are mainly found in hard plant organs such as wood, bast, and roots, offer protection in nature against feeding (bitter and astringent taste), and have antimicrobial and insecticidal effects. Technically speaking, tanning is an important industrial process, the end product of which, leather, is of great importance. Tanning is chemically-physically understood as the penetration of polyphenols into the collagen of fresh animal skin, which make the skin water-resistant and stabilize it through covalent and ionic bonding and the formation of stable hydrogen bridges to the proteins.

Tannins are complex and polymeric natural substances that are not obtained as pure substances, but as mixtures or extracts. Chemically, a distinction is made between condensed and hydrolyzable tannins. Without going into detail here, it can be said that the hydrolyzable tannins consist of sugars and gallic acid(derivatives) and the condensed tannins mainly consist of proanthocyanidins. The latter are condensed catechins. The ecological significance is complex: Tannins are probably important plant defense substances against herbivores because of their bitter taste. In addition, it is suspected that the loss of dietary protein makes the consumption of the plants unattractive.

12.8.1 Hydrolyzable Tannins

As the name suggests, these tannins, which are glycosidically built up from sugars (glucose, rhamnose, sugar alcohols) and gallic acids, can decompose under the influence of acid, i.e., be hydrolyzed. Hydrolyzable tannins like the ellagic acid (◘ Fig. 12.17) have no special technical applications, they are rather found in medicine (dentistry, tannins), in dermatological products and in certain foods such as red wine and tea (determining the taste).

Biosynthesis Although the biosynthesis of this class of substances is intensively researched, the exact formation of the oligomeric and polymeric structures is still poorly understood due to their high complexity. The simplest structure is the linkage of gallic acid with glucose to pentagalloyl glucose. The linkage of gallic acid occurs with UDP-glucose or other sugars. As with proanthocyanidins, C-C bonds can also form between the free positions C2 and C5 of gallic acid to ellagitannins, consisting of ellagic acid (◘ Fig. 12.17). The biosynthesis of gallic acid is based on the shikimic acid biosynthesis (C6C1). In addition to purely hydrolyzable tannins,

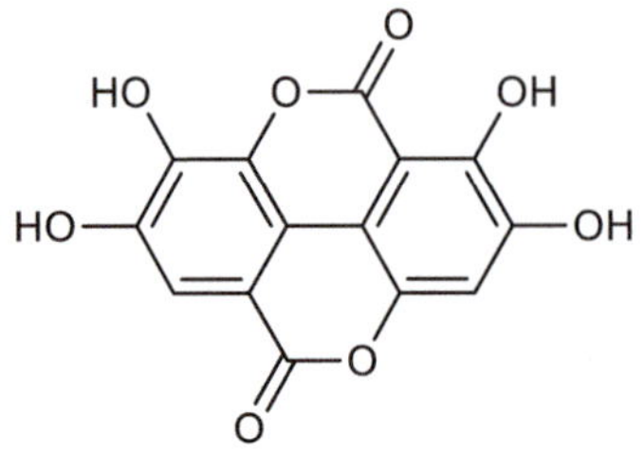

◘ **Fig. 12.17** Ellagic acid

mixed forms are also known, in which the gallic acid is bound to the phenolic hydroxyl groups of a condensed tannin. A typical example is galloylgallocatechin, which is known as a tumor-protective natural substance from green tea.

12.8.2 Condensed Tannins

This type of tannin consists of proanthocyanidins (PA) (e.g., catechins) as monomer units, which originate from flavonoid biosynthesis and polymerize via C-C bonds. The condensed tannins are divided into oligomeric and polymeric proanthocyanidins according to their size. Since depolymerization by hydrolysis is not possible, this group of tannins is referred to as "condensed". The basic building block is flavan-4-ol, whose biosynthesis via the shikimate pathway and the shortened polyketide pathway has already been explained above (◘ Fig. 12.11). The linkage of the two units to form a C-C bond is interesting. The linkage of the two proanthocyanidins often occurs oxidatively in C4AC8 or C6AC8 position. The biosynthesis is largely unclear in terms of type and number; it is unknown how highly complex structures with unusually high molecular masses of over 20,000 are biosynthesized. Derived from the proanthocyanidins, anthocyanins play a role as color pigments in plants and also in the food industry. Depending on the pH value and hydroxylation pattern, they carry a positive charge on the hetero-oxygen (flavylium cation) and have a red, violet, or blue color. The function of these dyes in the plant is UV-B protection on the one hand, and on the other hand, they serve as attractants for insects for pollination. Anthocyanins also play an important role in microbiological degradation reactions, e.g., in the fermentation of tea. The formation of the yellow and red dyes is based on complex reactions in which quinones are formed by oxidation reactions on the aromatic B-ring. These are very reactive and react to form high molecular weight, amorphous pigments, the phlobaphenes. Phlobaphene were first observed in the tanning of hides into leather.

Unlike today, tannins were used extensively in tanning in the past. The tanning process is based on the interaction between the water-soluble and swellable proteins of the skin and the polymeric tannins. In this process, the tannins are converted into quinones by oxidation, which are covalently bound to the proteins via sulfhydryl and amino groups. During the subsequent drying, the protein fibers no longer stick together and remain separate. The animal skin dries up like leather. The tannins bound to the protein surface reduce the solubility and swellability of the proteins, so that the leather does not dissolve and remains water-resistant.

❓ Self-check Questions
1. Identify the central metabolic pathways that lead into the shikimate pathway.
2. What central chemical structure characterizes products of the shikimate pathway?
3. Outline the main steps in the synthesis of tryptophan.
4. What aromatic natural substances are derived from the shikimate pathway? Name at least five.

5. Compare the natural biosynthesis of vanillin with the synthetic bioprocess of the company Evolva. Name important enzymes and metabolites.
6. Name and explain the function of the ammonium lyases in the biosynthesis of secondary natural substances, starting from the aromatic amino acids.
7. Find out how artificial biosynthesis pathways can be created using metabolic engineering.

Terpenes

Contents

13.1 Terpene Metabolism – 148

13.2 Biosynthesis of Terpenoids – 149
13.2.1 Mevalonate Pathway and Cholesterol Biosynthesis – 150

13.3 Monoterpenes and Essential Oils – 151
13.3.1 Essential Oils – 151

13.4 Sesquiterpenes – 156

13.5 Diterpenes – 160

13.6 Triterpenes – 163

13.7 Tetraterpenes – 169

13.8 Polyterpenes: Rubber – 171

© The Author(s), under exclusive license to Springer Fachmedien Wiesbaden GmbH, part of Springer Nature 2025
O. Kayser and N. J. H. Averesch, *Technical Biochemistry*, https://doi.org/10.1007/978-3-658-47121-7_13

> **Learning Objectives and Key Topics**
> - Terpenoid Nomenclature
> - Mevalonate Biosynthesis Pathway
> - DXP Biosynthesis Pathway
> - Cholesterol Biosynthesis
> - Chemical Diversity of Terpenoids
>
> **Technical Applications**
> - Enfleurage Process and Steam Distillation
> - Production of Standardized Extracts
> - Synthetic Biotechnological Production of Artemisinin

13.1 Terpene Metabolism

Terpenes are natural substances, the name of which is derived from the history of natural product chemistry. The naming is partly different: In addition to the now correct designation as terpenoids, the terms isoprenoids or terpenes are also used. This has historical reasons, as the term "terpenes" is derived from the proposal of the chemist Friedrich A. Kekulé to name this group of natural substances after the resin of the pine tree (*Pinus sylvestris*), from which turpentine oil is obtained by distillation. Turpentine oil is not a petrochemical, but a natural oil, which mainly contains 2-pinen, 2,10-pinen and 3-caren. Terpenes consist exclusively of carbon. Biosynthetically, terpenoids are clearly defined. In contrast to terpenes, they also contain functional groups such as hydroxyl groups and consist of a multiple of the C5 monomer. This monomer is isopentenyl diphosphate (IPP), which represents the "active isoprene" as the end product of the mevalonate pathway or the methylerythritol phosphate pathway (MEP). Both biosynthesis pathways will be discussed in detail below. Through secondary modification and extension to C15 to C30 building blocks, this very universal building block can be considered one of the most interesting in natural biosynthesis, as it establishes the group of natural substances of the terpenoids with more than 30,000 known structures (Hosseini and Pereira 2023). Isoprenoids or terpenoids occur in both plants and animals and microorganisms. The terpenoids are divided according to the multiple of the C5-IPP building block into

- Monoterpenoids: C10 or 2 × C5
- Sesquiterpenoids: C15 or 3 × C5
- Diterpenoids: C20 or 2 × C10
- Triterpenoids: C30 or 2 × C15
- Carotenoids: C40 or 2 × C20
- Polyterpenoids: n × C5

The linkage of the C5-IPP monomers follows the isoprene rule or "head-tail" rule (◘ Fig. 13.1). If you imagine the IPP molecule as a fish, the branched end is the tail, into which a second fish bites with its head. This simplified image is known as the so-called isoprene rule with a C1→C4 coupling. But here too, exceptions

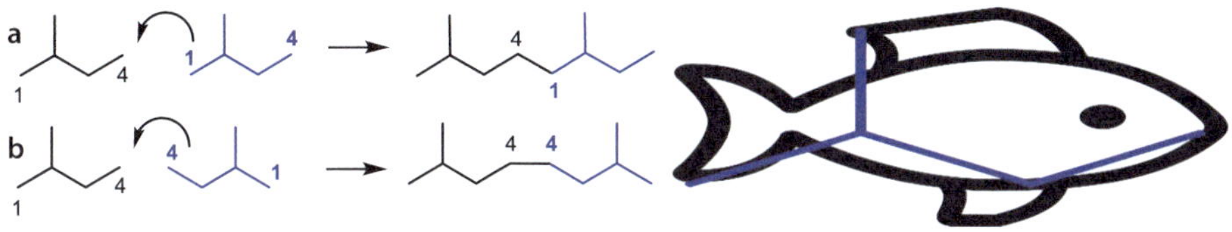

Fig. 13.1 Head-tail- **a** and head-head linkage **b** of the IPP-DMAPP condensation according to Friedel-Craft and the image of the fish as an example for the head-tail isoprene rule (right)

are known, which we will discuss further below. A prominent exception should be mentioned in advance: The coupling of IPP as a C5 building block to non-terpenoids such as proteins, the direct prenylation of hydroxyl groups or directly to carbons leads to biologically active hemiterpenes. Hemiterpenes are very rare; they are chemically modified C5 building blocks, e.g. by oxidation to hydroxides (prenol) or carboxylic acids (tiglic acid, isovaleric acid).

All other terpenes with two or more linked building blocks play an important role as natural substances in the food, cosmetics and flavor industry. Fragrances, for example, have accompanied humans throughout their cultural history. Today, the fragrance industry is a "big business" with a turnover of 11 billion €. However, about 75% of these aroma or fragrance substances are synthetically produced and only 25% are obtained by extraction from mostly plant-based raw materials. In addition to plant-based, animal fragrances such as musk and amber are also in high demand. To meet the customers' desire for organic products, the flavor industry has shown a strong interest in the increased use of biotechnological processes in recent years, as the example of patchouli shows.

13.2 Biosynthesis of Terpenoids

Active isoprene (isopentenyl diphosphate = IPP) can, as already mentioned, be formed via two biosynthetic pathways. The mevalonate pathway, known for several decades, decarboxylates three acetate units via mevalonic acid (C6 unit) to IPP as a C5 unit. In the mevalonate pathway, three ATP are consumed, and IPP is formed via a carbenium ion, which is in resonance isomerism with dimethylallyl diphosphate (DMAPP) due to the rearrangement of the double bond. This biosynthetic pathway, presented below, occurs in many eukaryotes including humans, in fungi such as baker's yeast, and in plants, and mainly serves the construction of cholesterol, ergosterol or sitosterol, which are important components of the cell membrane (Table 13.1). Cholesterol is a typical animal steroid found in the cell walls of mammalian cells. In contrast, ergosterol is typical for fungi, and sitosterol is characteristic for plants. Reasons for the differences in the side chain can be seen in the influence on fluidity. Fungi and plants live under colder climatic conditions and have a significantly lower maximum temperature in the environment than mammalian cells at 37 °C. A lower temperature means less fluidity in the cell walls, which is countered by structural chemical changes. In plants, the mevalonate pathway is also important for the biosynthesis of sesquiterpenes and triterpenes. In the animal

Table 13.1 Important sterols in mammals, plants, and fungi

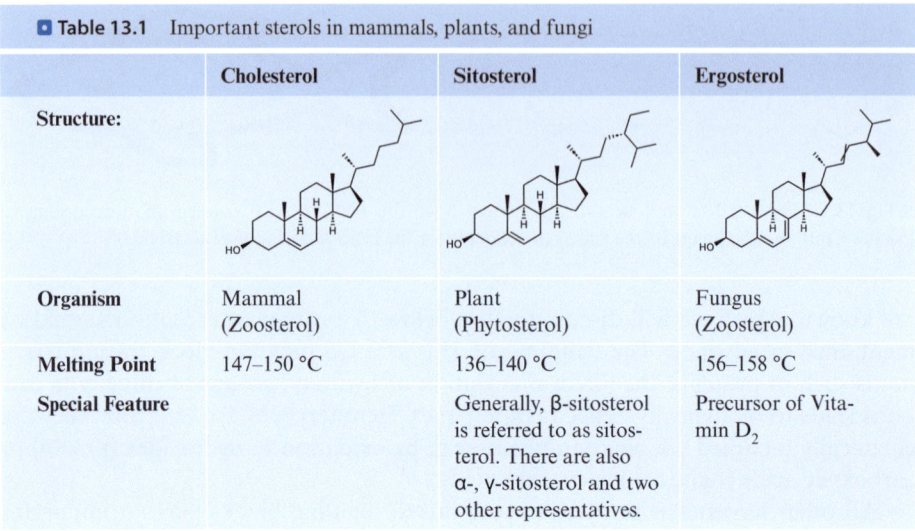

	Cholesterol	Sitosterol	Ergosterol
Structure:			
Organism	Mammal (Zoosterol)	Plant (Phytosterol)	Fungus (Zoosterol)
Melting Point	147–150 °C	136–140 °C	156–158 °C
Special Feature		Generally, β-sitosterol is referred to as sitosterol. There are also α-, γ-sitosterol and two other representatives.	Precursor of Vitamin D_2

organism, cholesterol is an important branching point to bile acids, gluco- and steroid hormones (testosterone, estrogen, vitamin D) like bile acids.

13.2.1 Mevalonate Pathway and Cholesterol Biosynthesis

At the end of the 1970s, a second biosynthesis pathway to IPP/DMAPP was found, which surprisingly did not consume acetate units. It turned out that this biosynthesis pathway begins with the linkage of pyruvate and glyceraldehyde-3-phosphate (G3P). Decarboxylation results in deoxy-D-xylulose-5-phosphate (DXP), which was formerly the name for the biosynthesis pathway (◘ Fig. 13.3 right), but was changed to methylerythritol phosphate and thus to the MEP pathway. The two biosynthesis pathways are spatially separated in the cell. The mevalonate pathway (◘ Fig. 13.3 left) is located in the cytosol and the MEP pathway in the plastid. However, the focus of this chapter is not the biosynthesis of the various terpenoid secondary substance classes. Of particular interest is cholesterol biosynthesis, which is found in many eukaryotic organisms. Interestingly, the MEP pathway (◘ Fig. 13.3 left) is found in many prokaryotic microorganisms, but rarely in higher animal organisms such as humans. This finding gave rise to the idea of developing a drug against plasmodia, the causative agents of malaria. In plasmodia, in contrast to humans, the biosynthesis of methylerythritol phosphate (MEP) is important for sterols. Since these biosynthetic enzymes do not occur in humans, the MEP pathway was considered as a drug target for specific drug development. With fosmidomycin (◘ Fig. 13.2), such a substance was also found, but did not make it to approval for financial and toxicological reasons. There are also approaches to use the MEP pathway in biotechnology for the production of isoprenoids by replacing the mevalonate pathway in *S. cerevisiae* with the MEP pathway. This offers the advantage of lower energy consumption and thus higher productivity. In addition, the product yield of the mevalonate pathway in baker's yeast is higher than with

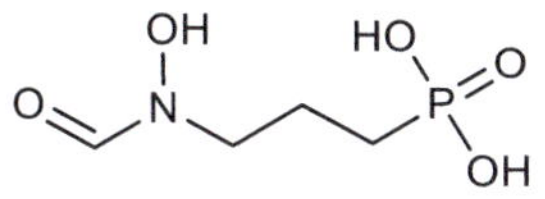

Fig. 13.2 Fosmidomycin

the MVA pathway. Thus, it is possible to completely replace isoprenoid biosynthesis and dispense with the MVA pathway.

13.3 Monoterpenes and Essential Oils

The basic reaction to monoterpenes is the linkage of IPP and DMAPP according to the isoprene rule. The term "mono" is misleading if one assumes that a C5 building block is meant. Rather, the linkage of two C5 units to a monomeric building block (C10) for the further terpenes is meant, because only with the C10 building block, the geranyl diphosphate (GDP), a real terpene is created. The analogous terpene, which lacks the diphosphate group and the oxygen function, is 2,6-dimethyloctane, which, like most monoterpenes, is volatile. Although the number of carbon atoms and substituents is small, a large structural diversity results from various biosynthesis variants. Responsible for this are terpene cyclases and terpene transferases, which can fold the carbon chain differently and link the ends, so that in addition to linear, also ring-shaped mono-, bi- and tricyclic monoterpenes are formed (■ Fig. 13.3).

In further biotransformations, additional substitutions, usually oxygen substitutions, can be carried out by monooxygenases, which alter the chemical, physical, and biological properties. A special group of monoterpenoids are the iridoids. Their basic structure is 2,6-dimethyloctane, which is linked from C3 to C7 (nepetalactone type). By oxidizing the terminal methyl groups C1 and C9, a second ring closure is made possible, and the resulting hemiacetal can be stabilized with glucose in a glycosidic bond. These glycosidic iridoids are referred to as secoiridoids, which are sensitive to hydrolysis and represent important building blocks in further biosynthesis pathways (e.g., in alkaloid biosynthesis, see ▶ Chap. 14).

13.3.1 Essential Oils

Many plants can be identified by their characteristic smell, such as peppermint, sage, or roses. Often, volatile substances that are perceptible to our nose are responsible for this. They do not occur in the plant as individual substances, but in a mixture, the essential oil (■ Fig. 13.4). We have already read in the chapter on fats and fatty oils that essential oils fundamentally differ from classical oil (see ■ Table 13.2). When essential oils are distilled, a solid, highly viscous residue remains, which is referred to as resin or balsam.

Although it is a mixture of substances that can consist of up to 3,000 substances, essential oils have common properties:

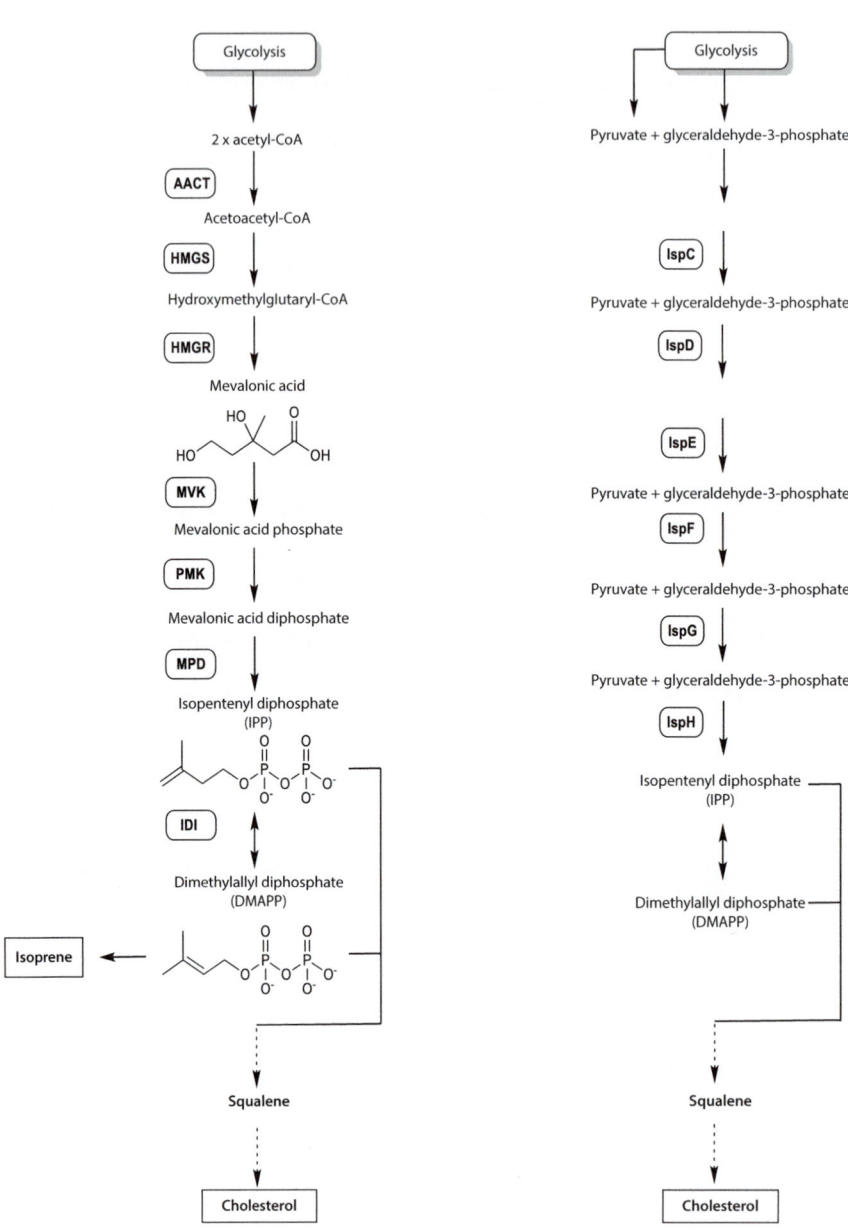

Fig. 13.3 Simplified Mevalonate (MVA) biosynthesis pathway (l) and biosynthesis of Methylerythritol phosphate (MEP) (r). DXS: 1-Deoxy-D-xylulose 5-phosphate synthase, DXP: 1-Deoxy-D-xylulose 5-phosphate, DXR (IspC): DXP reductoisomerase, GPP: Geranyl diphosphate, HMG-CoA: 3-Hydroxy-3-methyl-glutaryl-CoA, HMGR: 3-Hydroxy-3-methyl-glutaryl-CoA reductase, IDI: Isopentenyl-diphosphate:dimethylallyl-diphosphate isomerase, IPP: Isopentenyl diphosphate, IspG: 2-C-Methyl-D-erythritol-2,4-cyclodiphosphate reductase, IspH: 4-Hydroxyl-3-methylbut-2-enyl diphosphate reductase, MEcPP: 2-C-Methyl-D-erythritol-2,4-cyclodiphosphate, MEP: 2-C-Methyl-D-erythritol-4-phosphate (Methylerythritol phosphate), MVA: Mevalonate (Mevalonic acid)

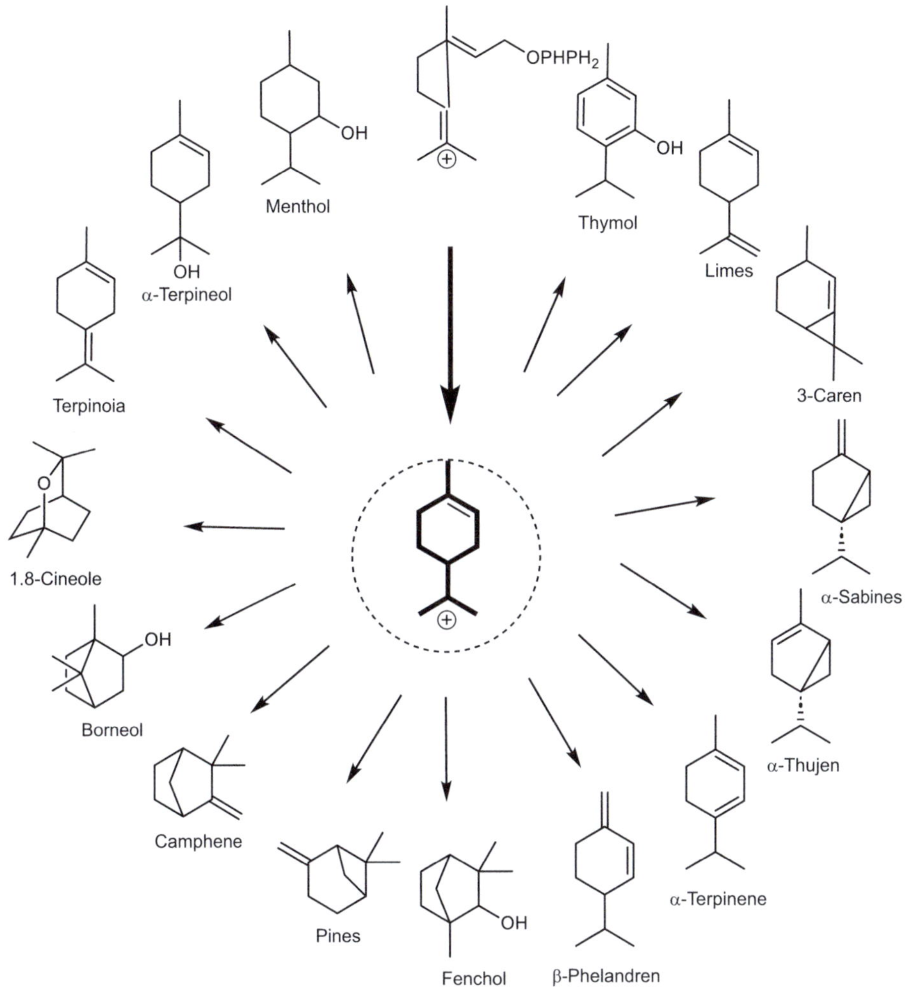

13.3 · Monoterpenes and Essential Oils

Fig. 13.4 Chemical diversity of monoterpenes based on the terpene building block in essential oils

- The majority are colorless or slightly yellowish liquids (exceptions are chamomile oil (blue) and *Hypericum* oils (red)).
- They have a characteristic smell and taste.
- They are volatile at room temperature.
- They have a lower density (p<1 g/mL) than water (exceptions are clove oil and cinnamon oil due to the very high aromatic content).
- They are often optically active.
- They have a high refractive index (depending on the number of double bonds).
- They show low solubility in water (e.g., rose water).
- They dissolve well in organic solvents.

Various methods are used to extract essential oils in technology and especially in the cosmetics industry. The classic and physically simplest method is the steam

Table 13.2 Differences between fatty and essential oils

	Fatty Oil	Essential Oil
Chemical Structure	Fatty acid esters with glycerol	Terpene mixture
Biosynthesis	Fatty acid biosynthesis	MEV and MEP biosynthesis
Localization	Oil cell	Trichome
Biological Function	Storage substance	Diverse (defense against herbivores, attraction of pollinators, antibiosis, fragrance)
Physical Properties	High melting point High evaporation temperature	Low melting point Low evaporation temperature

distillation of plant material such as flowers or leaves, as has been described for centuries for rose oil, chamomile oil, and peppermint oil (▶ Infobox 13.1). Another method is extraction with organic solvents such as ethanol, acetone, or isopropanol, which is used for less volatile compounds or to extract non-distillable substances.

Essential oils can oxidize and resinify when stored in the air (autoxidation, polymerization, hydrolysis), which can significantly change their chemical and physical properties (consistency, pH value, smell, taste). Quality control to verify identity and composition is carried out analytically using GC-MS, less often using HPLC-MS. The complex composition is used analytically as a metabolic fingerprint.

The extraction of essential oils in technology and especially in the cosmetics industry is carried out by various methods. A particularly gentle process is extraction with organic solvents, such as the enfleurage process, which produces very high-quality essential oils.

Chemistry and biological effect Essential oils are complex mixtures that exhibit a large variation of long-chain, cyclic mono-, di-, and sesquiterpenes and their derivatives. In addition to the aliphatic terpenoids, aromatic compounds are also present, which are derived from the shikimate pathway. Other organic compounds are alcohols, aldehydes, organic hydrocarbons, and various epoxides, carboxylic acids, ethers, and esters. Only a few essential oils consist of a single compound. A very good example is clove oil, which is used in dentistry for anesthesia. This clove oil consists of 95% eugenol; however, the quality is determined by the minor components that make up the smell and taste. The vast majority of essential oils are multi-component mixtures with up to 5,000 different substances, the description of which would go beyond the scope of this book.

Volatile substances can be isolated from almost all plants by steam distillation. The content of essential oils is usually not very high and typically fluctuates between 0.5 and 2%. However, from a content of 0.01%, we already speak of essential oil plants. The essential oil accumulates in the plants at special storage locations. Often these are trichomes on the leaf surface. In some plants, the essential oil is also located in specialized oil channels, which run through the fruit or the trunk of the

13.3 · Monoterpenes and Essential Oils

> **Infobox 13.1: Steam Distillation**
>
> Volatile substances form compounds with steam due to their hydrophilic properties. Some substances that have a higher boiling point than water can be extracted with water at about 100 °C (at 1 atm) through evaporation. This makes them significantly more volatile than the respective pure substance. This behavior, which can be observed, for example, in predominantly lipophilic terpenes and related chemical compounds, can now be used to extract substances from plant material using so-called steam distillation.
>
> In steam distillation, water and plant material are heated in a closed system. The steam penetrates, destroys the cell structures with the lipophilic essential oils, and the rising steam carries these with it. The mixture separates again in the condenser (see below). The composition and quality of the essential oils obtained depend heavily on the distillation conditions, i.e., the pressure and temperature of the steam, the design of the distillation apparatus, and the duration of exposure to the plant material.
>
> It is much more effective if the steam is passed through the plant material, e.g., on a grate. The distillate is collected in a receiver, with most essential oils (except for clove oil, sassafras oil, and cinnamon oil) having a lower density than water and therefore floating on the water surface, so that the phases can be easily separated from each other using a separating funnel. For this purpose, so-called Venetian bottles (conical collecting vessels) were used in the past.
>
> Different qualities can be obtained or desired ingredients can be specifically enriched through fractional distillation. A classic example is the extraction of blue chamomile oil, which is colored blue by a chemical degradation reaction to chamazulene. An exotic example is Ylang-Ylang (*Cananga odorata,* Annonaceae), a tropical tree species from Southeast Asia. It is harvested only once a year in the early morning and must be immediately processed into Ylang-Ylang oil in a several-hour steam distillation. In a further distillation process, Cananga oil can also be obtained from the same flowers.
>
>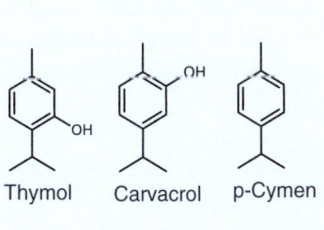
> Thymol   Carvacrol   p-Cymen
>
>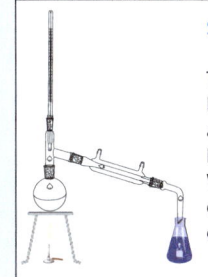
>
> **Steam distillation**
>
> The plant material is located in the piston and is connected to Diluted with water. The Water vapor drives the essential oil enters the cooler and is collected.

plant. Biologically particularly interesting are the trichomes, which look like small mushrooms and whose head cells produce the oil.

Many essential oils are used in the pharmaceutical, cosmetic, and food industries. In pharmacy, essential oils are used in so-called OTC drugs (OTC = over the counter, i.e., available without a prescription in the pharmacy) for mild colds of the respiratory tract. Other indications are joint pain (pine oil), skin inflammations (witch hazel oil), sprains *(Calendula officinalis),* and bruises *(Arnica montana).*

Most essential oils can also be found in the cosmetics industry, which is more due to marketing experts than to rational application (e.g., peppermint and calendula in toothpaste). In the food industry, essential oils such as peppermint oils are processed in confectionery or used as flavor correctors. However, essential oils can also be dangerous: The administration of camphor oils to children under 12 years and especially to infants is prohibited, as camphor can cause reflexive respiratory arrest. Essential oils are also gaining technical importance: The US Environmental Protection Agency (EPA) recommends limonene as a replacement for CFC-containing solvents. It has also been shown that eucalyptus oil is suitable as a diesel substitute for trucks and that essential oils can be good solvents for paints.

13.4 Sesquiterpenes

Sesquiterpenes are widespread in the plant kingdom. They often have a bitter taste and frequently serve in the plant for the defense against predators. Volatile sesquiterpenes also play an ecological role by inhibiting the seed maturation of surrounding plants and thus keeping competing neighboring plants at a distance. The structures of the sesquiterpenes have an odd number of 15 carbon atoms in the basic skeleton. Due to methyl group migrations and rearrangements (see Wagner-Meerwein rearrangement), sesquiterpenes are often not immediately recognizable in their structure. This also explains why the sesquiterpenes are the most structurally rich (5,000 are known) and have the most ring skeletons in the realm of terpenes. They occur in almost all plants, especially in those that contain essential oils.

Sesquiterpenes are known in medicine as antimicrobial, anti-inflammatory, anti-neoplastic, but also allergy-triggering natural substances. Their alkylating potential, similar to a Michael addition, particularly on thiol groups of proteins, is responsible for all these reactions (◘ Fig. 13.5). The figure shows the chemical reaction of helenalin, which is based on the same chemical reaction known for contact dermatitis.

Sesquiterpenes are formed by the condensation of IPP and GPP, resulting in the C15 building block farnesyl diphosphate (FPP) (2,6,10-trimethyldodecane, farnesol type). This long chain can be structurally modified by various terpene cyclases

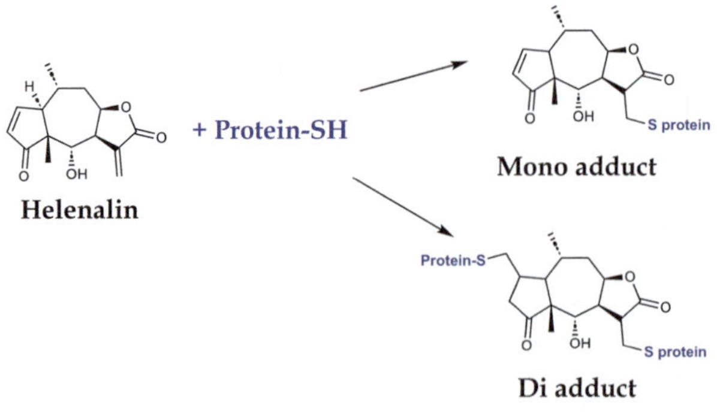

◘ **Fig. 13.5** Alkylation of thiol groups by sesquiterpene lactones with exocyclic double bond

13.4 · Sesquiterpenes

and transferases, so that different ring systems (mono-, di-, tri- and tetracyclic) are formed through internal cyclization. Oxidations at all points of the carbon skeleton can change the physicochemical and biological properties. Typical sesquiterpenes include, for example, artemisinin (◘ Fig. 13.6) as an antimalarial agent, phytotoxins, and phytohormones. Of the many known sesquiterpenes, artemisinin (Quinhaosu) as an antimalarial agent is of particular interest. In addition to the many known nitrogen-containing antimalarial agents, artemisinin is remarkable as a ring system with a peroxide substructure. The biosynthesis is not fully understood and ends with dihydroartemisinic acid. It is now assumed that peroxidation and the formation of a trioxane ring occur through photochemical reactions in the trichomes of the plant *Artemisia annua*.

Artemisinin This natural substance (◘ Fig. 13.6) against plasmodia, the causative agents of malaria, is an important tool in the fight against this tropical infectious disease. At the beginning of the millennium, attempts were made to reproduce the biosynthesis biotechnologically in yeasts. The biotechnological biosynthesis is a milestone in synthetic biology (B. Liu et al. 2011; Turconi et al. 2014). Artemisinin plays an important role as an antimalarial drug, as it is considered the last important drug against which the parasite from the genus *Plasmodium*, transmitted by the bite of the Anopheles mosquito, has so far developed hardly any resistance. A look at the structure shows that artemisinin is very difficult to synthesize: seven stereocenters and a peroxide group in a trioxane ring make the synthetic production of artemisinin expensive and uneconomical. On the other hand, extraction and isolation from plant extracts are not easy, as a continuous supply cannot be guaranteed due to crop failures. To ensure a sustainable supply of the drug to patients, who are mostly among the poorest in the world in Asia, Africa, and South America, a Bill Gates Foundation-funded initiative to biotechnologically produce artemisinin was launched at the beginning of this millennium. However, this was only successful up to the metabolite dihydroartemisinic acid. Subsequent chemical reactions were sufficient to successfully produce artemisinin, but the economy was lower than with isolation from plants. For these reasons, the project was discontinued by the pharmaceutical industry. Although the goal was not achieved and today a semi-synthetic process leads to artemisinin, the results are impressive and show an interesting path in synthetic biology.

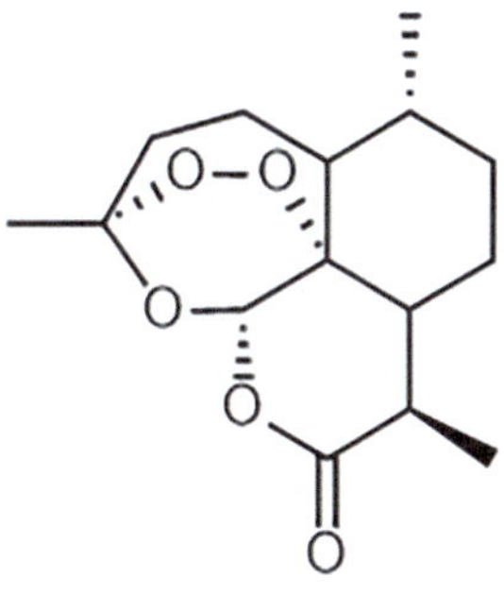

◘ **Fig. 13.6** Artemisinin

Gossypol Gossypol (Fig. 13.7) is a dimeric sesquiterpene from cotton *(Gossypium arboreum)* and other species that inhibits spermatogenesis. In the 1970s, clinical trials were conducted in China with men to develop gossypol as a "pill for men". However, these trials were discontinued because a large proportion of men remained permanently infertile after discontinuation of the natural substance (Coutinho 2002).

Cannabinoids No other plant is currently at the center of heated discussions as much as *Cannabis sativa* L. For some, it is simply a gateway drug that should be banned, for others it is an important medicinal plant, the potential of which is not yet known because its use is still prohibited in Germany (as of 03/2025). It is undisputed that *C. sativa* can lead to health damage like any drug when improperly used after illegal procurement. In particular, smoking ("getting high") in adolescents and very young adults can lead to a delay in mental development. Nevertheless, the ingredients of *Cannabis sativa* are highly valued in the treatment of multiple sclerosis, chemotherapy with vomiting, glaucoma, Huntington's disease, Tourette's syndrome with compulsive actions, Parkinson's disease, and chronic pain. The main ingredient is $\Delta 9$-tetrahydrocannabinolic acid (THCA) (Fig. 13.8), the content of which in the "grass" of "grandpa's good joint" was still at 5–7%, but in today's highly bred plants can already be up to 25%. This certainly raises the question of whether it is actually still a soft drug.

Due to the great interest in cannabinoids, a biotechnological process for the production of THC and cannabidiol (CBD, Fig. 13.9) in yeasts has recently been

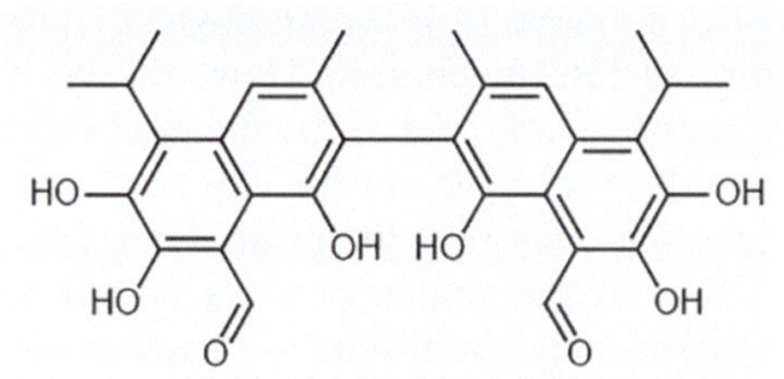

Fig. 13.7 Gossypol

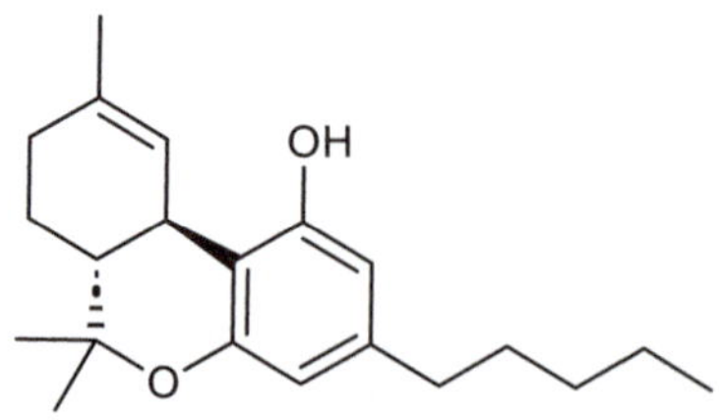

Fig. 13.8 $\Delta 9$-Tetradydrocannabinol

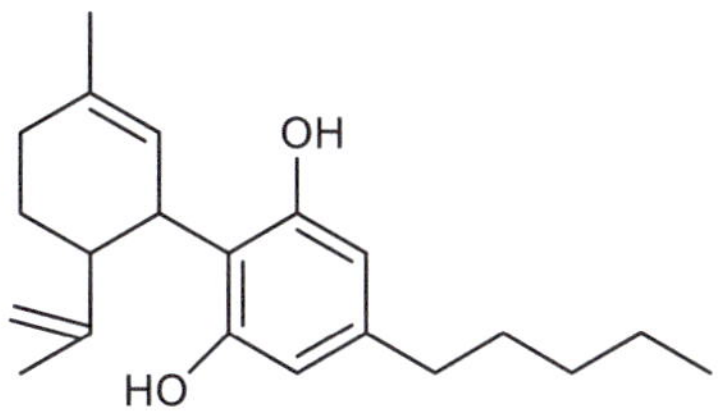

13.4 · Sesquiterpenes

Fig. 13.9 Cannabidiol (CBD)

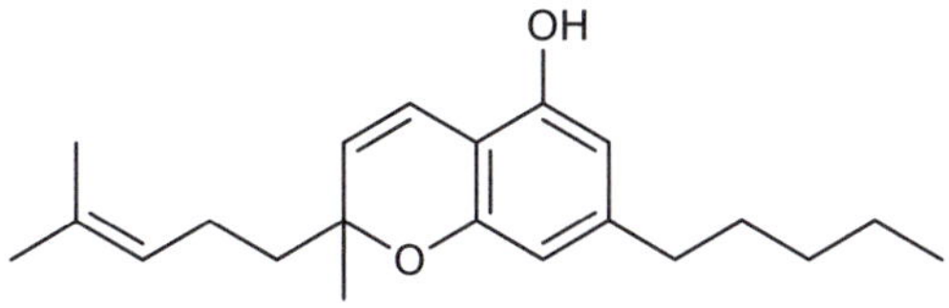

Fig. 13.10 Cannabichromene (CBC)

developed (Stehle et al. 2017; Thomas et al. 2020). Based on the polyketide pathway, which provides the aromatic precursor olivetol, a second biosynthetic building block is provided via the mevalonate pathway. This pathway, which takes place in the mitochondria, provides geranyl diphosphate (GPP), which is linked with olivetolic acid via a Friedel-Crafts alkylation. The resulting cannabigerolic acid (CBG) can now be converted to one of the three main cannabinoids by specific enzymes. The main metabolite is tetrahydrocannabinolic acid (THCA), which is converted by a THCA synthase. In addition to THCA, CBD and cannabichromene (CBC, Fig. 13.10) are produced in small amounts.

The biotechnological production (Fig. 13.11) is problematic, as the plant enzyme CsPT4 from *C. sativa* can only be poorly expressed as a membrane-bound catalyst in yeast. The misfolding leads to a significant loss of activity. An alternative is the cytosol-soluble enzyme NphB from *Streptomyces* spp. In addition to this problem, it also shows that the cellular provision of olivetolic acid or its direct conversion in the genetically modified yeast is not optimal. The linear precursor accumulates and is probably chemically converted to a lactone, which is no longer available for further reactions to THCA. The developed biotechnological process works, but a large-scale implementation is still pending.

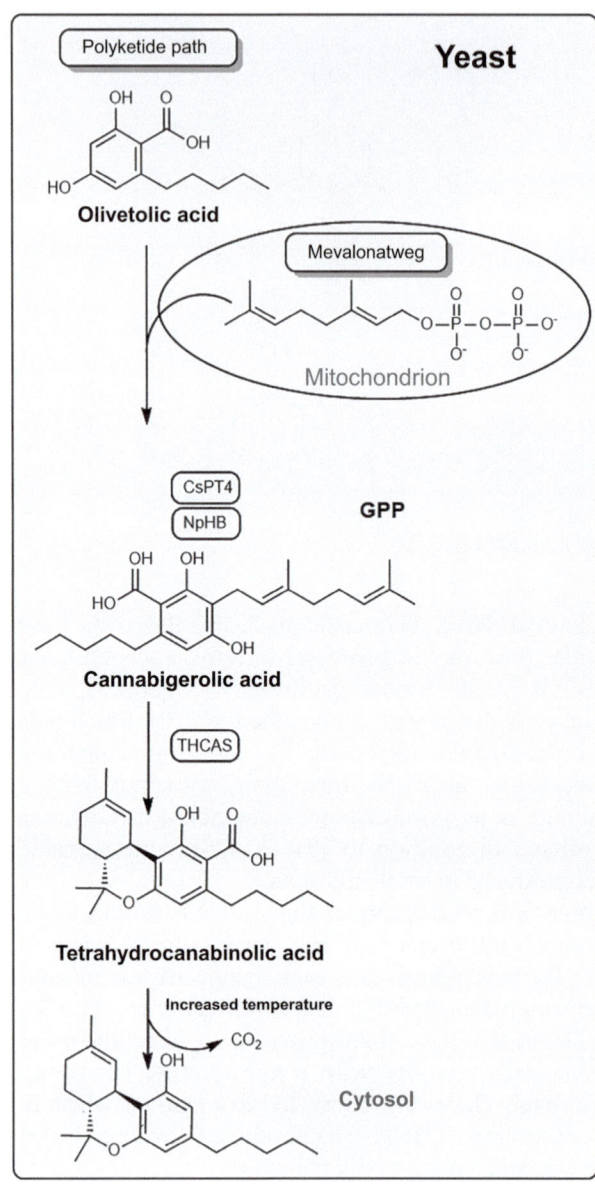

☐ **Fig. 13.11** Biotechnical synthesis of tetrahydrocannabinolic acid in genetically modified yeast

13.5 Diterpenes

Diterpenes are built from isoprene units. In total, there are four C5 building blocks that are linked to monoterpenes. Diterpenes are widespread in the plant kingdom and usually serve as a defense mechanism against predators due to their bitter taste. However, they are also popular with humans as flavoring agents, as various alcoholic beverages such as vermouth or gentian schnapps taste bitter due to diterpenes.

13.5 · Diterpenes

Other diterpenes like gibberellins play an important role as phytohormones in the plants themselves, as they regulate cell growth and cell division. Lipophilic diterpenes include phytol, a component of chlorophyll, and paclitaxel, a very lipophilic diterpene and important drug in cancer therapy.

The basic structure of diterpenes consists of two monoterpene units, which are formed by the coupling via double bond of geranylgeranyl diphosphates. Depending on the species-specific terpene cyclase, the ring closure to a macro ring system takes place, which is built up in further reactions into bi-, tri- and tetracyclic rings and supplemented by monooxygenases with oxygen functions (hydroxy, epoxy groups).

A large number of diterpenes are structurally known today, even if the level of information about their biological function is not correspondingly high. Prominent representatives are, for example, phorbol esters from spurge plants like *Jatropha curcas*, which has great technical importance as an energy plant. From the seeds of this plant, a vegetable oil is already being produced on a ton scale, which is used as biodiesel in trucks in Southeast Asia. During the technical extraction, phorbol esters are also pressed out, which can be problematic for the consumer who refuels biodiesel and thus comes into contact with them due to their pro-carcinogenic effect.

The biological significance of diterpenes is great, as important natural substances such as ginkgolides, paclitaxel (cancer drug), stevioside (sweetener) or the plant growth substances of the gibberellins have a diterpene skeleton.

Ginkgolides are found in the leaves of the ginkgo tree (*Ginkgo biloba*). The ginkgo tree is considered a living fossil and is the only surviving representative of its family (Ginkgoaceae). Ginkgolides are very complex compounds that cannot be easily synthesized. Therefore, ginkgo extracts are used for the prevention and treatment of neurodegenerative diseases such as Alzheimer's or senile dementia. They are supposed to improve the blood circulation of the brain and inhibit the formation and spread of free radicals. The production of standardized extracts for phytopharmaceuticals is an interesting and complex technical process and an example of the complexity of production and process control.

Stevioside is a sweet-tasting natural substance from the leaves of *Stevia rebaudiana*. The plant is a ground-level herb from the highlands of Paraguay, but is now also economically cultivated in Brazil and various Asian countries. The extraction of stevioside (◘ Fig. 13.12) is a technically relevant process and is carried out by crystallization from a methanolic extract. The approval of stevioside was critical in the EU and the USA, as there were not enough data on toxicology. From 2021, stevioside will be available in stores with the E-number E960.

Paclitaxel, is an important natural substance that was developed as a cancer drug in an academic research project with substantial tax funds. Particularly noteworthy for its pharmacological effect is that it carries no nitrogen as a diterpene. The mechanism of action of the cancer drug against cervical, ovarian, and breast cancer is also remarkable, as the built-up microtubule apparatus of the dividing spindle is stabilized so strongly during cell division that it cannot disintegrate again and cell division is halted. From the perspective of Technical Biochemistry, Paclitaxel (◘ Fig. 13.13) is of interest because the semi-synthetic derivative Docetaxel is used in therapy rather than the natural substance. Docetaxel (◘ Fig. 13.14) is obtained by the partial synthesis of Desacetylbaccatin III, which is obtained from

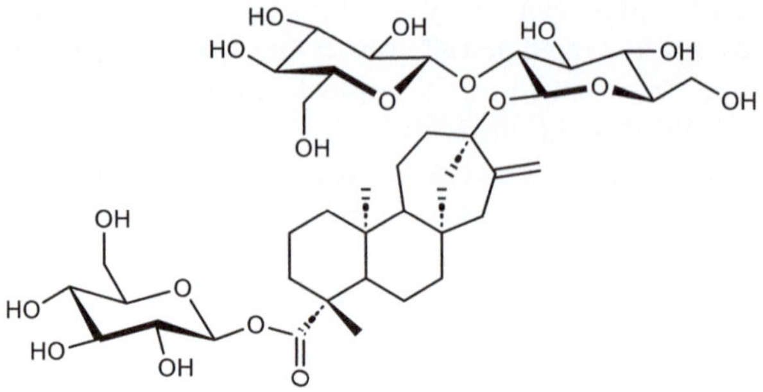

Fig. 13.12 Stevioside

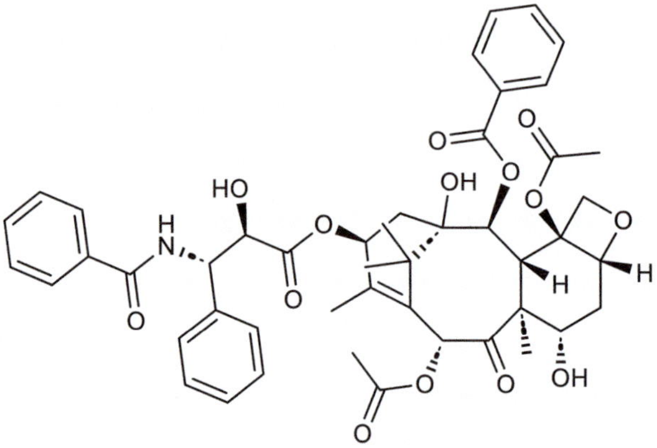

Fig. 13.13 Paclitaxel

plant cell cultures (Renneberg 2007; Zhong 2002). Technically, the provision of paclitaxel is problematic because it only occurs in traces in the plant. An example may illustrate this: To meet the needs of cancer-stricken women in the US alone for one year, all yews in North America would have to be felled. About 2 g are needed for the treatment of one woman, which are obtained from the bark of three at least 100-year-old yews. It is obvious that alternative production methods must be found. One such way is the isolation of the precursor 10-Deacetyl-Baccatin-III from the green needles of the yew (content 0.2%), which constantly renew during the vegetation period. This is converted into docetaxel (Taxotere®) in a multi-stage process (Fig. 13.14).

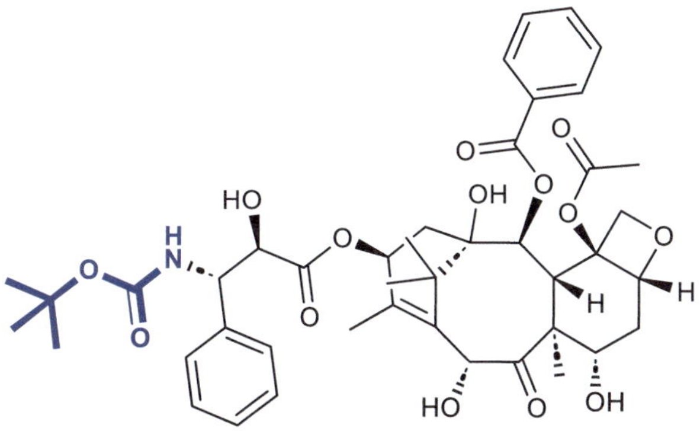

13.6 · Triterpenes

Fig. 13.14 Docetaxel, in blue the introduced chemical derivatisation

13.6 Triterpenes

The basic structure of the triterpenes and steranes (Fig. 13.15) is squalene (C30). We have already encountered cholesterol and ergosterol as important representatives in the chapter on the mevalonate pathway and in the context of cholesterol biosynthesis. As far as known, the biosynthesis is strictly conserved and can be found in all organisms that rely on triterpenes or steroids in their metabolism. Squalene is an important intermediate that is formed by reductive dimerization of two farnesyl diphosphates (C15) under the influence of NADPH. The linkage of the two farnesyl diphosphate building blocks occurs at the terminal C12 atoms, resulting in a symmetrical molecule. In anticipation of further cholesterol biosynthesis, the resulting *all-trans* squalene is converted into the so-called SSSWg and SWSWg (S = chair-, W = boat form) after specific folding with the help of further enzymes. These chair or boat forms refer to the conformation of the cyclohexanes in the tetracycle.

All organisms possess the enzymatic equipment to carry out the SWSWg folding, which leads to the steroids via lanosterol or cycloartenol (Fig. 13.16). Green plants also possess the enzymatic equipment to carry out the SSSWg folding, which is the basic prerequisite for the biosynthesis of triterpenes. Despite numerous secondary modifications in many triterpenes, the original configuration at the chirality centers at atoms C13 and C14 can be read.

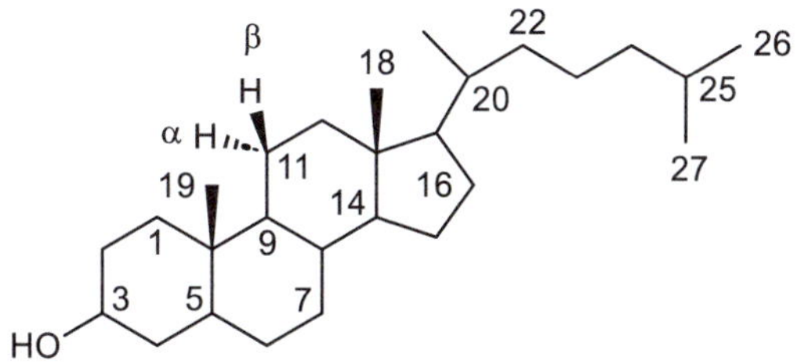

Fig. 13.15 Counting method for steranes

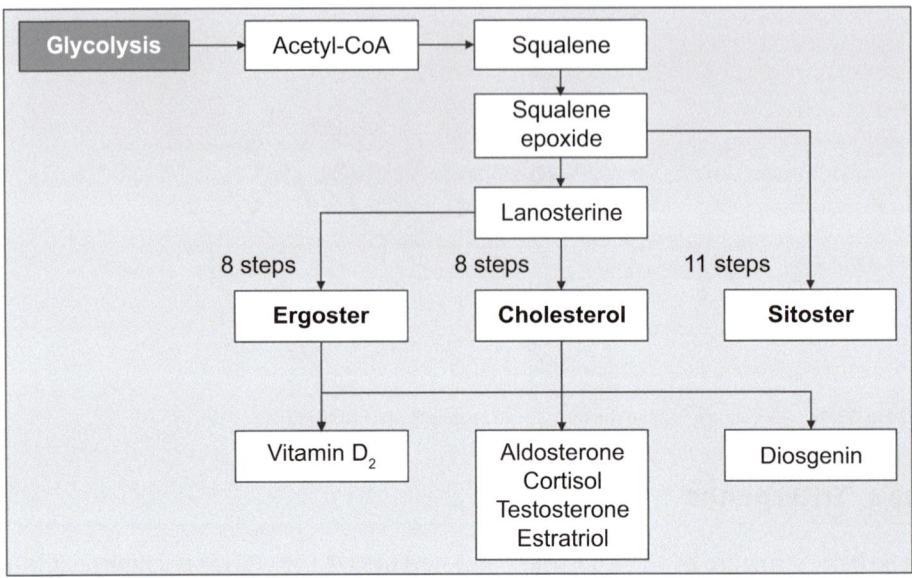

■ **Fig. 13.16** Simplified overview of the biosynthesis origins of ergosterol, cholesterol, and sitosterol

The cyclization of the folded chains in the SSSWg conformation occurs through the introduction of an epoxide ring between C2 and C3. The subsequent opening of the epoxide ring leads to the formation of a carbenium ion at C2 and the formation of a tetra- or pentacyclic triterpene, if in the latter case the side chains also cyclize. Alkylation (especially by SAM) of the double bond in the side chains of cholesterol results in a C28 sterol (mycosterol) in fungi and a C29 sterol (phytosterol) in plants. Further oxidative shortening of the side chains leads to pregnenolone, a C21 steroid with a C2 side chain at C17, which represents a biosynthetic key position for further hormonally active steroids (sex hormones and glucocorticoids). Squalene in the SWSWg conformation is converted to 2,3-oxidosqualene by molecular oxygen with the help of squalene monooxygenase. Opening of the epoxide ring with an electrophilic reagent induces ring closures that lead to a tetracyclic carbenium ion, which is stabilized by methyl group shift with the elimination of a proton. The primary cyclization product is a C30 steroid like lanosterol (animals, mold fungi) or the isomeric cycloartenol in green plants. Both compounds are converted into the same C27 steroid cholesterol by demethylation (2 × C4, 1 × C14) (■ Fig. 13.17).

Diosgenin is an important triterpene used for the semi-synthesis of steroids, especially for the synthesis of estrogen in contraceptives ("birth control pills"), steroid hormones and other corticosteroids. The peculiarity of diosgenin is that the substance has the same stereochemistry in the important four rings as lanosterol, which is the basic structure for biosynthesis in mammalian organisms. Approximately half of all drugs with a steroid skeleton are synthesized today starting from diosgenin. The chemist Russell Marker (1902–1995) discovered diosgenin in Mexican *Dioscera* species and developed a chemical degradation reaction to split the spiroketal ring to progesterone (Rudo et al. 2015). Starting from progesterone, chemical processes could be developed to produce cortisone and the semi-synthetic

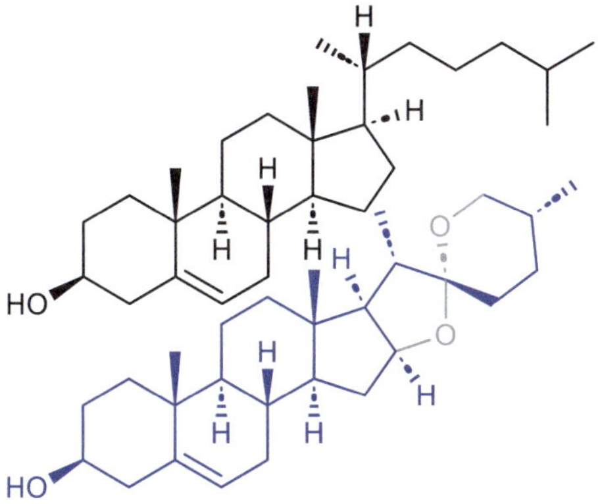

13.6 · Triterpenes

Fig. 13.17 Structural similarity between diosgenin and cholesterol (for didactic reasons, some atoms are shown in gray)

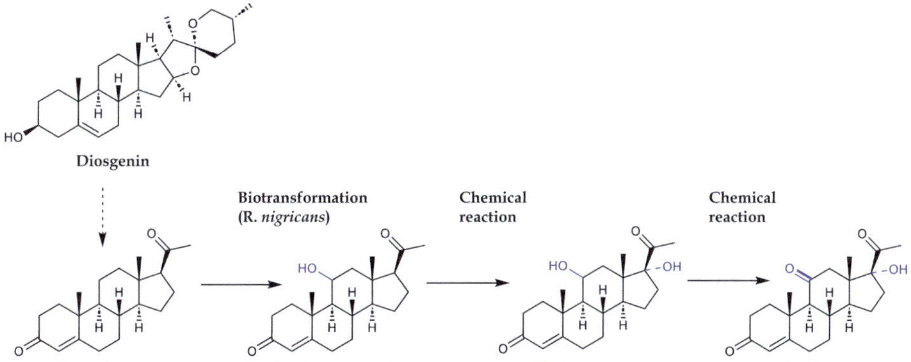

Fig. 13.18 Semi-biosynthesis of cortisone from diosgenin through enzymatic 11α-hydroxylation (*Rhizopus nigricans*) and chemical 17β-hydroxylation

drug estradiol. The main source for the extraction of diosgenin are the roots of *Dioscera* species, which can weigh between 4 and 5 kg. *Dioscera* plants are grown for 4 to 5 years before they can be harvested. The annual requirement is about 20,000 t.

The synthesis of cortisone is achieved through a semi-biotechnical process, in which *Rhizopus nigricans* allows selective hydroxylation in the 11-position (Fig. 13.18). This site-specific catalysis was developed by Upjohn in the 1950s. Through the use of biotechnology, the price of cortisone fell from 200 USD to less than 0.15 USD per gram. Screening of further *Rhizopus* strains showed that hydroxylations at other positions such as 7, 11β and 14α are also possible (Fig. 13.19). Cytochromes are responsible for the biocatalysis in *Rhizopus arrhizus*, and the strain yields 90% at room temperature in aqueous phase and neutral pH value. Alternatively, a second cortisone synthesis was established at Glaxo with

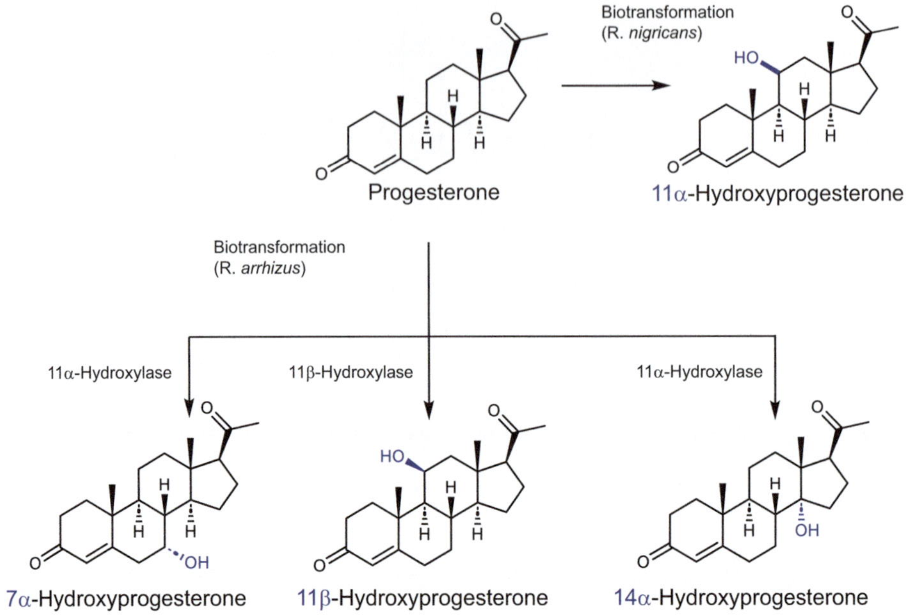

Fig. 13.19 Hydroxylation by *Rhizopus arrhhizus*

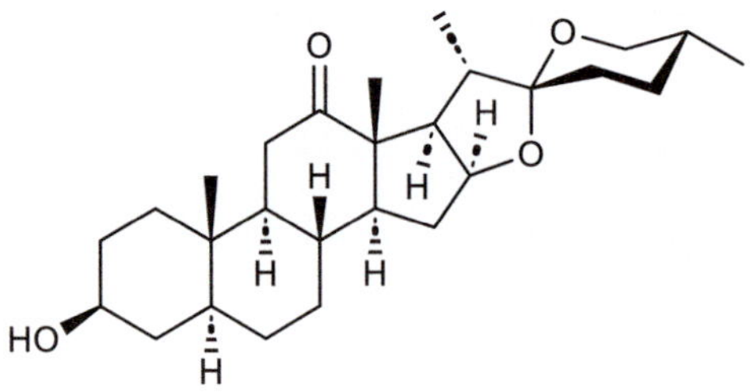

Fig. 13.20 Hecogenin

hecogenin (Fig. 13.20) from Agava sisalana. However, this could not displace the existing syntheses from diosgenin.

Today, several hundred enzymatic catalyses for the degradation of phytosterols are known. In summary, it can be said that pharmacologically active substances (indicated in brackets) were obtained through the following main reactions:
- Degradation of the side chain by oxidations (Estrogens, Progestogens)
- Hydroxylations throughout the sterane basic structure (Glucocorticoids, Progestogens)
- Introductions of double bonds by dehydrogenations (Glucocorticoids)
- Reductions of keto functions (Androgens, Progestogens)

Upon closer examination, it becomes apparent that the predominant biocatalytic reaction in the side chain (C20–C27) is carried out by bacteria and the oxidation reactions in the ring structure by fungi. The resulting chemical diversity ena-

13.6 · Triterpenes

bles the synthesis of steroids with equally diverse biological effects. This development can be impressively demonstrated in the production of hydrocortisone. While 29 chemical reactions were necessary for production in 1940, this number was reduced to 19 in 1950 with the introduction of a biotechnological step. In the 1980s, the number of chemical reactions was reduced to 9 with the introduction of recombinant producers, and today, the complete biosynthesis of hydrocortisone in a microbial producer is possible.

Sitosterol (see ◻ Table 13.1) is the most common sterol in the plant kingdom and is present in high concentrations in seeds (pumpkin, saw palmetto), nuts (cashew) and extracted vegetable oils such as sunflower oil or corn germ oil. Sitosterol does not have a simple structure, as secondary reductions and the introduction of substituents have made the basic structure very complex and usually represent a mixture (e.g., α-, β-, γ-sitosterol, stigmasterol, and sitostanol). No significant applications in medicine or technology are known for sitosterol itself. It is worth mentioning the biotechnological conversion as a starting material for pharmacologically relevant steroids (◻ Fig. 13.21).

Androstenedione (AD) already has a great structural similarity to the male sex hormone testosterone. In a final NADPH-dependent reduction, AD can be very efficiently converted to the end product with the help of a 17-ketoreductase (◻ Fig. 13.22).

Ergosterol (see ◻ Table 13.1) is found as a sterol in fungi, non-chlorophyll cryptogams, and ergot (*Secale cornutum*), a parasitic persistent form of rye that infects it as a fungus and gave the sterol its name. Ergosterol has an important function as a precursor of ergocalciferol, which has a similar structure to provitamin D_2. It differs from real vitamin D_2 by an additional methyl group in position C24. Despite the methyl group, it is active at the vitamin D receptor, which is why its formation is promoted by the UV irradiation of the fungi (◻ Fig. 13.23). The levels of vitamin D average 9.6 micrograms per 100 g, which is significantly above the levels of ordinary cultivated mushrooms.

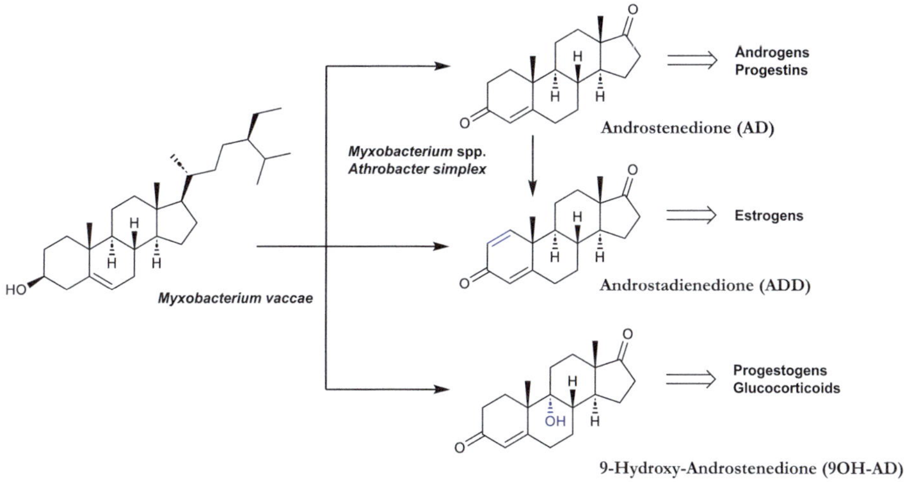

◻ **Fig. 13.21** Sitosterol, exemplary microbial biotransformations

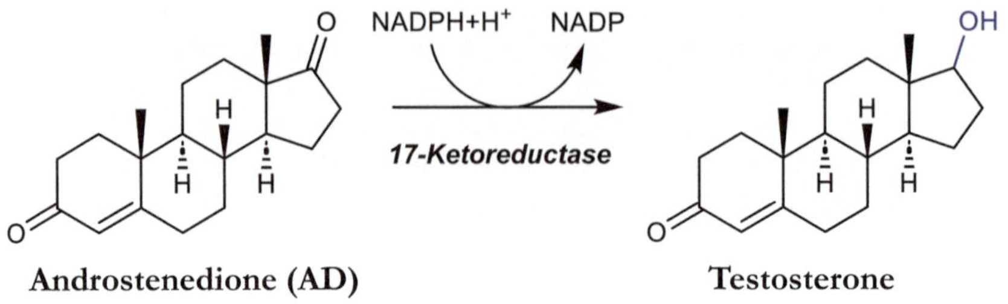

Fig. 13.22 Reduction of androstenedione to testosterone

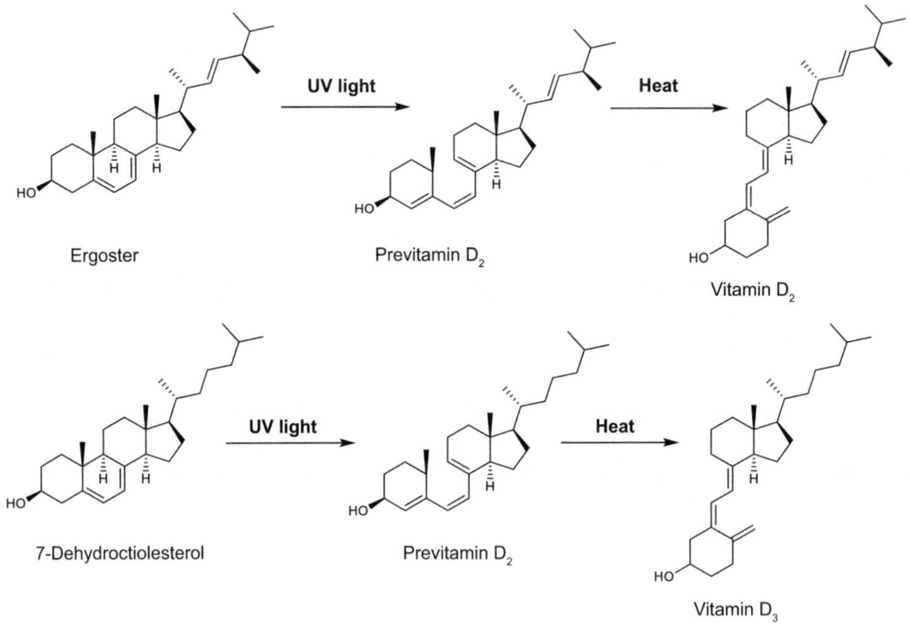

Fig. 13.23 Biosynthesis of Vitamin D_3 from Ergosterol (top) and Cholesterol (bottom)

Infobox 13.2: Is Licorice Dangerous?

Media often warns against excessive consumption of licorice. What's the truth, and how much do I need to eat to not put myself at risk?

Licorice is the thickened juice of the plant *Glycyrrhiza glabra,* which provides the mass for licorice production. In this mass, the triterpene glycoside glycyrrhizin (Fig. 13.24) can be detected, which has a desoxycortisone-like effect. In vitro and in vivo in rats, it has been shown that glycyrrhizin affects the mineral balance of corticosteroids, which has effects on blood pressure: It inhibits the effect of cortisone on inflammation. Although this is not pleasant and should be avoided by conscious consumption. However, the risk is rather low, as at least 2 bags per day would have to be consumed for at least 1 week to cause the mentioned symptoms.

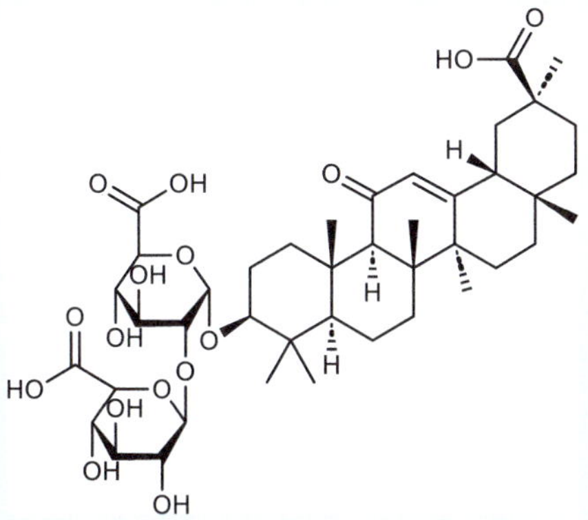

Fig. 13.24 Glycyrrhizin

13.7 Tetraterpenes

Many are familiar with the yellow, orange, and red colors of plants, which are composed of eight isoprene units (Fig. 13.25). By stringing together conjugated double bonds (9–11) as the smallest unit of chromophores, almost all color nuances can be created through chemical variations of the long carbon chain and by oxygen attachment in nature as well as by genetic modification. Carotenoids are poorly water-soluble and react quickly with air and light. Dyes consisting only of carbon atoms are referred to as carotenoids, those containing one or more oxygen atoms as xanthophylls. The best-known carotenoid is the red dye lycopene in tomatoes. Of great medical and nutritional physiological interest is β-carotene, which is predominantly synthetically produced. Physiologically, it is vital for humans as a precursor of vitamin A and is mainly absorbed through all reddish types of vegetables.

Carotenoids are of medical importance. Physiologically, they are significant because the α- and β-carotenoids are oxidatively split into two C20 parts by β-caro-

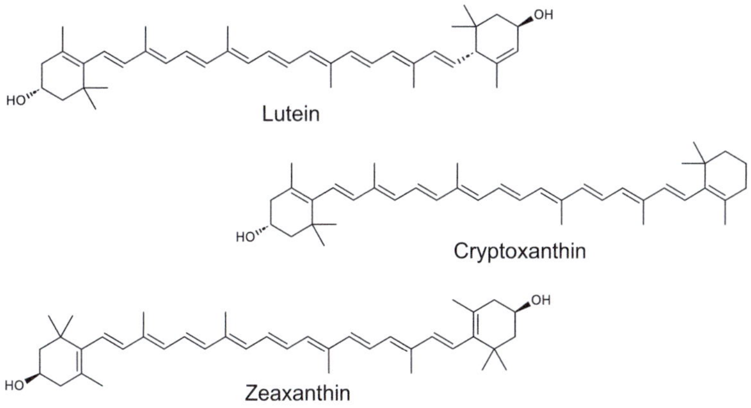

Lutein

Cryptoxanthin

Zeaxanthin

Fig. 13.25 Economically and medically important carotenoids

tene-15,15'-monooxygenase, resulting in vitamin A via retinal. Retinal is reduced to the alcohol retinol, which is transported as a fatty acid ester in chylomicrons to the visual receptors. Even if she did not know this molecular process, Grandma knew: "Eat your carrots, they are good for your eyes." In addition to these economically relevant carotenoids, lutein as well as ionones and irones should be mentioned as metabolic degradation products, which are obtained through microbial fermentation.

β-Carotene   The dye extraction (◘ Fig. 13.26) can traditionally be done by extraction and concentration from red beets or carrots. Newer technical production routes are also well documented with genetically modified microorganisms such as *Erwinia herbicola*.

Damascenone and Jonone   Strictly speaking, both representatives are not tetraterpenes, but their degradation products, which are linked to a ring with a chain length of 13 carbon atoms. They are components of essential oils and, for example, closely associated with the scent of roses. Chemically, they are degradation products of carotenoids. Four structures of damascenone are well known and are valued as important flavorings in strawberries, currants, but also in cognac, rum, or scotch. Due to their popularity, they also play an important role in the perfume industry. The main representatives are β-damascenone (◘ Fig. 13.27) and β-jonone (◘ Fig. 13.28).

Astaxanthin   This oxidized carotenoid is structurally related to β-carotene and has similar metabolic and physiological functions. Technically, the production of astaxanthin (◘ Fig. 13.29) plays a major role, which is used not only because of its orange color (colorant E161j) in salmon farming. Thus, organic and conventional salmon can be distinguished by their pale (organic) or strong (conventional) orange coloration. It is also interesting that astaxanthin has a vitamin-like effect on salmon. It increases fertility and immune defense in breeding. The (3R,3'R)-stereoisomer (◘ Fig. 13.29) is produced on a ton scale in the fungus *Xanthophyllomyces dendrorhous*. Synthetically, a racemate is produced, while salmon biosynthesize exclusively (3S,3'S)-astaxanthin.

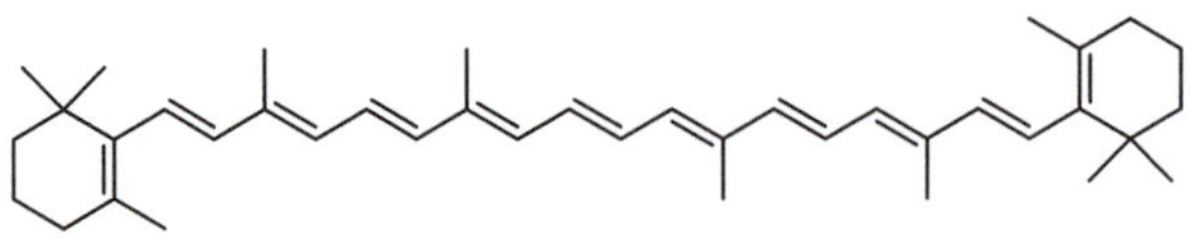

◘ Fig. 13.26   β-Carotene

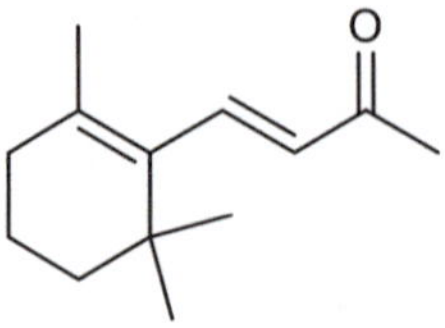

◘ Fig. 13.27   β-Damascenone

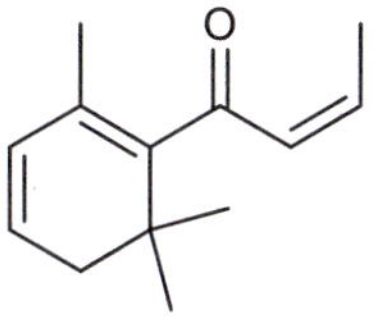

Fig. 13.28 β-Jonone

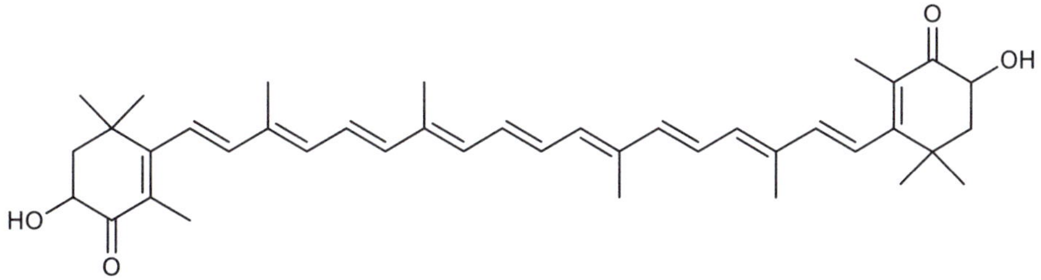

Fig. 13.29 Astaxanthin (without consideration of stereochemistry)

13.8 Polyterpenes: Rubber

In contrast to the terpene groups discussed so far, the polyterpenes are a chemically not exactly determinable natural substance. Due to the polymerization, an approximate molecular weight is obtained, which is indicated in chemistry with a minimum and maximum size. A section of a possible polyterpene is shown in ◘ Table 13.3. The building block of the polymer is the *cis*-linked 2-methyl-1,3-butadiene with a molecular mass of 500,000 to 1 million per mol. Many of the polymeric terpenes exhibit a rubber-like physical property that strongly depends on temperature, humidity, light exposure, and storage time. The physiological function is assumed to be the sealing of the tree after injury to prevent the penetration of infections. Biosynthesis occurs through a very frequently repeated Friedel-Crafts alkylation of the IPP monomer unit with the help of *cis*-prenyltransferases. Biosynthesis and elongation occur in micellar globules, in which the polyisoprene chain lies and the C1 end to be extended protrudes into the phospholipid layer (Men et al. 2019). In addition to the prenyltransferase, other stabilizing proteins are found here, such as the Rubber Elongation Factor (REF). Their task is to fix the monomeric IPP building block and feed it to the *cis*-prenyltransferase (◘ Fig. 13.30). The extended chain is continuously pushed further into the interior of the micelle. This rubber is referred to in the literature as latex, although other plants besides the rubber tree *(Hevea brasiliensis)* can be interesting sources of latex (Nair 2010). Initial approaches to the microbiological production of rubber are documented. The synthetic-biological approach is characterized by the cloning of the relevant prenyltransferase and stabilizing proteins in *E. coli,* albeit with modest success. A problem for successful biosynthesis is the lack of micellar structures that enable long-chain biosynthesis. Although the biosynthesis of short-chain polyterpenes could be demonstrated, transferring the biosynthetic machinery to more efficient production organisms such as

Table 13.3 Comparison of natural polyisoprene polymers

	Rubber	Gutta-percha	Balata	Chicle
Parent plant	*Hevea brasiliensis*	*Palaquium gutta*	*Manilkara bidentata*	*Manilkara zapota*
Structure				
Stereochemistry	cis-1,4, all-Z	trans-1,4, all-E	trans-1,4, all-E	trans-1,4, all-E cis-1,4, all-Z
Molecular weight	8,000–30,000	1,500		

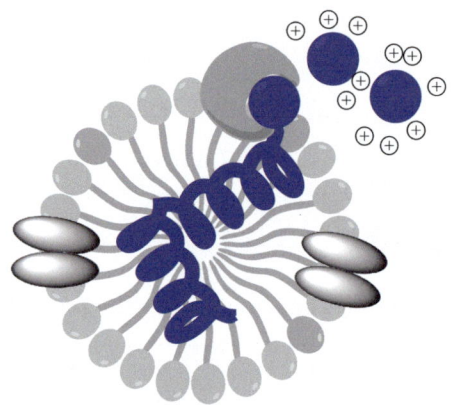

Fig. 13.30 Micelles as the site of polyterpene biosynthesis, ● IPP, ☾ cis-Prenyltransferase, ◯ REF, ⊕ Mg^{2+}/Ca^{2+}-ions

Yarrowia lipolytica or *Cryptococcus curvatus,* which as oleaginous microorganisms possess micellar fat droplets, could solve the efficiency problem. Heterologous biosynthesis in plants is also not without problems (Salehi et al. 2022). It has been suggested to disrupt the squalene synthase gene (SQS/ERG9) of the isoprenoid branch to significantly increase the substrate flow into rubber biosynthesis at the expense of steroid biosynthesis. This approach seems to be successful in plants. However, the temporary coexpression of the presumed natural rubber biosynthesis complex in the leaves of *Nicotiana benthamiana* did not lead to the formation of rubber particles. This suggests that other components are involved in the biosynthesis of natural rubber or the biogenesis of rubber particles.

Rubber is a plant product obtained by tapping the bark of a South American tree. Known as "tree tears" to the Mayans and Aztecs, they used the emulsion referred to as latex or milky sap to impregnate materials to make them waterproof. These waterproof items were mostly drinking pouches and rain tarps, but it is also known that the first football was made of rubber. The Spanish conquistadors brought rubber to the Spanish court for the first time in the mid-16th century, where it met with little interest due to its stickiness and melting in heat. With

13.8 · Polyterpenes: Rubber

its second discovery by the Frenchman François Fresneau de La Gataudière (1703–1770), who was tasked with the technical exploration of the colony as an engineer in French Guiana, rubber reached France, where it was examined at the Academy of Sciences and first simple products were developed. It was not until 1854 that Charles Goodyear (1800–1860) solved the basic problem of poor durability when he accidentally invented vulcanization in his kitchen. By adding sulfur and heating, a chemical linkage of the polymer chains occurred, leading to the desired permanent structure and durability.

Vulcanization, named after Vulcanus, the ancient Roman god of fire and metalworking, is not a biochemical process. It is a chemical reaction in which thermoplastic natural rubber is converted into elastomers. The raw rubber is mixed with sulfur-releasing chemicals such as S_2Cl_2 in a concentration of 1.5 to 2.5% and catalysts and heated to 120 to 160 °C. In this process, the long-chain rubber molecules are linked together via short-chain sulfur bridges. The number of sulfur bridges determines the hardness of the elastomer. Over time, rubbers become brittle and fragile as the sulfur bridges are replaced by oxygen bridges during aging.

Technical Importance Rubber finds a wide range of applications in medicine and technology. Typical areas of use include the production of latex gloves (Proskauer 1958), condoms, hoses, seals, and car tires. Its use in medicine is severely limited, as rubber can cause allergies in workers. The reason is the presence of proteins that are leached out by sweat and provoke the immune system. Therefore, synthetic elastomers are used in medicine. A non-allergenic rubber as natural rubber is obtained from the Russian dandelion and is in prototype development.

Gutta-percha Chemically speaking, gutta-percha is a poly-(1,4-*trans*-isoprene) natural rubber and physically belongs to the thermoplastics. Gutta-percha is similar to natural rubber, but has a lower elasticity and a lower molecular mass at room temperature (◘ Table 13.3). Gutta-percha is the dried latex of the Southeast Asian tree *Palaquium gutta*. However, since natural rubber is only available in limited quantities, synthetic analogs are often used. These include 1,3-butadiene and 2-methyl-1,3-butadiene, also known as isoprene. There is now also BioIsoprene®. Together with DuPont, the companies Goodyear and Genecor have genetically modified *E. coli* to produce isoprene. The mevalonate pathway for the production of IPP/DMAPP was built into the genome of *E. coli*, so that the C5 building block is biosynthesized as a precursor of the polymeric rubber. Isoprene is gaseous at room temperature, escapes, and rises in the bioreactor without damaging the cells. The isoprene gas has a purity of 99%, which is important because impurities inhibit the catalysts that polymerize isoprene into rubber.

In addition to these two mainly used biological elastomers, two other polyterpenes should be mentioned that today play no major role in technology (drive belts, shoe soles) or medicine. One is **Balata,** which is obtained from the latex of the South American balata tree. It has a wood content of up to 40% and is similar to natural rubber, with the linking of the monomeric building blocks occurring in *trans* configuration. Another polymer is **Chicle rubber,** a mixture of *cis* and *trans* polyisoprenes. It has a low degree of polymerization and only becomes plastic at a temperature of 50 °C. Today, it is used as a plastic-free biological raw material for chewing gum (◘ Table 13.3).

❓ Self-check Questions

1. Distinguish between terpenes and terpenoids. Which monomer underlies both substance classes?
2. Name the biosynthetic pathways that provide the basic building block of terpenoids.
3. Provide the subdivision of terpenoids.
4. You have learned about fatty oils in the previous chapter. Now explain the chemical, biosynthetic, and physical differences to essential oils.
5. Explain the physical and chemical differences between gutta-percha and natural rubber.
6. Explain why a fully biotechnological production of artemisinin is currently not possible.
7. Explain the chemical group of steroids. Elucidate the differences that exist in biological activities between sexual steroids, glucosteroids, and bile salts despite similar structures.
8. Elucidate the sterically specific oxidations in the C11 and C17 positions of the sterane ring framework.

Alkaloids

Contents

14.1 Definition – 177

14.2 Chemistry – 179

14.3 Technical Importance – 179

14.4 Biosynthesis – 180

14.5 Functions in the Plant – 181

14.6 Alkaloids Derived from Phenylalanine and Tyrosine – 183
14.6.1 Biosynthesis of Mescaline – 184
14.6.2 Opium Alkaloids – 184
14.6.3 Ingredients of Opium and Their Effects – 188
14.6.4 Colchicine – 190

14.7 Alkaloids derived from Tryptophan – 191
14.7.1 Biosynthesis – 191
14.7.2 Psilocin and Psilocybin – 191
14.7.3 Physostigmine – 193
14.7.4 Melatonin – 195
14.7.5 Lysergic Acid Alkaloids – 196
14.7.6 Vinca Alkaloids – 199

14.8 Quinoline Alkaloids – 201
14.8.1 Camptothecin – 202

14.9 Alkaloids Derived from Ornithine – 205
14.9.1 Tropane Alkaloids – 206

© The Author(s), under exclusive license to Springer Fachmedien Wiesbaden GmbH, part of Springer Nature 2025
O. Kayser and N. J. H. Averesch, *Technical Biochemistry*, https://doi.org/10.1007/978-3-658-47121-7_14

14.10   Alkaloids Derived from Histidine – 210

14.11   Purine Alkaloids – 211

14.12   Alkaloids Derived from Arginine – 212

Learning Objectives and Key Topics
- Nomenclature of Alkaloids
- Chemistry and Chemical Diversity
- Biological and Physiological Effects in Plants and Humans

Technical Applications
- Synthetic biotechnological production of alkaloids such as Thebaine and Scopolamine
- Extraction methods in the pharmaceutical industry

Important Alkaloids
- Opiates
- Tropane Alkaloids
- Purine Alkaloids

In 1805, the pharmacist F. W. Sertürner isolated the alkaloid morphine from opium in Paderborn, which he tried to characterize the following year. He discovered that the compound reacts basic and induces a numbing, sleep-like state. He dedicated the substance to the god of sleep "Morpheus" and gave it the name morphine, which is still known today. Gomez isolated a similarly basic reacting mixture from cinchona bark in 1810, from which Pelletier and Caventou isolated quinine and cinchonine in 1920 and separated them from each other. The pharmacist W. Meissner (1792–1853) already suggested in 1819 to call these basic reacting nitrogenous natural substances "alkaloids" because they reacted "like alkali".

14.1 Definition

It is very difficult to find a uniform definition for alkaloids, and the discussion of the last almost 50 years shows that none of the involved researchers is really satisfied (Wink 1998). The problem is that many very different substances with very heterogeneous structural features from very different biological sources with also very different biological effects had to be summarized. Initial attempts to order the alkaloids according to their chemistry and structural features (◘ Fig. 14.1) led rather to a very large confusion. Later, only nitrogen-containing aromatic compounds were recorded, until aliphatic alkaloids were also found in pepper, paprika and some corals, which were then summarized under the term pseudoalkaloids. The reference to the biological origin, e.g. Solanacea alkaloids, did not help further, as alkaloids with the same structural types were also found in other families (e.g. Convolvulaceae). It is difficult to formulate a perfect or at least clear definition, which is why alkaloids today are simply considered as nitrogenous natural substances with strong physiological effects (except for antibiotics).

The current classification into groups or classes is done based on biosynthesis (from which amino acid the alkaloids can be derived). About 10–15% of all plants

Chapter 14 · Alkaloids

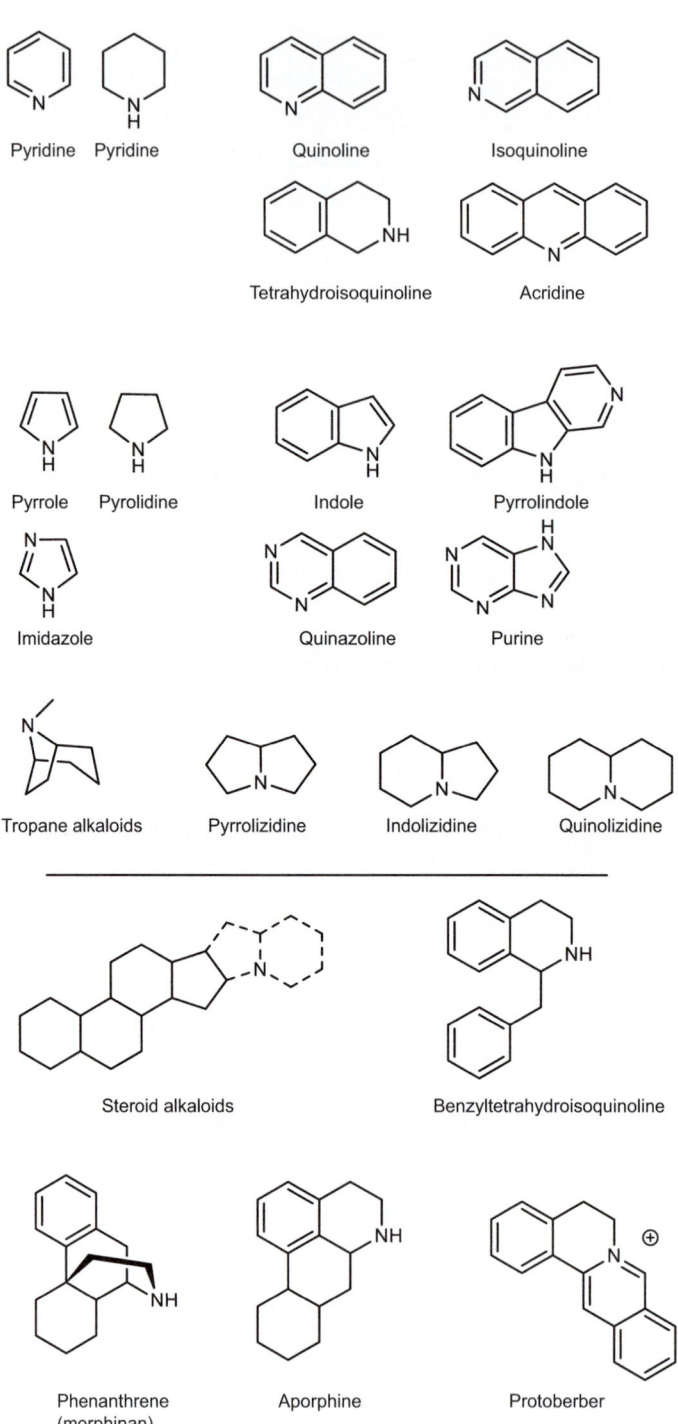

Fig. 14.1 Basic heterocycles of significant alkaloids (above) and typical structures of important alkaloid classes (below)

contain alkaloids, and to date, approximately 9,000 alkaloids have been isolated and described from 5,000 plants. Typical alkaloid-bearing plant families are Apocynaceae, Loganiaceae, Papaveraceae, Rubiaceae, and Solanaceae.

The occurrence of alkaloids in microorganisms has been known for a long time. St. Anthony's fire (Ergotism) is a disease feared in Europe since the Middle Ages. It is a poisoning caused by the sclerotia of ergot *(Claviceps purpurea)*, which parasitizes rye in damp weather and produces dangerous alkaloids. In humans, who ingested ergot alkaloids (see also lysergic acid alkaloids) with bread, they led to a permanent narrowing of the blood vessels in fingers and toes, so that these died off due to lack of blood circulation (Hofmann 1978).

14.2 Chemistry

Alkaloids feature at least one nitrogen atom as a structural characteristic; some alkaloids such as ergotamine contain up to five nitrogen atoms. The nitrogen can be present as a primary amine ($R\text{-}NH_2$) in pseudoalkaloids like capsaicin, as a secondary amine (R-NH-R) as in most alkaloids, or as a tertiary amine (R_3N as in galantamine or tubocurarine). In some cases, alkaloids also exist as quaternary ammonium salts ($[R_4N^+X^-]$), such as pancuronium, which is contained in the arrow poison curare. A very good overview of the chemistry and biology of alkaloids is provided by the book by E. Breitmaier (Breitmaier 2008). The degree of basicity of alkaloids can vary and depends on the molecular structure, the bonding of the carbon atoms to the nitrogen, and the functional groups in structural proximity. At acidic pH, the nitrogen is protonated: The alkaloid becomes more hydrophilic and dissolves as a salt in water. At alkaline pH, deprotonation occurs: The base becomes more lipophilic, and the alkaloid only dissolves in organic solvents.

This basic principle of pH-dependent solubility is exploited in isolation in downstream processes or in medicine to increase bioavailability and uptake in the body. A very vivid and drastic example is the smoking of Crack, the free base of cocaine. By smoking, the cocaine base is absorbed within seconds through the lungs, and the drug immediately reaches the brain, as it can easily overcome the blood-brain barrier. As we will see, most alkaloids have a complex structure. For organic chemistry, this means that their synthesis is very difficult or often uneconomical in industry. From a process engineering point of view, the question often arises: How can alkaloids be effectively isolated from plants? As a mental exercise, one can ask this question for almost all alkaloids discussed in this chapter. A real challenge is posed by the vincristine alkaloids, which only occur in traces in the plant.

14.3 Technical Importance

Only a few alkaloids are still used as pure substances in medicine (see box). Compared to the 19th century, when the pharmaceutical industry knew no or only a few highly effective synthetic drugs, the structures of the alkaloids have been systematically examined for their pharmacophoric substructures over the last almost 100 years. Based on these findings, drug design prevailed and accelerated the discovery of new highly active substances with often fewer side effects. Today's pharmacopoeia includes alkaloids such as codeine as a cough suppressant, colchicine for gout attacks, pilocarpine in ophthalmology, and camptothecin as well as *Vinca* alkaloids as anticancer agents. Quantitatively, two alkaloids such as scopolamine and thebaine play an important role as starting materials for partial syntheses. Thebaine is an important starting material for the synthesis of very strong opiates (naturally occurring morphinan alkaloids in opium) and opioids such as buprenorphine, etorphine and oxycodone and opioid antagonists like naloxone. The term opioid is not clearly defined, as all morphine-like alkaloids are summarized, but in a narrower sense it refers to the (semi-)synthetic painkillers.

> **Alkaloids with current technical importance**
>
> **Starting materials for syntheses**
> – Thebaine: Starting material for the partial synthesis of buprenorphine, etorphine, naloxone, and oxycodone
> – Scopolamine: Starting material for the synthesis of tiotropium
>
> **Pure substances in application**
> – Codeine
> – Colchicine
> – Camptothecin
> – Pilocarpine
> – *Vinca* alkaloids

14.4 Biosynthesis

The biosynthesis of alkaloids does not proceed uniformly, as one might suspect from the example of the natural substances already discussed, such as cinnamic acids, polyketides, or fatty acids. Due to the great structural diversity, it is suspected that each alkaloid is constructed in its own specific way. This is only partially correct, as structurally similar alkaloids are usually biosynthesized from the same amino acid. For example, morphine is structurally a benzylquinoline, which biosynthetically belongs to the family of tyrosine alkaloids. Other structures such as codeine, papaverine, and thebaine are structurally similar to morphine, and large parts of the synthesis can be derived from the same biosynthetic pathway, but they have different biological effects and functions in the plant. The main alkaloids discussed in this chapter can be derived from the amino acids summarized in ◘ Table 14.1.

An overview of the compounds of the individual biosynthetic pathways and the amino acids underlying the alkaloids is given in ◘ Fig. 14.2. Acetate units (C2) (e.g., in cocaine or coniine) or IPP units (C5) (e.g., in terpene indole alkaloids such as vincristine) can be involved in the construction. This leads to alkaloids like coniine and capsain, which are mainly built from acetate units, to γ-coniceine, which

Table 14.1 Alkaloids biosynthetically derived from amino acids

Tyrosine/Phenylalanine	Morphine, Apomorphine, Thebaine, Papaverine, Mescaline, Adrenaline, Aristolochic Acid, Chelidonine, Noscapine, Colchicine, Emetine, Galantamine
Ornithine	Cocaine, Scopolamine, Hyoscyamine, Senecio Alkaloids
Lysine	Lobeline, Piperine, Sparteine
Histidine	Epibatidine, Nicotine, Anabasine, Arecoline, Pilocarpine
Tryptophan	Lysergic Acid, Ergot Alkaloids, Ergotamine, Psilocybin, Harmane, Catharanthine, Vindoline, Vinblastine, Vincristine, Ajmalicine, Yohimbine, Reserpine, Strychnine, Camptothecin, Quinine, Physostigmine

is mixed from acetate and malonate, and to terpene alkaloids, which are built from the iridoid secologanin or via the triterpene biosynthesis. Examples of terpene alkaloids are soladinine, aconitine, or tomatine. General reactions in the formation of alkaloids are:

- Decarboxylation and transamination of amino acids via reactive aldehydes as intermediates
- Formation of Schiff bases (= azomethines) for the formation of C=N bonds from an aldehyde and an amine
- Mannich condensation (construction of C-C-N bonds) from carbonyl compounds, amines (or from Schiff bases) with a -C-H group
- Aldol condensation (construction of C-C bonds)
- Incorporation of OH- and CH_3 groups
- Methylations of OH groups

14.5 Functions in the Plant

Alkaloids can occur in all organs of the plant, but are mainly found in certain parts such as bark, bast and leaves (*Cinchona* bark, *Papaver* fruits, Belladonna leaves). This means that alkaloids do not have to be present or biosynthesized in certain tissues or organs of the plant. The biosynthesis takes place in the young organs or tissues of the plant, which does not exclude an internal transport of the alkaloids from the biosynthesis tissue to the accumulation organs such as leaf or fruit. During transport, chemical changes can occur (e.g., hyoscyamine in the root as a precursor of scopolamine in the leaf). Alkaloids can also be converted into a water-soluble transport form by glycosylation. Opinions differ on the function of alkaloids in the plant. Some properties clearly indicate the benefit for the plant, others are rather unclear:

- Alkaloids are non-toxic excretion products for otherwise harmful metabolic intermediates for the plant. But: In other plants (e.g., *Conium, Nicotiana, Papaver*) alkaloids are reabsorbed into the metabolism.

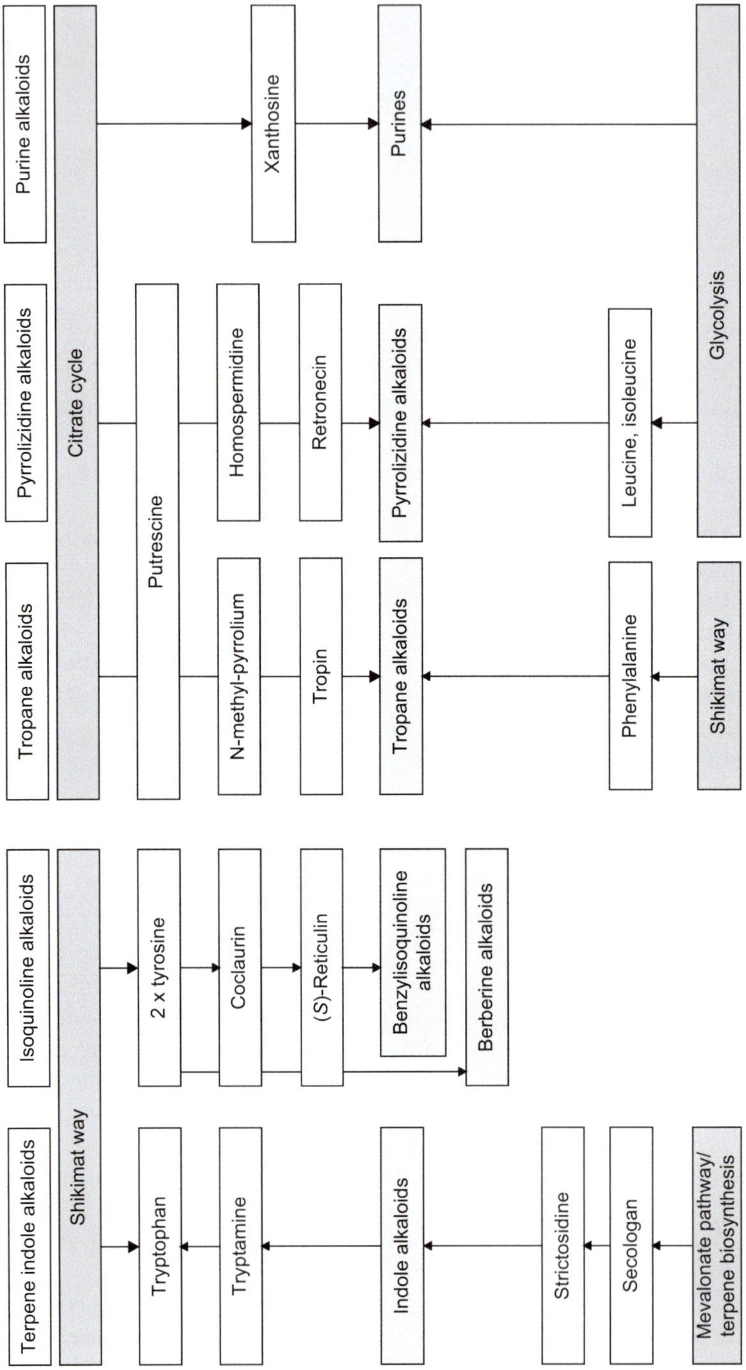

Fig. 14.2 Biochemical origin of the alkaloids discussed in the book

- Alkaloids can function as carriers of nitrogen and thus accumulate it. This can be a storage function, but also a metabolization of the cell poison ammonia (NH_3).
- Alkaloids influence the growth of plants, so they could have a function similar to hormones.
- Alkaloids are defensive substances for plants against microorganisms or herbivores. This thesis is supported by the fact that some alkaloids are antimicrobial or often toxins.

14.6 Alkaloids Derived from Phenylalanine and Tyrosine

Most microorganisms and plants build their proteins from 20 proteinogenic amino acids, which they synthesize themselves. These amino acids include L-phenylalanine, L-tyrosine and L-tryptophan, which belong to the aromatic representatives and derive from the shikimate pathway. The C6C2N backbone of these alkaloids is derived from phenylalanine or tyrosine, but can be broken up or further modified by secondary modifications (N-hydroxylation, decarboxylation, ring substitutions). Instead of phenylalanine, structurally similar compounds such as L-DOPA or dopamine can also be introduced into the biosynthesis. The family of phenylalanine alkaloids also includes the ipecacuanha alkaloids, whose biosynthesis deviates and will be discussed further below. However, two alkaloid groups can generally be built up via phenylalanine or tyrosine:
- Benzylisoquinoline alkaloids, e.g. morphine, codeine, tubocurarine, papaverine: They are formed by Mannich condensation from phenylethylamine derivatives, derived from phenylacetaldehyde. For example, papaverine is formed from two molecules of L-DOPA. Benzylisoquinoline alkaloids consist predominantly of phenylalanine, but can also originate from tyrosine as the original amino acid.
- Phenylethylamine alkaloids, e.g. mescaline, ephedrine, L-DOPA, dopamine: These alkaloids are built up from tyrosine.

Benzylisoquinoline alkaloids include opium alkaloids such as morphine or codeine (methylated morphine). The biosynthesis is complex and involves approximately 20 steps, which will not be described in detail here. More important is to gain a basic understanding, therefore the most important steps are explained, which can also be of technical importance. In the case of benzylisoquinoline alkaloids, reticulin is an important intermediate stage, as two amino acids form a central ring system through Mannich condensation. Reticulin is formed from dopamine and 3,4-dihydroxyphenylacetaldehyde, which together form (S)-coclaurin and (R)-reticulin. These simple structural variants show that different alkaloid classes can be biosynthetically justified despite the same chemical building blocks. Thus, the proximity of the benzylisoquinolines to the morphinans can be explained by rearrangement along the drawn axis. Other alkaloids can also be built from reticulin, which will be discussed later. However, the variations can not only be explained by the linkage of the two (modified) tyrosine building blocks, but also by different linkage of the ether bridges, direct C-C bonds and hydroxylations of the aromatics:
- Through C-C linkage: Alkaloids of the aporphine type (aporphine, boldine)

- Morphinan alkaloids
- Through rearrangement alkaloids of the phthalidoisoquinoline type (Noscapine, Hydrastine)
- By linking two molecules of reticulin via ether bridge formation to alkaloids of the bisbenzylisoquinoline type (tubocurarine)

As heterogeneous as this class of alkaloids may appear, they all have a uniform principle, namely the linkage of a C6C3 building block (dopamine) as an amine component and a C6C2 building block (*p*-hydroxyphenylacetaldehyde) as a non-amine component. In the alkaloids of berberine and isoquinoline phthalides, an additional formaldehyde equivalent (C1) is incorporated.

14.6.1 Biosynthesis of Mescaline

The precursors of mescaline biosynthesis (tyrosine, phenylalanine) originate from the shikimate pathway. However, the direct biosynthetic starting material in the narrower sense is dopamine. Mescaline is the main alkaloid of the peyote cactus Lophophora williamsii, Cactaceae, which occurs in northern Mexico and the southwestern USA. The biosynthesis of mescaline is very simple (◘ Fig. 14.3). First, a decarboxylation of phenylalanine takes place, in further steps the third hydroxylation on the ring is carried out by monooxygenases. The methylation of the hydroxy groups to methoxy groups is carried out by SAM. In addition to mescaline, the cactus contains 60 other minor alkaloids, which however have only a minor effect. The peyote cactus was formerly used by indigenous peoples of North America for its hallucinogenic effect in ritual actions. The complete synthesis was achieved by Späth as early as 1919, and today's synthetic derivatives are 20 to 50 times more hallucinogenic than mescaline. Synthetic analogues include 4-methyl-2,5-dimethoxy-α-methylphenylethylamine (STP = Serenity-Tranquility-Peace) or dimethoxymethylamphetamine (DOM) with high addiction potential. The hallucinogenic effect of mescaline and amphetamines is probably based on the binding and activation of the serotonin receptor 5-HT_{2a}.

14.6.2 Opium Alkaloids

This group of alkaloids accumulates in the latex of *Papaver somniferum,* Papaveraceae, which can be found in the schizogenous latex tubes of all parts, especially the capsule fruit. Morphinan alkaloids are formed by R-reticulin and thebaine. More than 30 alkaloids have been found in the genus *Papaver,* most of which belong to the benzylisoquinoline type. The natural biosynthesis of morphinan alkaloids is complex and proceeds through about 20 intermediate stages. Here too, the biosynthesis will not be explained in detail, but reference will be made to literature on natural product biosynthesis and pharmaceutical biology (Nuhn 2006). The aim is to promote a better understanding to solve technical questions. As already explained, the biosynthesis of (*S*)-reticulin plays an important role (◘ Fig. 14.4). The reticulin structure is split into the dienone salutaridin by oxidation, so that with

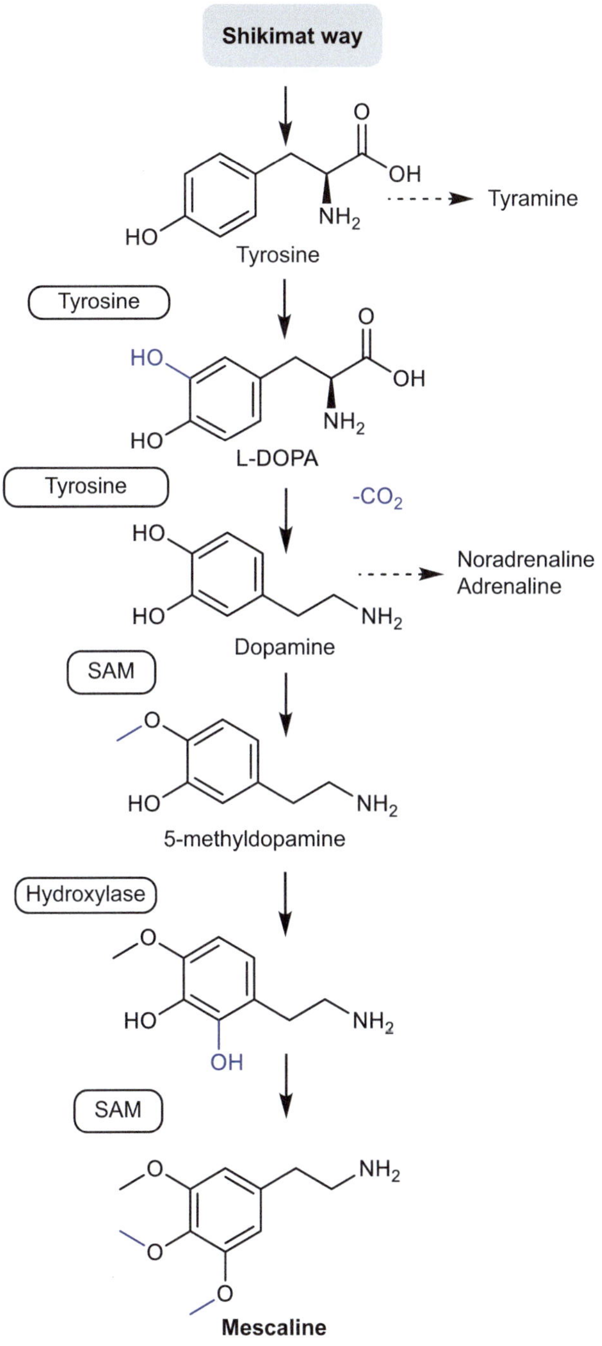

14.6 · Alkaloids Derived from Phenylalanine and Tyrosine

Fig. 14.3 Mescaline biosynthesis

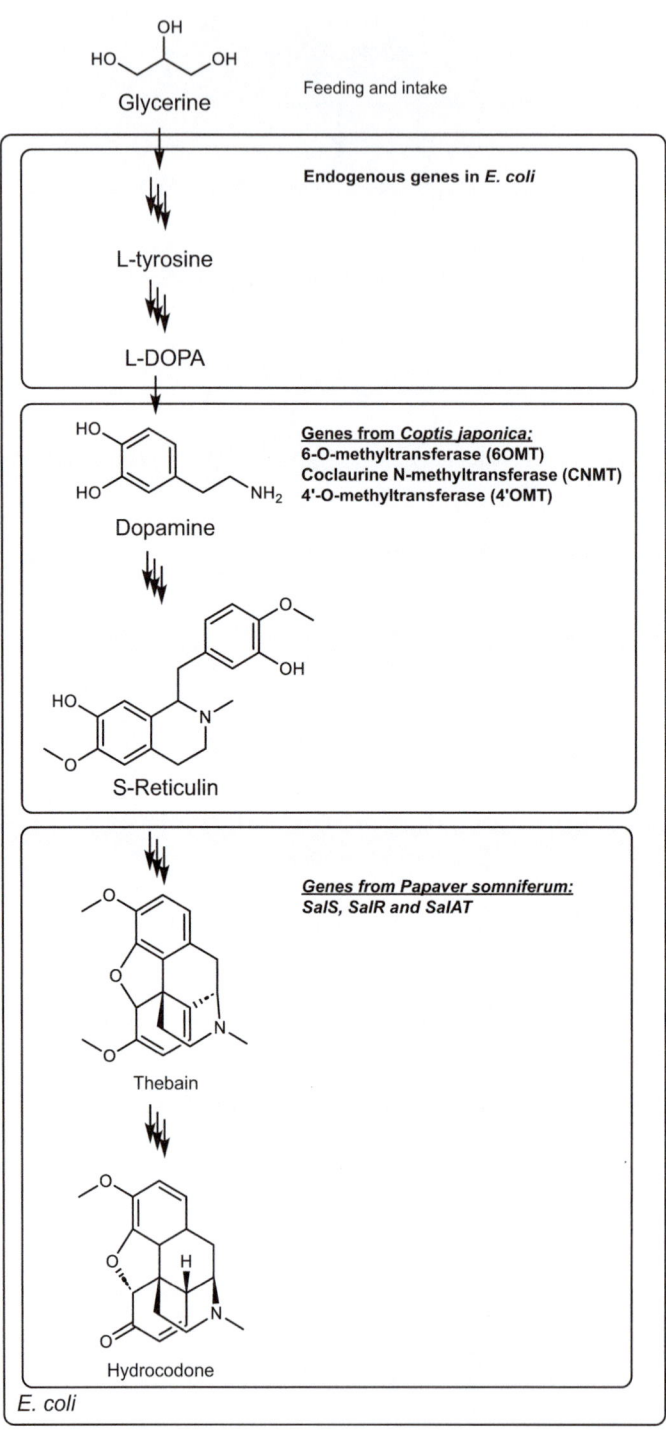

Fig. 14.4 Heterologous biosynthesis of thebaine

salutaridinol the aromatic structure, which is completely removed to form an epoxide as a reactive intermediate to thebaine, is completely removed. In further steps, morphine, the precursor of codeine, which is methylated at the free hydroxyl group of morphine, is formed. Morphine is the natural starting material for heroin, which is synthetically obtained by double acetylation of the free hydroxyl groups. Heroin is a semi-synthetic drug of Bayer AG. All other semi-synthetic opiates are based on thebaine.

The heterologous biosynthesis to thebaine was recently established in yeast. Starting from the biosynthesis of reticulin, it leads to thebaine and hydrocodone. The complete pathway to morphine has not yet been constructed. The heterologous biosynthesis was established in *E. coli* and starts with glycerin as a precursor, which is supplied to the medium and the microorganism. This trihydric alcohol is metabolized by glycolysis, and with the help of specific enzymes, the first important intermediate product dopamine is formed from tyrosine and L-DOPA. The reticulin biosynthesis is borrowed from Captos japonica; the further biosynthesis to hydrocodone is based on genes from Papaver somniferum (◘ Fig. 14.4). The quantities of the precursors formed and the hydrocodone produced so far are in the lower microgram range and are not yet economically interesting.

Papaver somniferum, **Opium Poppy**   There are various varieties of *P. somniferum*. It is suspected that the species originated from a culture of *P. setigerum* which still grows wild in the Mediterranean today. The genus *Papaver* can grow in all temperate latitudes, with the opium alkaloid content being higher in warmer areas. Due to the increasing illegal cultivation, other morphine-poor species such as *P. bracteatum* are also cultivated, which mainly produce thebaine.

Naloxone

Naloxone is a competitive morphine antagonist that displaces morphine, heroin, and other derivatives from the enkephalin receptor and immediately cancels their effect. Naloxone is administered for complications after operations, for overdoses of heroin, and as a "sobering aid" for drug addicts in withdrawal and aftercare. The duration of action is about two hours, shorter than most opiates and opioids.

Etorphine

Etorphine is more lipophilic than morphine due to the ethyl group on the two oxygen atoms, therefore it can more easily cross the blood-brain barrier and is up to 100 times stronger. Due to the extreme effect, the therapeutic range is very narrow (high risk of respiratory depression), which is why etorphine is not used in human medicine. In veterinary medicine, etorphine is used for the euthanasia of large animals (injection or tranquilizer darts).

Opium is the dried latex obtained by incising the immature seed pods of *Papaver somniferum*, Papaveraceae (Opium poppy). Each capsule, as the fruit of *P. somniferum* is called, yields 20–50 mg of morphine. To obtain 1 kg, about 2,000 plants must be cultivated and the same number of capsules incised. In addition to the capsules, morphine is also obtained from opium straw, which consists of the stems of the plant and still contains about 1% of morphine, codeine, and thebaine. The extraction is carried out using the Kabay method: After the addition of $Ca(OH)_2$, $CaSO_4$ precipitates, and the alkaloids are converted into the base. After the addition of NH_4Cl, morphine can be precipitated as a salt in the aqueous phase. Legal cultivation is heavily regulated by the United Nations (Narcotic Control Board) and is only allowed in Hungary and Turkey in Europe. Another important legal cultivation area is, curiously, Tasmania in Australia. The actual demand is about 3,000 tons of opium per year, of which 90% is used to produce codeine. Illegal cultivation makes up the largest part of the known total cultivation area and is primarily used to extract morphine, which is converted into heroin by acetylation. To produce one ton of heroin, seven to ten tons of opium and up to two and a half tons of acetic anhydride are needed. According to UNODC, an estimated 8,200 tons of opium are illegally produced in Afghanistan each year. Thus, opium poppy cultivation brings about ten times the income compared to wheat cultivation. Other major illegal cultivation areas are southern Afghanistan and the so-called Golden Triangle between Thailand, Laos, and Myanmar.

14.6.3 Ingredients of Opium and Their Effects

In addition to morphine, opium contains other alkaloids, as well as other natural substances such as flavonoids, sugars, and terpenes. The most important alkaloids and their content are given in ◘ Table 14.2, but heroin, chemically diacetylmorphine, is not found in opium. It is a synthetic substance that has been synthesized and traded as a cough remedy under the brand name Heroin® by Bayer since

14.6 · Alkaloids Derived from Phenylalanine and Tyrosine

Table 14.2 Main alkaloids of opium

Alkaloid	min./max. content	Average content
Morphine	3–23%	12%
Noscapine	2–10%	5%
Papaverine	0.5–1.3%	1%
Codeine	0.2–3%	1%
Thebaine	0.2–1%	0.5%
Narceine	0.2–0.7%	0.5%

1896. Despite its very high potential for addiction, it was not removed from the market until 1971 and has been subject to the Narcotics Act in Germany ever since. In the UK, the drug is still available on prescription in palliative medicine. Under the trade name Diamorphine®, heroin has been available again in Germany since 2009 as a medication for diamorphine-supported substitution treatment. Codeine is the monomethyl ether of morphine, which is obtained by semi-synthesis from morphine. Codeine is less analgesic than morphine and suppresses the cough reflex, which is why it is used in medicine as a cough remedy. Codeine is partially demethylated to morphine in the liver, which is excreted in the urine. Codeine has virtually no analgesic effect, and the risk of addiction is low. Thebaine does not have a numbing effect, but rather an antispasmodic effect. Thebaine does not play a role as an active ingredient or drug, but is used as a technical base material for semi-synthetic opiates. Drugs synthetically derived from thebaine such as oxycodone, etorphine, buprenorphine, and naloxone are clinically important opiates.

Buprenorphine

Buprenorphine is a derivative of etorphine and is semi-synthetically produced from thebaine. It has a cyclopropylmethyl ring on the nitrogen, which increases the analgesic effect compared to morphine by a factor of 100. In addition, the effect lasts longer (6 to 8 h), which is why buprenorphine is administered for severe operations with strong pain. Buprenorphine is also used as a substitution agent in opioid dependence.

Starting from the opiate derivatives presented here, further potent analgesics such as methadone and pethidine have been developed over the last 70 years, which barely recognize the original morphine as a lead structure. The goal of extending the half-life of morphine to two to four hours and minimizing the risk of dependency has not been achieved to this day. Methadone is used in heroin substitution because it is available orally and has a half-life of eight hours.

14.6.4 Colchicine

Colchicine (◘ Fig. 14.5) also belongs to the protoalkaloids and reacts to weak acids. Because of the exocyclic nitrogen, colchicine is formally not a real alkaloid. Interesting is the presence of an unusually large ring with seven carbon atoms, in whose biosynthesis phenylalanine and tyrosine are involved as starting amino acids, but are not immediately visible in the finished molecule. Colchicine has an interesting structure with its 7-membered tropolone ring. A special feature of the structure of colchicine is that the two aromatic rings are not coplanar, but are twisted against each other at an angle of 54°, which is important for the pharmacological effect and the binding to tubulin. The main alkaloid colchicine is contained at about 0.8% in the seed and 0.6% in the tuber of *Colchicum autumnale*. The nitrogen in colchicine is part of an amide function, therefore has no significant basicity and does not form defined salts. In addition to 20 other contained alkaloids, colchicine is the main alkaloid and is used in medicine to treat acute gout attacks. The mechanism of action is not known and the indication is surprising, as colchicine is a strong cell poison. Colchicine inhibits mitosis and prevents the formation of the tubulin apparatus. In addition to colchicine, the natural substances podophyllotoxin, noscapine, and combretastatin also bind as ligands to the so-called colchicine site of tubulin. The biosynthesis begins with the rearrangement of autumnalin, which is formed in a Mannich reaction from dopamine and cinnamaldehyde. Rearrangements of the cyclohexane ring and the aromatic ring of the 4-cinnamaldehyde lead to an expansion to the 7-ring. In this process, the nitrogen of the former dopamine is arranged exocyclically.

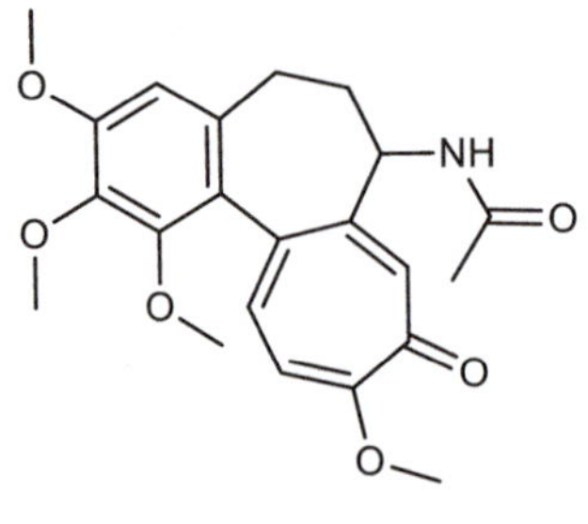

◘ Fig. 14.5 Colchicine

14.7 Alkaloids derived from Tryptophan

14.7.1 Biosynthesis

The biosynthesis starts from the amino acid tryptophan, which is characterized by an indole structure. Since this substructure (indole C2N) occurs in many alkaloids, it is also referred to as the class of indole alkaloids. Decarboxylation of tryptophan produces tryptamine, which is the actual starting substance for the biosynthesis of these alkaloids, but also of various physiological mediators such as serotonin, in our body. Most of the alkaloids discussed here are therefore directly derived from tryptamine.

To explain the structural diversity, one must consider the bioorganic reaction mechanism of tryptamine in further biosyntheses. Simple ring closure from tryptamine produces the alkaloid psilocybin or physostigmine. Mannich condensation of tryptamine with a C2 carbonyl compound produces harman as a further intermediate for more complex biosynthetic branches. A large group of alkaloids that can be derived from harman are the *Rauwolfia* alkaloids, the best-known representative of which is yohimbine. *China* alkaloids such as quinine are further complex alkaloids in which tryptamine and a monoterpene are linked via a Mannich reaction. In the final structure, these reactions are no longer detectable due to complex rearrangements, as can be clearly seen in the structure of quinine. The incorporation of a C5 isoprene unit into tryptamine leads to further interesting alkaloids such as lysergic acid (◘ Fig. 14.9), which can be found in the alkaloid group of ergot alkaloids.

In addition to the already mentioned *Rauwolfia* and China alkaloids, the *Vinca*, *Catharantus* and *Strychnos* alkaloids also contain a monoterpene. This monoterpene is secologanin (C10), which is involved as a central intermediate in strictosidine. Strictosidine is of great importance for chemical and structural diversity. As (◘ Fig. 14.6) shows, the folding in the biosynthetic enzyme plays a crucial role in the formation of this alkaloid subtype. Further biosynthetic modifications of the strictosidine building block (rearrangements, conformational changes) result in the alkaloids of the Corynanthe/Strychnos, Aspidosperma and Iboga types, named after the plant genera that predominantly biosynthesize these alkaloid types.

14.7.2 Psilocin and Psilocybin

Psilocin and Psilocybin Mushrooms of the genus *Psilocybe, Panaeolus* and *Conocybe*, e.g. *P. mexicana* and *P. pelliculosa,* found in Mexico, contain the substance psilocybin (◘ Fig. 14.7), the phosphoric acid ester of psilocin. These alkaloids have a strong hallucinogenic effect and are subject to the narcotics regulation in Germany. The small mushrooms are revered as sacred in South and Central America and used for religious ceremonies. In the West, the mushrooms and their ingredients became known through a publication in "Life Magazine" in 1957; active ingredient isolation and structural elucidation were carried out in 1958 by Albert Hofmann. *Psilocybe* mushrooms are now illegal drugs and pose mainly mental health risks in

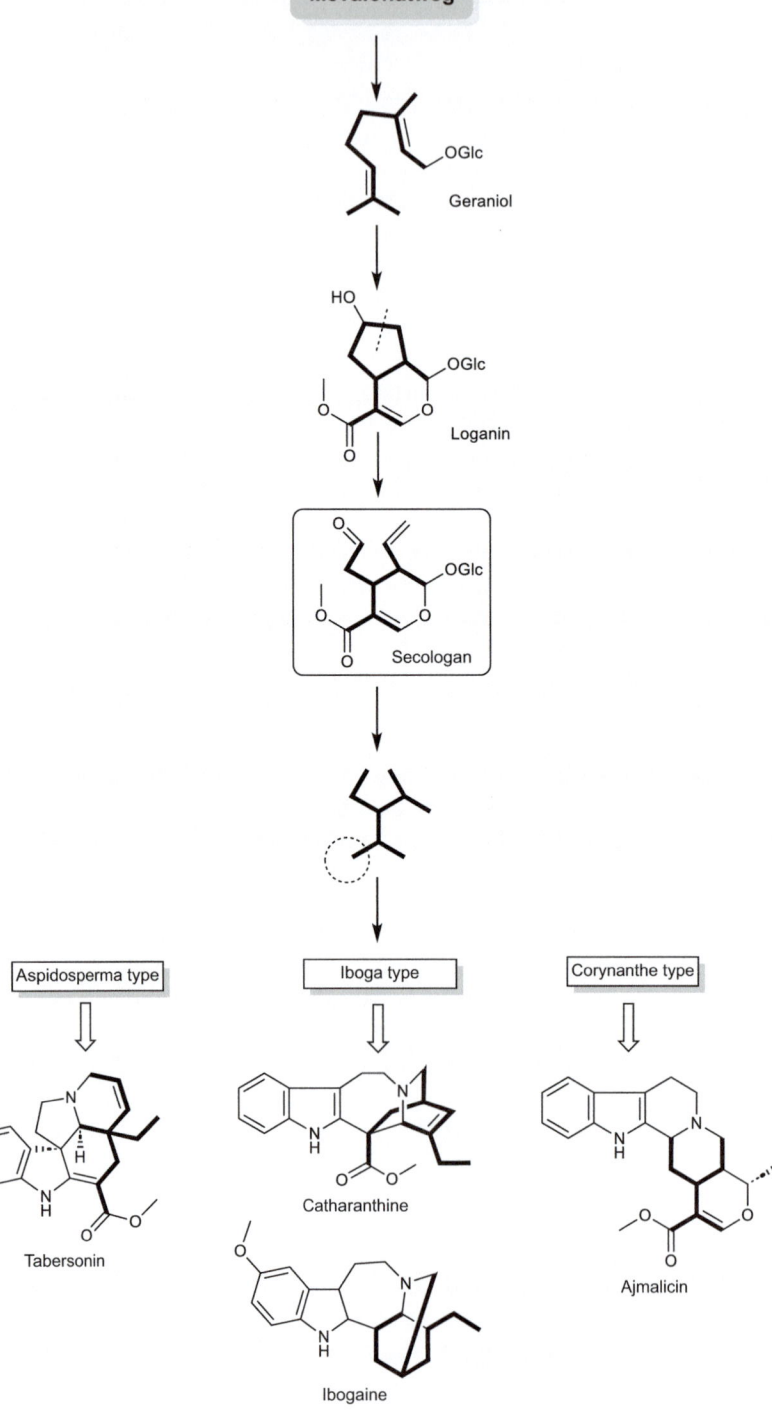

◘ Fig. 14.6 Alkaloids from the biosynthesis starting from secologanin

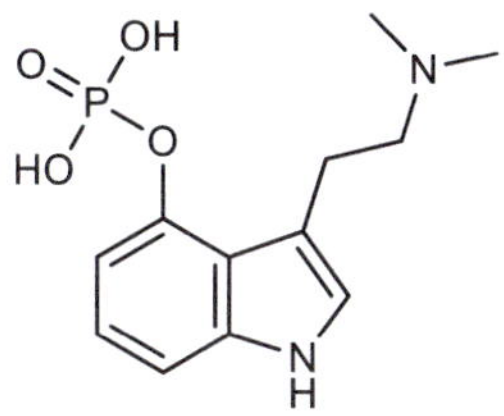

14.7 · Alkaloids derived from Tryptophan

Fig. 14.7 Psilocybin

case of overdose and misuse. Interestingly, the US Centers for Disease Control and Prevention (CDC) classify psilocybin as less toxic than aspirin. The mechanism of action of psilocybin is unknown; it is suspected that psilocybin acts as a precursor of psilocin and alters the serotonin metabolism in the brain.

More than 80 Psilocybe species have proven to be psychoactive, while more than 50 species have been classified as inactive. More than 30 hallucinogenic species have been identified in Mexico, but active species can occur anywhere in the world. Psilocybe mexicana has been used for many years by the indigenous peoples of Mexico in traditional ceremonies. The history of this plant goes back to the Aztecs. *Psilocybe semilanceata* is a widely distributed species in temperate latitudes with a similar effect. Consumption of the mushroom leads to optical hallucinations with rapidly changing shapes and colors and an altered perception of space and time, with the effect gradually subsiding and no lasting damage or dependencies occurring. The active hallucinogens, with a proportion of about 0.3%, are the tryptamine derivatives psilocybin and psilocin, which are structurally related to the neurotransmitter 5-HT, explaining their neurological effect.

14.7.3 Physostigmine

Physostigmine (Fig. 14.8) is formed by a ring closure of tryptamine, so that in addition to the indole ring, a pyrrolizidine ring is formally also formed. Physostigmine is still obtained from the Calabar bean, the fruit of *Psysostigma venenosum*, Fabaceae, a West African liana. The content is 0.1–0.2% based on the dry weight. The "beans", which are botanically classified as legumes, are used by the population to carry out divine judgments, because the content and thus toxicity vary. In medicine, physostigmine is used as a miotic for the treatment of cataract. Unlike the other alkaloids presented, no newer synthetic products have been developed from physostigmine; rather, physostigmine is still used directly as a natural substance. It acts as a parasympathomimetic and inhibits cholinesterase, thereby reducing the breakdown of acetylcholine.

Physostigmine is a reversible acetylcholinesterase inhibitor. It prevents the normal breakdown of acetylcholine and thereby enhances cholinergic activity. Although it is hardly used as a drug today, it was crucial for the research of the function of acetylcholine as a neurotransmitter. It was mainly used as a miotic to constrict the pupil, often to enhance the effect of mydriatics. It lowers intraocular

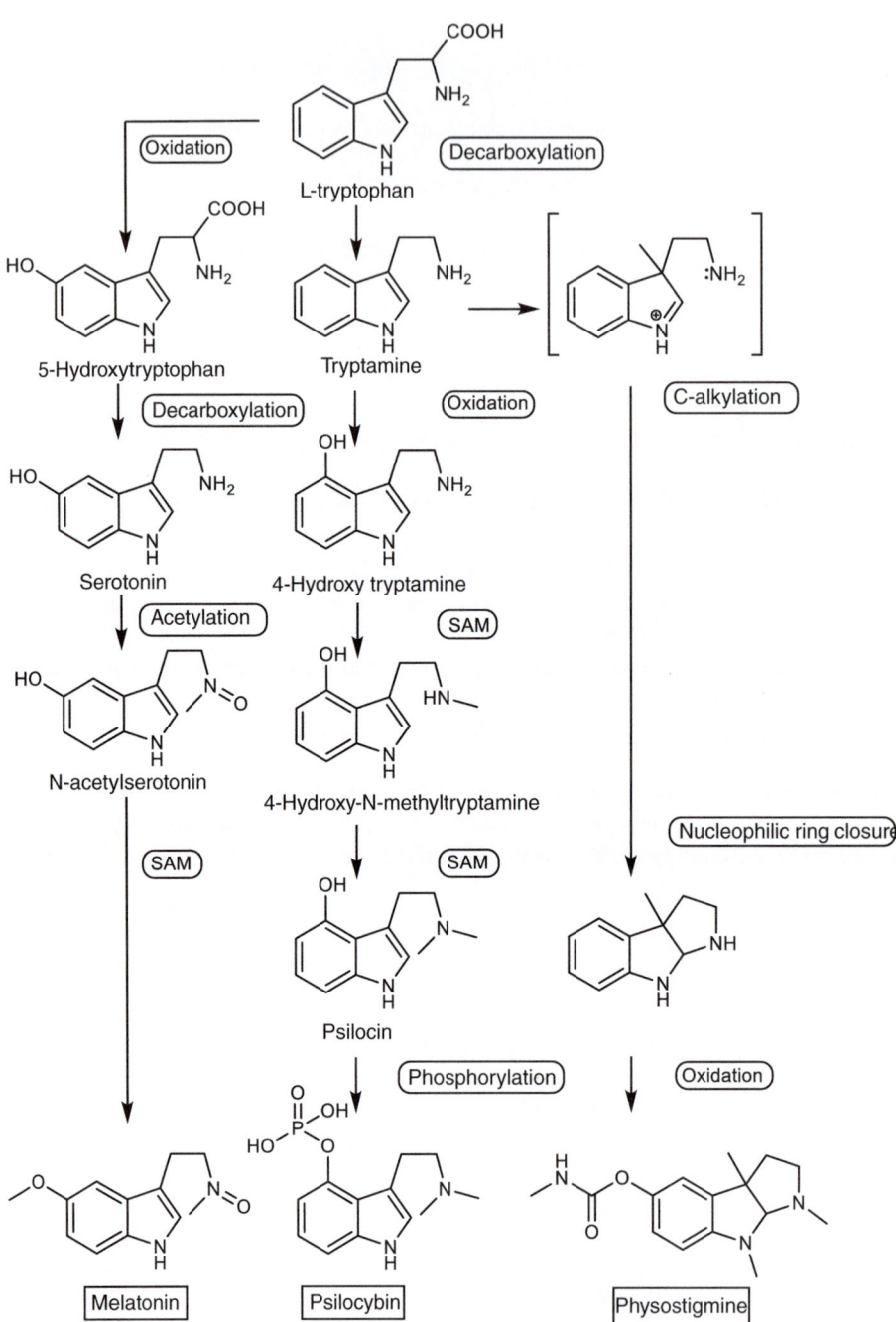

Fig. 14.8 Psilocybin and Physostigmine Biosynthesis

pressure by constricting the pupil and increasing aqueous humor outflow and is, along with pilocarpine, an important agent for the treatment of glaucoma. Since it prolongs the effect of the body's own acetylcholine, physostigmine can be used as an antidote against anticholinergic substances such as hyoscine. Physostigmine is interesting as an acetylcholinesterase inhibitor in the treatment of Alzheimer's disease, which is characterized by a dramatic loss of function of the central cholinergic system. Physostigmine analogs are already in use (e.g., rivastigmine) or are being tested in clinical trials (e.g., phenserine). These analogs have a longer duration of action, lower toxicity, and better bioavailability than the natural substance itself.

Biosynthesis   The biosynthesis (◘ Fig. 14.8) starts from tryptophan, which decarboxylates the terminal carboxy group to the alkaline-reacting tryptamine. This metabolite can itself exert physiological effects and is known as a neuromodulator. The terminal amino group can be converted by a monoamine oxidase to indole-3-acetaldehyde, which is important for further biosynthesis reactions in the group of alkaloids. From tryptamine, the neurotransmitter serotonin and the hormone melatonin involved in the regulation of the sleep-wake cycle are formed (◘ Fig. 14.8). For the formation of psilocybin and physostigmine, a hydroxylation is carried out by a Cyp450-4-hydroxylase. The terminal amino group is doubly methylated with the help of SAM to form psilocin. A psilocin kinase phosphorylates psilocin to psilocybin and physostigmine. Our knowledge of the biosynthesis is still incomplete, as it is not yet known whether a hydroxylation and then a methylation occurs first or vice versa. In the biosynthesis of physostigmine, a ring closure is observed under nucleophilic attack of the terminal amine. The aromatic hydroxyl group is introduced by a carbamate at the hydroxyl group, in contrast to psilocybin. The biosynthesis of physostigmine is located in a gene cluster that consists of seven genes or modules. The formation of a carbamate is rare in phytochemistry.

Biotechnology   After a large-scale screening for physostigmine in microorganisms, the actinomycetes Streptomyces griseofuscus and Streptomyces pseudogriseolus were discovered as physostigmine producers. Based on the physostigmine biosynthesis pathway, a new biosynthesis pathway was established to produce melatonin in *E. coli*. However, the physostigmine biosynthesis pathway is not a biosynthesis pathway that has been implemented with modern genetic engineering.

14.7.4 Melatonin

Chemistry and Biosynthesis   Melatonin (◘ Fig. 14.8) is an indoleamine derived from tryptophan. The amino acid is hydroxylated at position C5 by a tryptophan hydroxylase and decarboxylated to serotonin. The terminal amine group is acetylated by a serotonin-N-acetyl-transferase and the aromatic hydroxyl group is methylated with SAM (◘ Fig. 14.8).

Biological Effect   Melatonin is a hormone that naturally occurs in the human body and is primarily produced by the pineal gland (Epiphysis) in the brain during darkness. It plays a crucial role in regulating the sleep-wake cycle, also known as the circadian rhythm. The production and release of melatonin are influenced by light, with darkness stimulating production and light inhibiting it.

Biotechnology For the biosynthesis of melatonin in *E. coli* and yeast, a genetic biosynthesis pathway could not be relied upon. The gene cluster involved in the biosynthesis of melatonin in microorganisms has not yet been identified. Therefore, almost all genes necessary for establishing the melatonin biosynthesis pathway in genetically modified bacteria were cloned from animals, humans, and plants. The genes coding up to serotonin come from humans or from the blood fluke *Schistosoma mansoni*and from cattle Bos taurus. When these enzymes originating from mammals and plants are heterologously expressed in *E. coli*, the expression level is low or the expression product is inactive, which limits high melatonin production in prokaryotic cells. In addition, metabolic and enzymatic strategies were used to optimize the biosynthesis pathway and the melatonin production was increased 11-fold compared to the first-generation strain.

14.7.5 Lysergic Acid Alkaloids

In this group of alkaloids, IPP is inserted as a terpene building block through C-C bond formation on the aromatic ring (◘ Fig. 14.9). IPP and DMAPP from the mevalonate pathway are incorporated at position C4 by an aromatic terpene transferase. After ring closure, the tetracyclic ergoline skeleton is formed. Ergoline is methylated to the inactive methylergoline. By introducing a double bond between C8 and C9, the clavine alkaloids(chanoclavine I) are formed. If the methyl group at C8 remains free, it is agroclavine, and elymoclavines are formed by oxidation to the primary alcohol. Further oxidation of C8 to the carboxy function in β-position and rearrangement of the double bond to C9-C10 results in lysergic acid. These alkaloids are chemically labile, which is why the carboxy function easily rearranges to the α-position. This results in ineffective lysergic acids. By coupling amines, amino alcohols, or amino acids to the carboxyl function of lysergic acid, *iso*-lysergic acid and the secal alkaloids are formed. The biosynthesis of ergopeptin takes place non-ribosomally on a multi-enzyme complex, as is also known from the peptide antibiotics discussed in ▶ Chap. 15. By forming an amide of the carboxyl function in C8 position with amino acids, two groups of ergot alkaloids can be distinguished:
— Simple acid amides such as ergometrine. These ergot alkaloids are poorly soluble in water.
— Peptide alkaloide, represented by ergotamine (◘ Fig. 14.11), ergocristine, ergocryptine, and ergocornine (in mixture = ergotoxin) as ergopeptines. These alkaloids are insoluble in water. In addition, there is a subgroup, the ergopeptams, in which the oxazolidine ring of the first amino acid is open. Here, the number of representatives of the ergopeptines predominates.

Extraction Industrially, *Claviceps paspali* is used for the production of ergometrine and *C. purpurea* for the production of ergotamine, ergosin, ergocristine, and ergocryptine grown above ground. Asparagine is preferred as a nitrogen source over am-

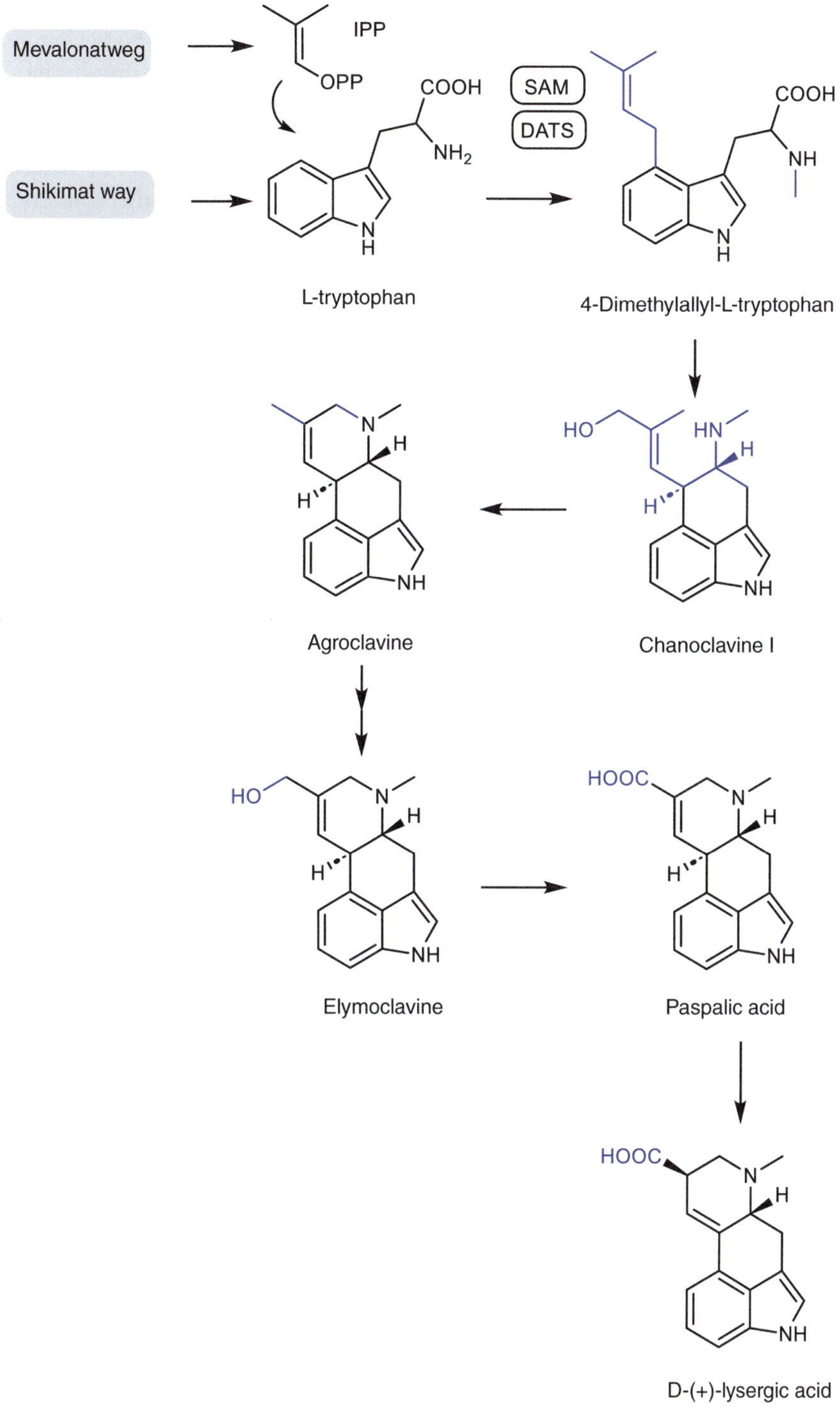

14.7 · Alkaloids derived from Tryptophan

Fig. 14.9 Lysergic acid, abbreviated biosynthesis

monium salts. Ergot is the sclerotium of the ascomycete *Claviceps purpurea*, which infects the rye(*Secale cereale*) and then grows out of the fruiting body. The infection can occur naturally (± 0.2% total alkaloid content) or be artificially induced (± 1% total alkaloid content), as is still technically done today for the extraction of sclerotia for the production of ergot alkaloids. The artificial infection has the advantage that high-performance strains of the mold can be used. An alternative to the not entirely safe cultivation in the open field is the direct cultivation of *Claviceps paspali*in suspension culture. New studies are examining the use of endophytes of the plant *Argyrea nervosa*, which is consumed by drug users as "natural LSD", although no LSD, but only LSA (lysergic acid amide) is detectable.

Physiological Effects   In the Middle Ages, rye infected with *Secale cornutum* was often consumed. This led to ergotism (death of limbs due to vascular constriction), also called St. Anthony's fire in religious terms. The last cases of ergotism in Germany occurred in the 1980s with the onset of organic farming in Australia and the USA. The renunciation of fungicides allowed the fungus to spread. Current screening methods and improved cultivation methods have sufficiently solved this problem. Secale alkaloids are used in gynecology for their labor-inducing effects and to treat uterine bleeding. In pain therapy, ergot alkaloids are used in the prophylaxis and treatment of severe migraine attacks. Semi-synthetically produced ergot alkaloids are methylergometrine and methysergide. Although they are largely structurally identical, they have different effects: methylergometrine is used for uterine bleeding, methysergide for the prophylaxis of migraine attacks (▶ Infobox 14.1).

Infobox 14.1 About natural and synthetic LSD – or what Woodstock and cluster headaches have in common

Lysergic acid diethylamide (LSD) was accidentally synthesized in 1943 by Albert Hoffmann at the then Ciba-Geigy (now Novartis) (the exact date is still celebrated as Bicycle Day by users). In a self-experiment, Hoffmann found that LSD has a very strong hallucinogenic effect. The physiologically active dose is comparatively low at 25 ng for an adult (compared to 500 mg of aspirin!). After World War II, LSD was introduced into the clinic, but it was not taken by patients, but by the doctor or psychiatrist, as they believed they could better empathize with the sick psyche of the patients. LSD (◘ Fig. 14.10) was also researched as a possible weapon, but with dubious success, as the loyalty of the soldiers was limited and it was not known how abnormal and unpredictable the enemy would behave. In the 1960s, LSD became a cult in the hippie scene, and it didn't take long before it

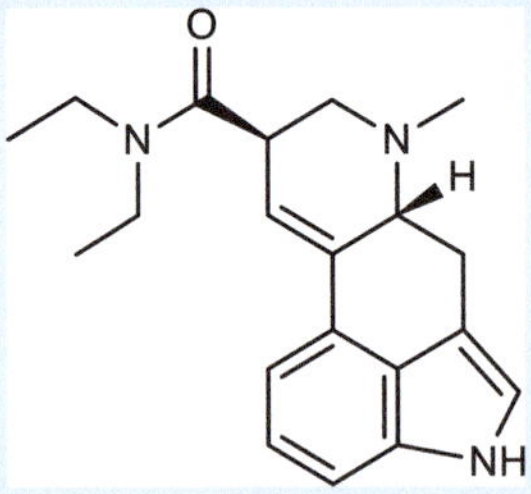

◘ **Fig. 14.10**   Lysergic acid diethylamide (LSD)

was banned due to alleged murders under the influence of LSD. From then on, LSD was only used in medical research, especially because of its mechanism of action (constriction of blood vessels, where opioids fail) for the treatment of very painful cluster headaches. Interestingly, LSD does not cause physical or psychological dependence. For those interested in the discovery and history, the documentary on the television channel arte is highly recommended. Lysergic acid alkaloids also occur in plants, but LSD is not a natural substance! Whether lysergic acid derivatives are formed by endophytes that live symbiotically with and in the plant is currently being intensively discussed. The fact is that lysergic acids such as ergotamine (◯ Fig. 14.11) occur in *Secale cornutum* and some species of the family Convolvulacae such as *Argyrea nervosa* (Hawaiian baby woodrose) and *Rivea corymbosa* (Ololiuqui) as well as in *Ipomoea* species, which were used by the Aztecs in Mexico for religious purposes.

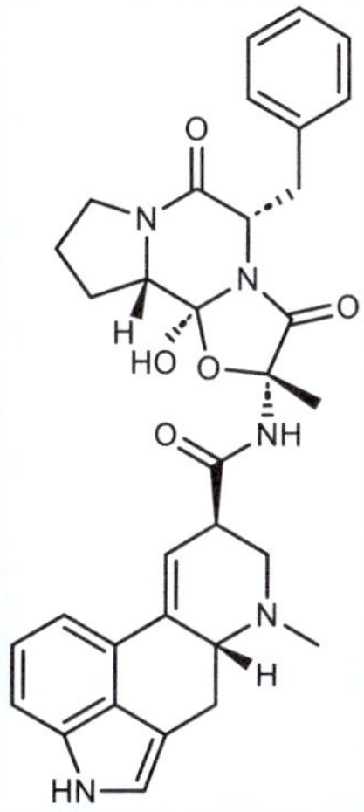

◯ **Fig. 14.11**   Ergotamine

14.7.6 Vinca Alkaloids

The Madagascar periwinkle Catharanthus roseus (*Vinca rosea*, Apocynaceae) is a shrub-like perennial plant native to Madagascar, now widespread in the tropics, which can also be found in our gardens as an ornamental plant due to its shiny, dark green leaves and pretty, five-lobed flowers. As it is easy to cultivate, it is grown in many parts of the world, e.g. in the USA, Australia, India, and Europe. Before its career as a cancer drug, the plant was originally investigated for its folkloric use as a tea for diabetics and its potential blood sugar-lowering effect. Animal experiments on rabbits showed a reduction in white blood cells (leukopenia), which led to the idea of using the plant and its ingredients as a cancer drug. Of the over 150 alkaloids in the extracts studied, the indole alkaloids proved to be particularly effective. In this group, it is the dimeric bisindole alkaloids such as vinblastine, vincristine, vinleurosin and vinrosidin. The prefix "Vin-" still points to the botanical origin (*Vinca rosea*) of the substances. Today, vinblastine and vincristine are used clinically.

Biosynthesis   From strictosidine, tabersonin is first built up as an important intermediate (◯ Fig. 14.12). This natural substance already shows the complex structure of the later Vinca alkaloids. The tabersonin, which belongs to the Aspidosperma structure type, is structurally transformed. On the one hand, a side path can lead to vinblastine as the target structure via the side path of catharandin or via six enzymes to vindoline and from there directly to vinblastine. The biochemical reactions to

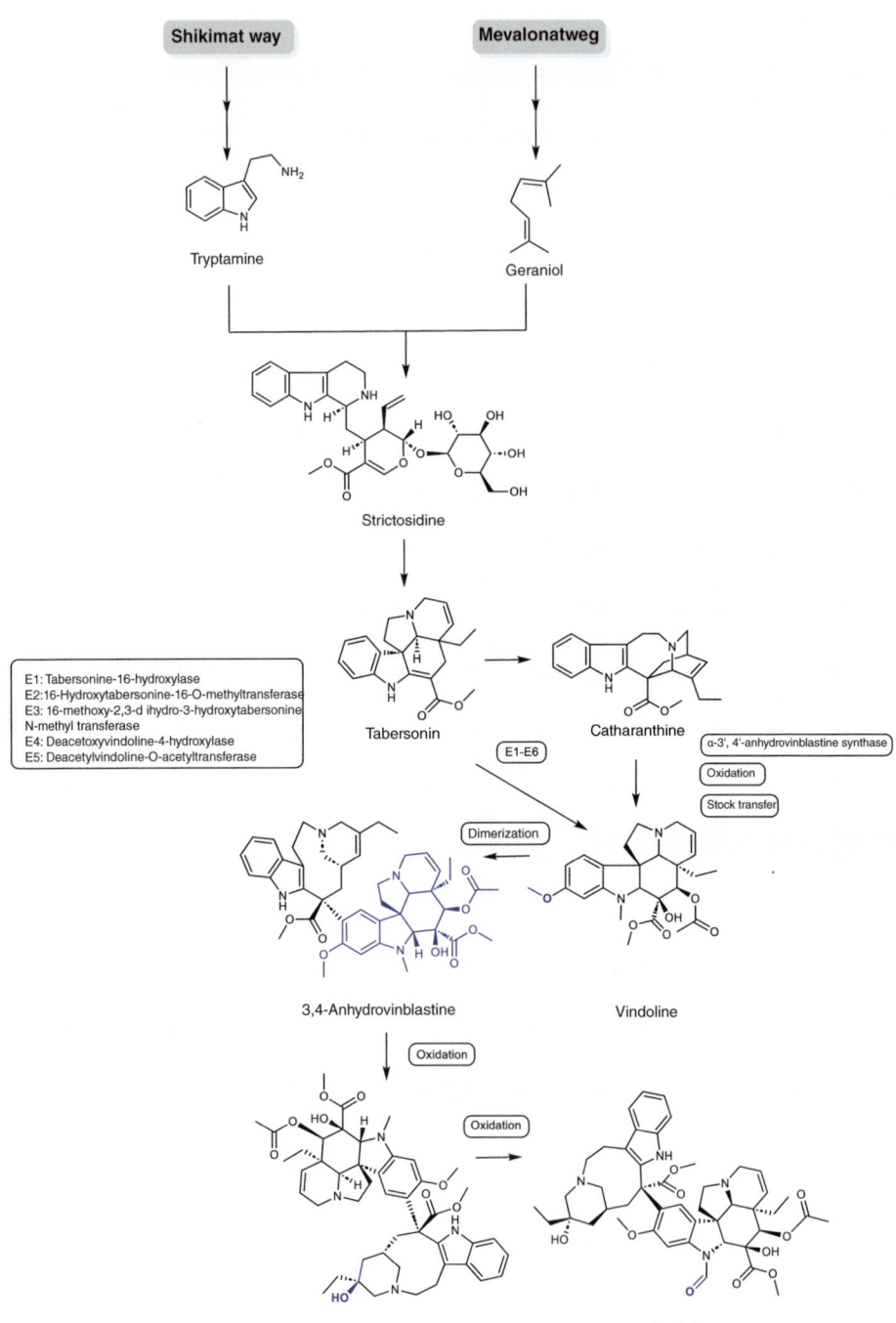

Fig. 14.12 Abbreviated biosynthesis of vinblastine and vincristine

vinblastine can be summarized as methylations and oxidation reactions. The dimerization takes place with vindoline. The conversion of vinblastine to vincristine occurs by forming an N-formyl substituent by oxidizing the N-methyl group of vinblastine.

Extraction   A major problem with the clinical use of vinblastine and vincristine is the very small amounts in the plant. Although the total alkaloid content of the leaf can be up to 1% or more, more than 500 kg of *Catharanthus* leaves are needed to obtain 1 g of vincristine. This yield (0.0002%) is the lowest among all medically relevant alkaloids that are commercially isolated. The extraction is costly and time-consuming, as it requires large amounts of raw material and extensive chromatographic fractionations. In addition, the plant produces a much higher proportion of the secondary alkaloid vinblastine than vincristine, with the latter natural substance being medically more valuable. Fortunately, it is possible to convert vinblastine to vincristine by controlled oxidation with chromic acid via demethylvinblastine. Demethylvinblastine occurs in nature and can be produced chemically or by microbiological N-demethylation with *Streptomyces albogriseolus* from vinblastine.

Considerable efforts have been made in the field of semi-synthesis of "dimeric" alkaloids to form dimers from monomers such as catharanthine and vindoline, which are produced in much larger quantities in *Catharanthus roseus*. It is now possible to convert catharanthine and vindoline into vinblastine with a yield of about 40%, although the dependence on natural monomers still limits commercial use. Excellent yields of anhydrovinblastine can also be achieved by electrochemical oxidation of a catharanthine-vindoline mixture. An enzymatic coupling with commercial horseradish peroxidase is also possible.

Semisynthetic Analogs   Vindesine is a semi-synthetic amide derivative of vinblastine, which was introduced for the treatment of acute lymphatic leukemia in children. Vinorelbine is a newer semi-synthetic modification of anhydrovinblastine. In this structure, the indole-C2N bridge in the unit derived from catharanthin was shortened by one carbon atom. The analogs have a broader anticancer effect, but less neurotoxic side effects than vinblastine or vincristine. They are used intravenously for the treatment of advanced breast cancer and non-small cell bronchial carcinoma or orally for the treatment of small cell bronchial carcinoma. These compounds inhibit cell mitosis by binding to the protein tubulin in the mitotic spindle and preventing polymerization into microtubules. They have a similar mechanism of action as other natural substances, e.g. colchicine and podophyllotoxin.

14.8 Quinoline Alkaloids

Quinoline alkaloids are obtained from the bark and branches of *Cinchona pubescens* and *Cinchona succirubra*, Rubiaceae, by peeling and extracting the bark. The tree originally comes from Peru and the northwest of South America and has a great cultural significance, which is also reflected in the country's flag. With the Spaniards, the tree came to Europe as Jesuit's bark and was used until the beginning of the 19th century as a fever remedy before it was replaced by salicin and later by acetylsalicylic acid. Because of its great importance as a malaria remedy in the emerging colonial empires, the demand for the treatment of malaria remained high.

To break Peru's cultivation monopoly, cinchona trees were also cultivated in Asia, especially by the Dutch in Indonesia in 1854. Both in India and in Indonesia, there are still partially wild plantations where cinchona bark is grown for the food industry. Starting from quinine (◘ Fig. 14.13), many antimalarial drugs have been developed, but unfortunately, they have strong resistance problems today. Quinine has remained the great exception, as no resistances have been reported to date. In addition to quinine, other alkaloids such as quinidine (◘ Fig. 14.13), cinchonine and cinchonidine were isolated from cinchona bark in 1829 by Pelletier and Caventou.

Biosynthesis These cinchona alkaloids are characterized by two amines in the aromatic quinoline ring and in the aliphatic piperidine ring (C5N) and belong to the corynanth type (◘ Fig. 14.13). Cinchona alkaloids occur in the bark as complexes, bound to tannins, or as salts of cinchonic acid. Cinchona alkaloids have two nitrogen atoms, one of which is in the aromatic quinoline ring system and the other in the aliphatic piperidine ring. The biosynthesis is known to start with the linkage of secologanin with tryptophan to form strictosidine, which then undergoes rearrangement reactions. In the first step, glucose is cleaved off by hydrolysis, the ring is opened, and the double bond becomes more reactive. In the further course, the carboxyl group is cleaved off and the C-N bond in the former side chain of the tryptophan is broken. The now present intermediate product corynantheal builds the bridged piperidine ring by reaction of the aldehyde with the nitrogen of the former β-amino group of the tryptophan to form cinchoninone. By reducing the carbonyl to alcohol, the quinine skeleton is complete and can be modified by secondary modifications of the aromatic compound such as hydroxylation (quinonidine) and methoxylation (quinine) or epimerization at C8 (cinchonine and quinidine).

Application Quinine, a bitter substance that most people appreciate for its taste in soft drinks (e.g., Bitter Lemon, Tonic Water), is a remarkable substance that fluoresces intensely bright blue under UV light and has a variety of properties. The idea of bitter tonic drinks originated from the former British colony of India, where the aim was to combine the pleasant with the useful. Along with the refreshing drink, quinine was to be taken for malaria prophylaxis. However, the amount of quinine in the drink is so small that it has no therapeutic effect.

Quinine stimulates the uterine muscles and was therefore used in the past as a labor-inducing agent. In this context, quinine was also misused as an abortifacient, which often led to the death of the expectant mother when taken in very high doses. Quinine also has analgesic, antipyretic, and local anesthetic effects. As a sulfate salt, it has antispasmodic effects and was used for the prevention and treatment of muscle cramps. Today, quinine is no longer used in medicine because of its side effects on the heart. Until the Second World War, the cinchona tree was intensively cultivated on plantations in Indonesia. From quinine, modern malaria drugs such as chloroquine and mefloquine (Lariam®) were developed.

14.8.1 Camptothecin

Camptothecin (◘ Fig. 14.14) is derived from the Chinese "Happy Tree" - a tree with the botanical species name *Camptotheca acuminata*, Nymphaceae, or *Nothapo-*

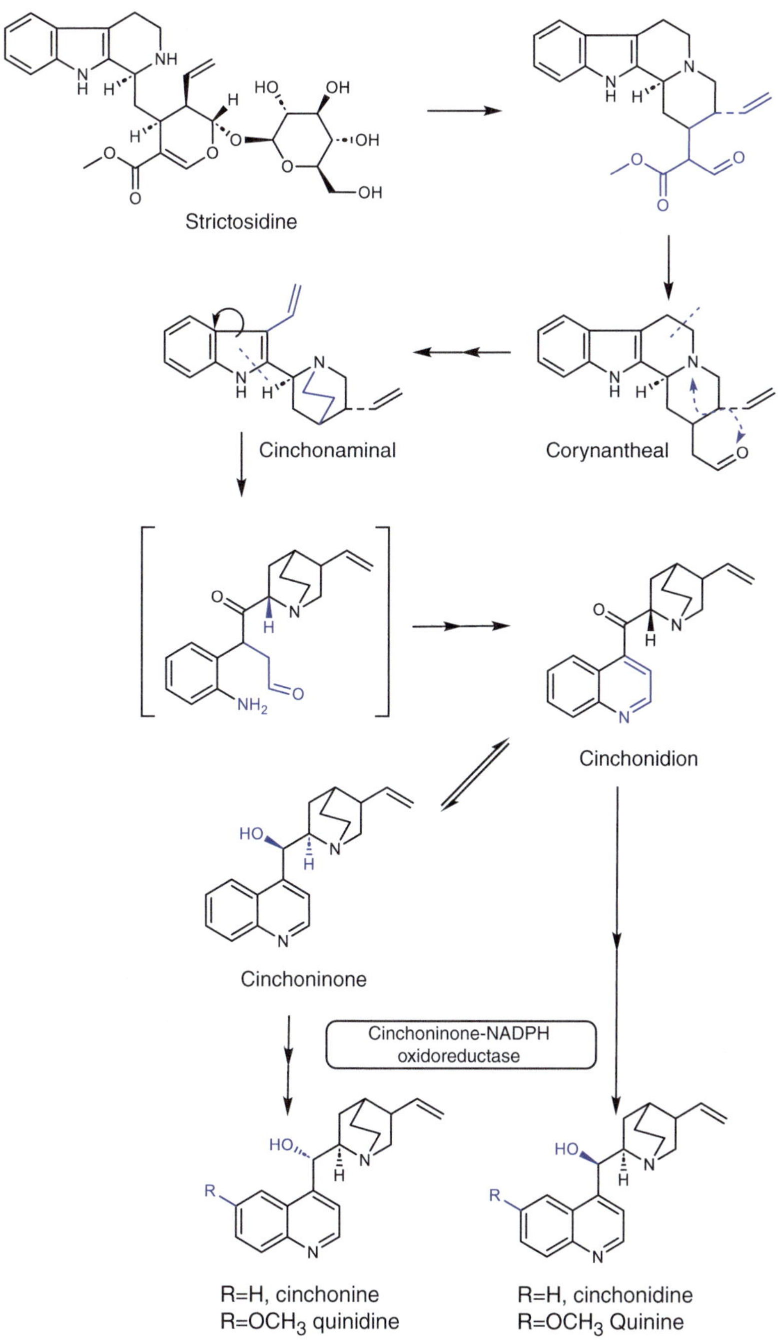

14.8 · Quinoline Alkaloids

Strictosidine

Cinchonaminal

Corynantheal

Cinchonidion

Cinchoninone

Cinchoninone-NADPH oxidoreductase

R=H, cinchonine
R=OCH₃ quinidine

R=H, cinchonidine
R=OCH₃ Quinine

Fig. 14.13 Biosynthesis of quinine and quinidine

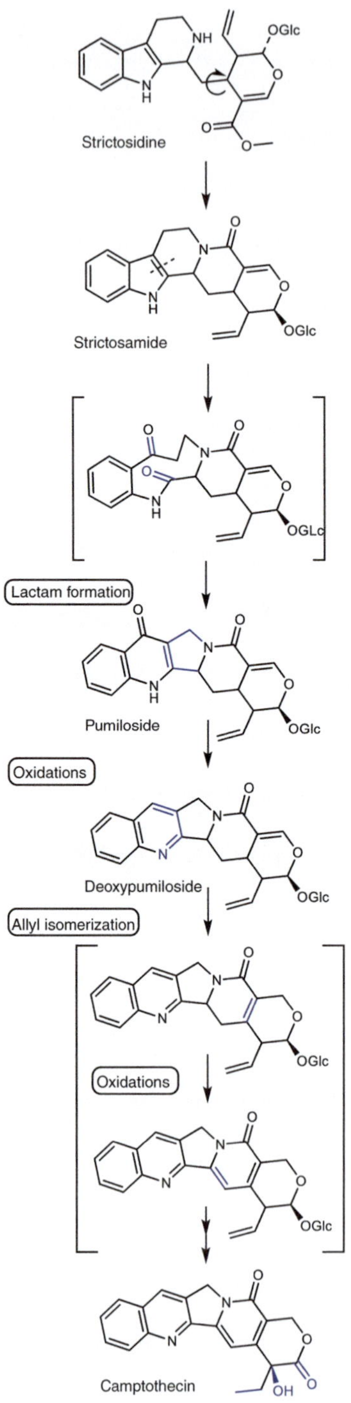

Fig. 14.14 Biosynthesis of camptothecin

dytes foetida, Icacinaceae. The seeds of *C. acuminata* contain 0.2 and the leaves 0.4% of the alkaloid and its by-product 10-hydroxycamptothecin (0.05%), which is 10 times more active than camptothecin. According to its chemical structure, camptothecin is an indolizidinequinoline system, whose "indole ring" (hence also abbreviated as indole alkaloids) was biosynthetically extended to a quinoline ring (◘ Fig. 14.14).

In Traditional Chinese Medicine (TCM), extracts are used to treat cancer. The first clinical trials were conducted in 1972, but severe hemolytic side effects led to a significant dose reduction. Since only 3 out of 61 patients responded at the end of the studies, the clinical trials were discontinued. After it was shown in 1985 that Camptothecin is a topoisomerase I inhibitor, studies were resumed to find better tolerated and water-soluble derivatives as cancer drugs. As a result of this research, the fully synthetic analogues topotecan (Hycamtin®) and irinotecan (Campto®) for the treatment of colon, lung and cervical cancer came into practice from development in 1995.

The current demand is about 1,000 kg for the synthesis of the above-mentioned partial synthetics. Camptothecin is obtained by extraction with dilute acid to transfer the protonated molecule into the aqueous phase. In a second step, the obtained extract is slightly alkalinized with a base (NaOH, KOH) and shaken with a water-insoluble solvent to transfer the basic alkaloid into the organic phase. If the purity is more than 90%, further chromatographic steps are carried out, leading to a purity of 99.9%. The problem is the contamination and separation of 20-vinylcamptothecin (20-Vinyl-CPT), which is not efficiently possible. By hydrogenation with a platinum catalyst and subsequent purification with activated carbon, the contamination can be degraded and the degradation product can be separated.

14.9 Alkaloids Derived from Ornithine

Within this alkaloid family, Ornithine is the only non-proteinogenic amino acid (C4N) that serves as a starting substance for another important alkaloid biosynthesis pathway. The biosynthesis of Ornithine is part of the urea cycle or results directly from L-glutamic acid via the citric acid cycle. Starting from Ornithine, a decarboxylation to putrescin takes place, which is rearranged in ring closure to N-Methyl-Δ1-pyrrolinium cation. This reaction is a typical Schiff base reaction of a carbonyl group with a primary amine. At this point, the biosynthesis pathway to nicotine also branches off, which is synthesized by combination with nicotinic acid from aspartate. In the biosynthesis of the tropane alkaloids, however, N-Methyl-Δ1-pyrrolinium is linked with two acetate units, so that the intermediate hygrin leads to Tropinone, forming a double ring system. To reach Tropinone, a decarboxylation and a stereospecific ring closure take place. If the carboxyl group is retained, cocaine is formed, if decarboxylation occurs, *Atropa* alkaloids are formed. The carbonyl group is reduced to alcohol and forms the basis for esterification with an aromatic acid from the shikimate pathway. In the tropane alkaloids, as the bridged ring system is called, the alcohol has the α-ol- (or *trans*-) configuration. The β-configuration is also called pseudo-configuration.

14.9.1 Tropane Alkaloids

These alkaloids are derived from tropane-3-ol (tropine). The tropane molecule consists of a pyrrolidine and a piperidine ring, which are connected by a ring. Esterification with tropic acid results in L- and D-hyoscyamine, with only the L-form occurring in the plant. After extraction and processing, a rearrangement of the stereoisomers from the L- to the D-form takes place. This mixture is referred to as atropine. Further tropane alkaloids can be produced through secondary modification. An important tropane alkaloid is scopolamine (◘ Fig. 14.15), which is of great importance as a starting material for the synthesis of other medicinal substances. For the synthesis of N-butylscopolamine (◘ Fig. 14.16), the plants *Duboisia myoporoides* and *D. leichhardtii* are cultivated in Australia. From the leaves, which can contain up to 5% scopolamine, scopolamine is technically complex to isolate. The dioxygenase hyoscyamine 6β-hydroxylase (H6H) hydroxylates hyoscyamine at the 6β-position, in the second step an epoxidation at C6 and C7 of the hyoscyamine takes place, resulting in scopolamine (◘ Fig. 14.18). This chemical change also alters the effect profile. While atropine is considered euphoric and stimulating to invigorating, scopolamine has a calming and soothing effect. In the past, scopolamine was also used by secret services as a truth serum; today it is also misused for criminal purposes.

14.9.1.1 Effects and Applications of Tropane Alkaloids

Tropane alkaloids belong to the pharmacological class of drugs known as parasympatholytics (anticholinergics). These natural substances, which act on the nervous system and the brain, cause the parasympathetic nervous system to respond not at all or only partially to the neurotransmitter acetylcholine. The effects of L-hyoscy-

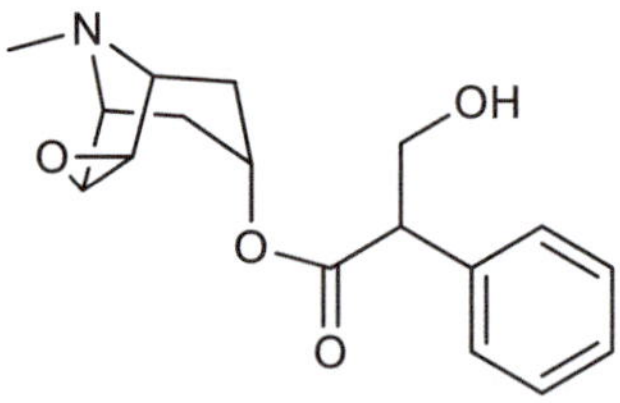

◘ **Fig. 14.15** Scopolamine

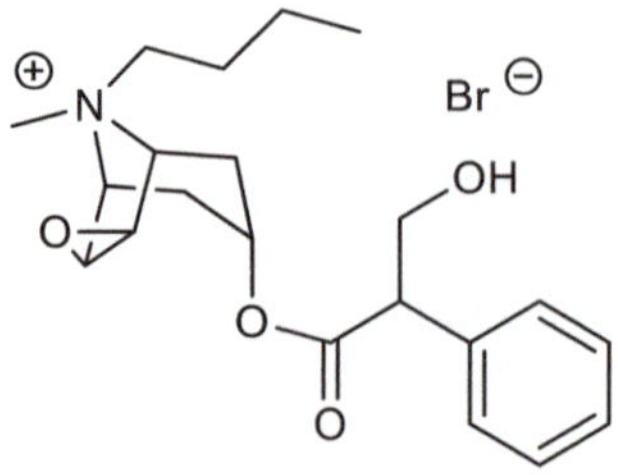

◘ **Fig. 14.16** N-Butylscopolamine

14.9 · Alkaloids Derived from Ornithine

amine can therefore be considered exemplary for parasympatholytic effects: mydriasis, inhibition of saliva flow and mucus formation in the gland cells, drying of the skin and suppression of sweat production, paralysis of intestinal peristalsis, muscle cramps, constriction of the bronchi, bile ducts and urinary tract. In case of overdose, strong cerebral excitation occurs first, followed by depression. It is used primarily for spasms, colics of the gastrointestinal tract and urogenital organs, to reduce gastric juice secretion (peptic ulcer), as a mydriatic and as an antidote for poisoning with parasympathomimetics (physostigmine, organic phosphorus compounds such as parathion). An important medication is Buscopan® with the active ingredient N-Butylscopolamine (◻ Fig. 14.16). The central effect of atropine and L-hyoscyamine is very pronounced due to its good CNS penetration and manifests itself in confusion and hallucinations in case of overdose. Plants containing tropane alkaloids are also misused. Extracts of Mandragora officinarum are used as an aphrodisiac, and smoking leaves of the angel's trumpet (*Brugmansia*-species) is popular among teenagers, but dangerous. According to the statistics of the poison control center in Baden-Württemberg, one third of the admitted consumers must be treated intensively immediately.

Scopolamine is still used today in transdermal patches and tablets (active ingredient name hyoscine) for the treatment of motion sickness (Scopoderm®, Scop® (USA), Travelcalm® (AUS)); otherwise, it mainly serves as a starting material for the synthesis of Buscopan®, for which tons of plant material must be technically extracted and the obtained scopolamine must be chemically processed into the quaternary ammonium salt N-butylscopolamine bromide. An important derivative of scopolamine is tiotropium bromide (◻ Fig. 14.17), which is prescribed for the treatment of obstructive lung diseases such as COPD (chronic obstructive pulmonary disease). Tiotropium is introduced into the lungs in very small doses using a special inhaler, the development of which goes back to nuclear energy. To concentrate the uranium, the uranium ores had to be very finely atomized, and a special piezotechnology was integrated into the nuclear physics development, which was later adapted for the treatment of COPD. In the past, COPD was diagnosed in heavy smokers, but today, due to high air pollution in many major cities, especially in Asia and Africa, it is also known among non-smokers (◻ Fig. 14.18).

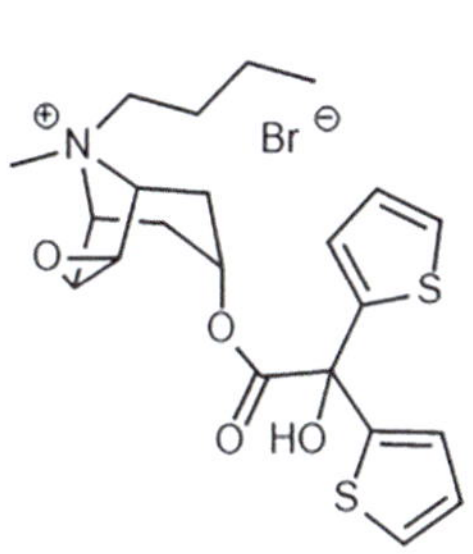

◻ **Fig. 14.17** Tiotropium bromide

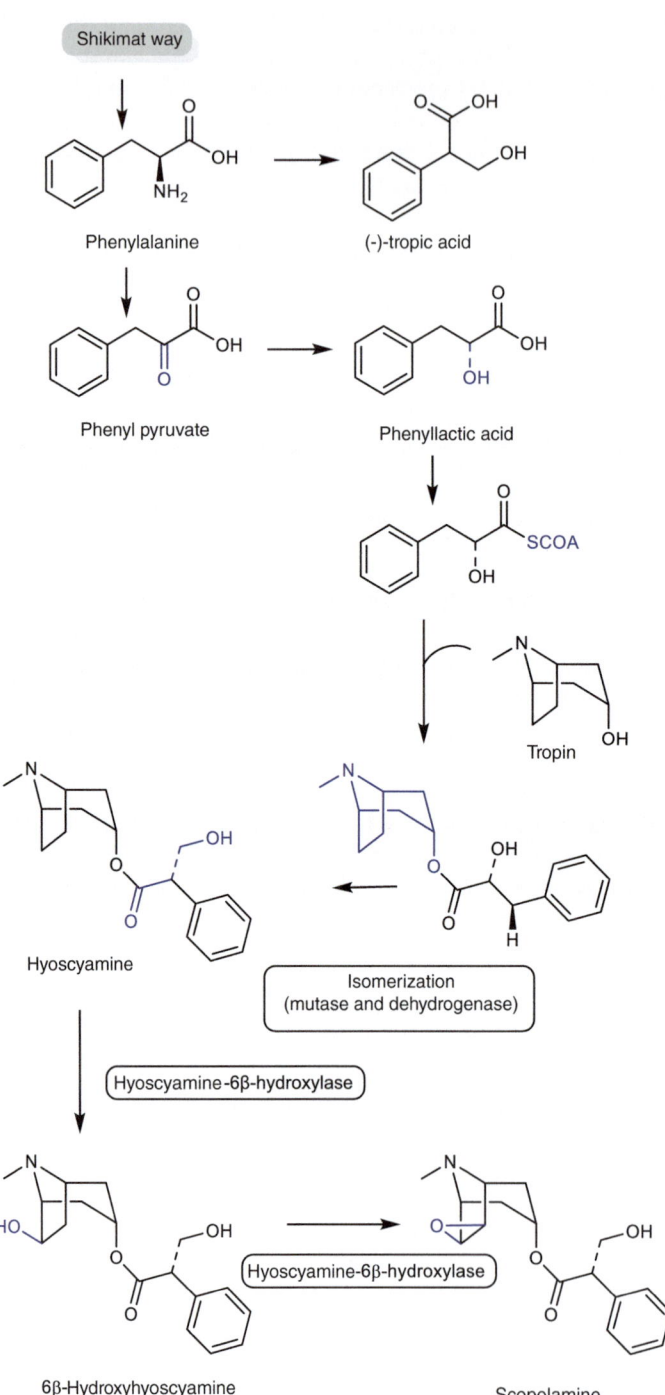

Fig. 14.18 Scopolamine biosynthesis

14.9 · Alkaloids Derived from Ornithine

14.9.1.2 Ecgonine Alkaloids

In contrast to the tropane alkaloids already discussed, the carboxyl group remains in the C2 position in cocaine, which can be esterified with methanol. Cocaine, as the first important representative, has a methyl ester, while the acid function remains free in the other secondary alkaloids. Another ester is located at C3, where the hydroxyl group in the α-position is esterified with benzoic acid. Cocaine (◘ Fig. 14.19) can also be synthetically produced by extracting other ecgonine alkaloids and hydrolyzing their aromatic esters (truxillic acid, cinnamic acid) and re-esterifying them with benzoic acid.

Cocaine has a strong local anesthetic effect and is very popular as an illegal drug due to its euphoric effect. The local anesthetic effect has been known since its discovery by Albert Niemann, who reported a numbness at the tip of his tongue in 1859 when he first isolated cocaine in its pure form. In medicine, cocaine was celebrated as a miracle drug, a myth later significantly contributed to by psychoanalyst Sigmund Freud. Its first application was as a local anesthetic in ophthalmology and later in surgery for spinal anesthesia, which, together with chloroform, enabled pain-free operations. In the 19th century, cocaine and extracts of the coca plant were considered health-promoting in refreshments such as Coca-Cola or wine (Vin Mariani). With the acceleration of industrial society around and after the First World War, cocaine became a widely used drug. In the German Reich, cocaine was banned in 1929 by the first Opium Act. Today, the legal consumption of coca leaves for traditional reasons is only allowed in Peru, Bolivia, and Colombia. In Colombia, possession of up to one gram of cocaine is not punishable. In South America, chewing coca is very popular, and old depictions and clay figures often show the Inca with thick cheeks because they took coca leaves with some lime into their cheeks to suppress hunger and fatigue. The lime, as a basic substance, converts the cocaine from the salt form to the base form, but also promotes the hydrolysis to ecgonine, which has a euphoric effect. The risk of addiction is surprisingly low with this application. About 50 g of leaves are consumed per day, which corresponds to about 1 to 2 g of cocaine. For comparison: About 5 g of cocaine is considered an overdose and is usually fatal for the first-time user.

The addiction potential of cocaine is very high, which is why it is ostracized as an illegal drug and the production, transport, and consumption (smoked, snorted) are severely punished in almost all countries. Cocaine no longer has any pharma-

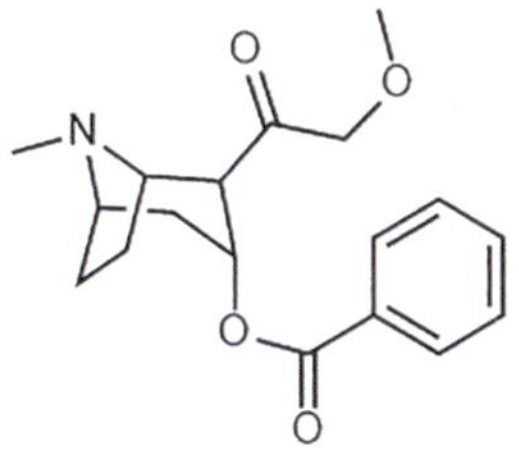

◘ **Fig. 14.19** Cocaine

ceutical or medical use and is obsolete as a drug. Derived from cocaine is the group of local anesthetics, which have revolutionized dentistry, as they (e.g., articaine, lidocaine, procaine) allowed for the first time pain-free operations. Cocaine was first used as a local anesthetic in 1884 by the New York dentist Morgan J. Howe during a tooth extraction.

14.10 Alkaloids Derived from Histidine

There is a structural relationship between the alkaloids of this family and histidine, the biosynthetic amino acid at the beginning of biosynthesis. Pilocarpine (◘ Fig. 14.20) is a typical representative of this alkaloid family and is also chemically referred to as an imidazole alkaloid. The eponymous imidazole ring in pilocarpine is connected with a methylene bridge and a five-membered lactone ring.

Pilocarpus jaborandus, Rutaceae, but also other species such as *P. microphyllus* and *P. pennatifolius* are small shrubs or trees that grow in South America. Phytogeographical distribution areas are Brazil and Paraguay, where the shrubs are also cultivated on farms and the leaves are traded. Pilocarpine is isolated from the leaves (content 0.7–0.8% w/w). Pilocarpine is a chemically unstable substance that degrades by more than half when stored at room temperature for a year. Pilocarpine is still used today as a muscarinic parasympathomimetic with effect on the acetylcholine receptor in ophthalmics against glaucoma. The natural substances muscarine, pilocarpine, and the biogenic neurotransmitter have structural similarities, which is why they act on the same muscarinic acetylcholine receptor and have similar physiological effects (◘ Fig. 14.21). Pilocarpine salts are of great value in ophthalmology and are used in eye drops as a miotic and for the treatment of glaucoma. Pilocarpine is a cholinergic agent that stimulates the muscarinic receptors in the eye. This leads to a narrowing of the pupil and an increased outflow of aqueous humor in glaucoma. Pilocarpine is effective in both narrow-angle glaucoma and open-angle glaucoma. However, the ocular bioavailability of pilocarpine is low. It is quickly excreted, resulting in a short duration of action. Pilocarpine is an atropine antagonist; it alleviates dry mouth, which is very common in patients undergoing radiation therapy for the treatment of oral and throat cancer. There are clear structural similarities between pilocarpine and acetylcholine (◘ Fig. 14.21). Specific drug carrier systems have been developed for pilocarpine in pharmaceutical technology, which enable dose-continuous drug release.

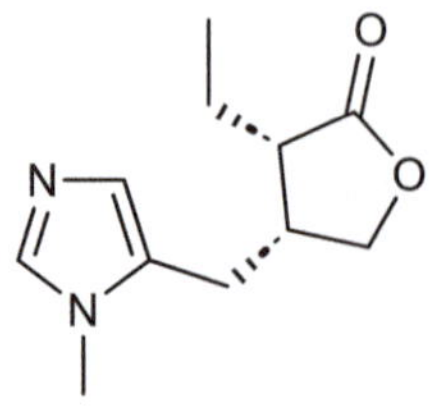

◘ **Fig. 14.20** Pilocarpine

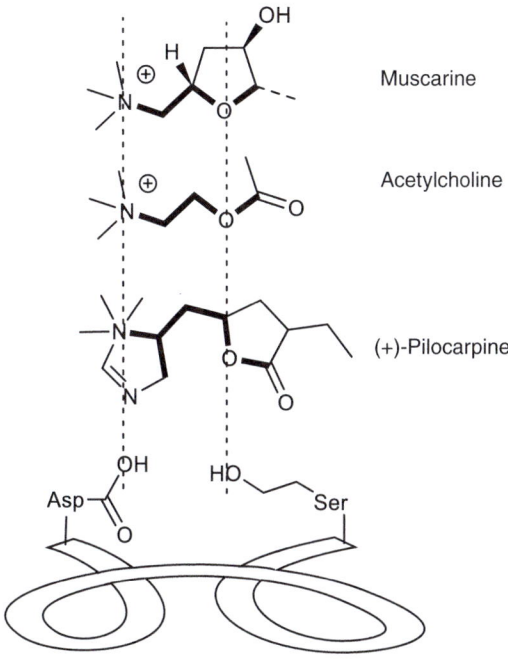

◘ **Fig. 14.21** Muscarinic acetylcholine receptor (excerpt)

14.11 Purine Alkaloids

Purines are found in all living cells, as two of the four bases of DNA and RNA are purines. The question arises whether these can also be alkaloids, as these, by definition, should also have a strong physiological effect. In fact, the two purines adenine and guanine as starting materials have none of these effects. The reason for this lies less in the structure than in the lack of methylation of the nitrogen atoms and the associated lack of effect. On the other hand, methylation also has advantages, because the caffeine contained in coffee, also a purine alkaloid, is metabolized differently and does not lead to gout. On the contrary: Up to four cups of coffee a day are said to reduce the risk of gout by about 40%.

Purines as secondary substances such as caffeine, theobromine, or theophylline are derived from xanthine (dioxypurine) (see ◘ Table 14.3). The second part of the original definition, namely that alkaloids must react basic, is also not true. Caffeine reacts neutrally, theobromine and theophylline even react slightly acidic. In the xanthine derivatives, the two keto groups (C=O) can enolize, i.e., they can switch between C-OH and C=O by shifting the double bond. In plants, caffeine is contained alongside plant acids such as chlorogenic acid in coffee or tannins in *Cola nitida* or *C. acuminata* and guarana. Caffeine is released through roasting (coffee) or fermentation (cola and guarana). Tea (*Camellia sinensis*) and cocoa (*Theobroma cacao*)

Table 14.3 Chemical structures of caffeine, theophylline, and theobromine

Purine alkaloid	Basic structure	R_1	R_2	R_3
Xanthine		H	H	H
Caffeine		CH_3	CH_3	CH_3
Theophylline		H	CH_3	CH_3
Theobromine		CH_3	CH_3	H

contain theobromine, which has a similarly stimulating effect as caffeine. All purine alkaloids share certain pharmacological properties:
- Stimulation of the CNS, especially the respiratory center and the heart by inhibiting the adenosine receptor
- Effect on the kidneys and increased stimulation of the heart muscle
- Spasmolytic effect on smooth muscle and bronchi (theobromine is used as a medication for asthma)

Biosynthesis In the biosynthesis of purine alkaloids (◘ Fig. 14.22), unlike other alkaloids, the imidazole and pyrimidine rings are constructed from several small building blocks. The biosynthesis appears difficult at first glance, but it is not. It is rather complex, as several biosyntheses converge. At this point, the formation of the building blocks is presented in a shortened and greatly simplified form. For those interested in the entire biosynthesis, the literature is referred to (Berg et al. 2018). The starting compound is ribose phosphate from the pentose phosphate pathway. The purine building block inosine is linked with ribose. The inosine biosynthesis is complex, and individual carbon and nitrogen atoms are supplied by the amino acids glycine (N7, C4, C5), glutamine (N3, N9) and asparagine (N1), formic acid (C2, C8) and carbon dioxide (C6). The purine alkaloids are further formed by oxidation (C=O groups) and SAM methylation to theobromine (3,7-dimethylxanthine) and caffeine (1,3,7-trimethylxanthine). Theophylline (1,3-dimethylxanthine) is biosynthesized via a side route through xanthosine and xanthine (◘ Fig. 14.22).

Theophylline Of the known purine alkaloids, theophylline still plays an important role in medicine as a drug for the treatment of bronchial asthma. The alkaloid causes a relaxation of the respiratory muscles, so that the pressure in the lung vessels is reduced and the air intake is improved through the transport of mucus by the cilia. Theophylline is not technically extracted from plants, but has been synthetically produced since 1900 through Traube synthesis. The access via organic synthesis led to numerous derivatives, which improved the effect and bioavailability of theophylline (Etofylline, Fenetylline).

14.12 Alkaloids Derived from Arginine

The structure of saxitoxin (◘ Fig. 14.23) consists, among other things, of a purine ring, which follows the same biosynthetic route as described above. The exact biosynthesis is unknown (◘ Fig. 14.25), but it is known that L-arginine is involved and

14.12 · Alkaloids Derived from Arginine

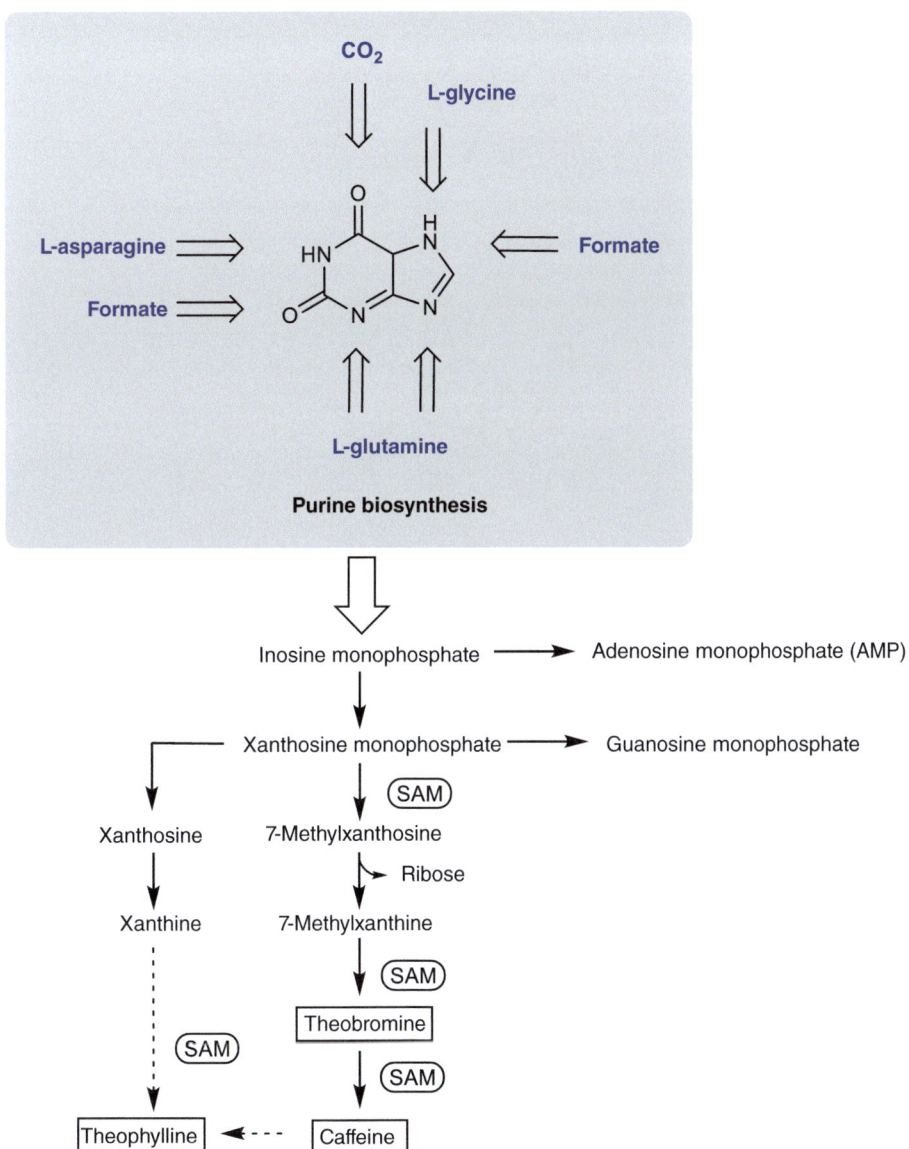

◘ **Fig. 14.22** Purine biosynthesis and formation of caffeine, theobromine and theophylline (simplified representation)

is coupled to acetate building blocks. Saxitoxin is a very strong nerve poison that blocks the sodium channels of nerve cells and depolarizes the membrane. The cause of poisoning is often the consumption of hygienically unsound mussels and shellfish. Even at a dose in the nanogram range per kg body weight, significant symptoms can be observed.

Another poison derived from this biosynthesis is tetrodotoxin (◘ Fig. 14.24), which is produced by *Vibrio* bacteria. The exact biosynthesis is also not fully un-

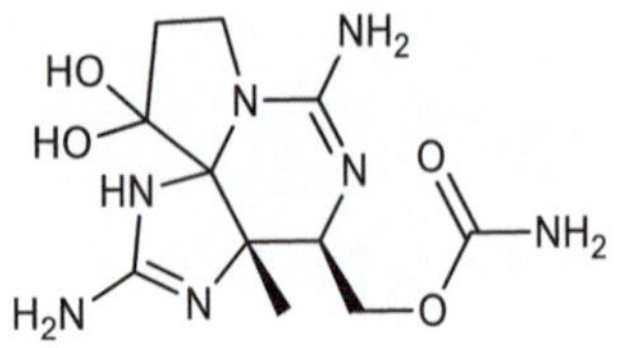

○ **Fig. 14.23** Saxitoxin

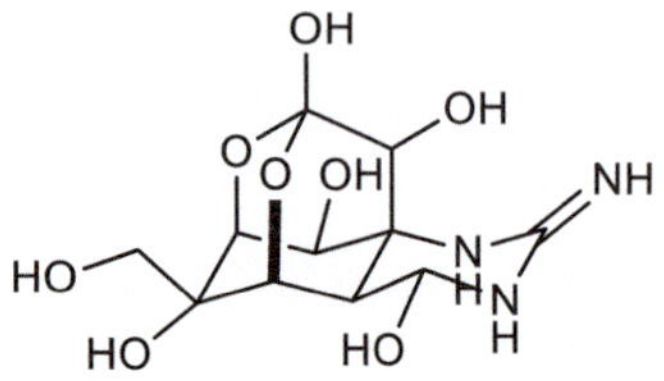

○ **Fig. 14.24** Tetrodotoxin

derstood. The poison and bacteria were first isolated from the Japanese pufferfish (Takifugu rubripes), which lives in symbiosis with these bacteria. The first poisoning was documented by James Cook (1728–1779) when he and his First Officer George Forster (1754–1794) ate an unknown fish in the South Seas in 1774. The Fugu prepared from the pufferfish is now considered a delicacy in Japan and can only be prepared by trained and tested chefs, as the liver and gallbladder of the pufferfish must be cleanly removed to eliminate the poison. Unfortunately, there are still several accidents in Japan every year when laypeople buy and prepare this fish without knowledge of the technique or out of overconfidence (○ Fig. 14.25).

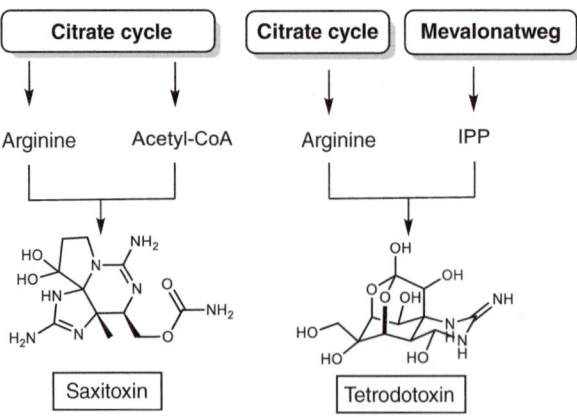

○ **Fig. 14.25** Saxitoxin and Tetrodotoxin biosyntheses

14.12 · Alkaloids Derived from Arginine

❓ Self-check questions

1. What property originally gave alkaloids their name? What is the current definition?
2. Explain the chemical systematics of alkaloids. Name five chemical structures from which alkaloids are derived.
3. Draw the structure of a catecholamine.
4. Show the general biosynthesis of morphine (structures and individual steps are not necessary).
5. Name at least three alkaloids that are derived from aromatic amino acids and have different biosynthetic pathways.
6. Describe the effect of LSD.
7. Draw the structure of two tropane alkaloids.
8. Show the biosynthetic pathway of the terpene alkaloids and indicate where the cross points are.
9. Outline and name the known biosynthetic building blocks in the structure formulas below.

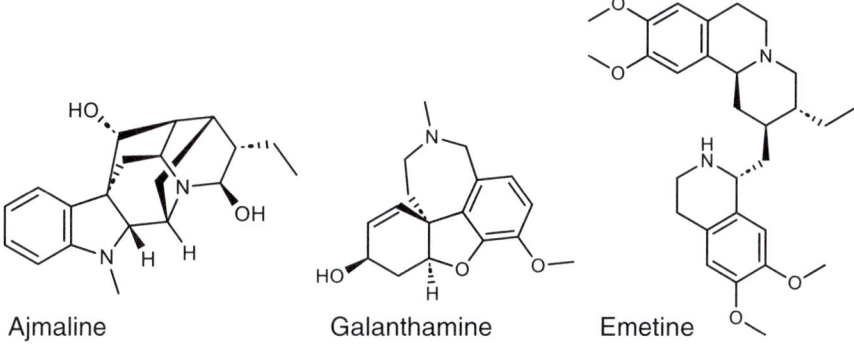

Ajmaline   Galanthamine   Emetine

Antibiotics

Contents

15.1 History – 218

15.2 Non-ribosomal Peptide Biosynthesis – 220

15.3 Penicillins and Cephalosporins – 222

15.4 Polyketide Antibiotics – 227
15.4.1 Avermectins – 228
15.4.2 Macrolides – 229

© The Author(s), under exclusive license to Springer Fachmedien Wiesbaden GmbH, part of Springer Nature 2025
O. Kayser and N. J. H. Averesch, *Technical Biochemistry*, https://doi.org/10.1007/978-3-658-47121-7_15

Learning Objectives and Key Topics
- Chemistry and chemical diversity
- Polyketide biosynthesis
- Non-ribosomal peptide biosynthesis
- Metabolic Engineering

Technical Applications
- Biotechnological production of antibiotics
    - Penicillins
    - Tetracyclines
    - Macrolides

Important Alkaloids
- Penicillins
- Tetracyclines
- Polyketide antibiotics
- Peptide antibiotics

15.1 History

The pharmacological group of antibiotics is dominated by natural substances. Only a few synthetic antibiotics are known and used in the clinic, and the vast majority of antibiotic structures currently researched and used are of non-plant origin. This may be surprising, but the chemical-ecological role of antibiotic natural substances in microorganisms such as bacteria and fungi, but also higher marine organisms, can be explained by the strong competition for food spaces. High division rates of microorganisms and rapid invasion into ecological niches are typical characteristics for ubiquitous colonization and rapid displacement. Humans take advantage of this microbial chemical defense. Due to evolution, microorganisms are capable of forming a chemically diverse repertoire of seemingly arbitrarily complex natural substance scaffolds, which can be highly specific and highly effective against invading microorganisms and higher predators.

> An **antibiotic** (Greek ἀντί- anti- "against" and βίος bios "life") in the original sense is a naturally formed, low-molecular metabolic product of fungi or bacteria, which inhibits the growth of other microorganisms or kills them. An antibiotic in a broader sense is also an antimicrobial substance that does not occur in nature and is produced semi-synthetically, fully synthetically or genetically, but is not a disinfectant.

In its cultural history, humans only knew microorganisms to a limited extent. Only with the discovery of the first microorganisms and the discovery of penicillin by Louis Pasteur, Paul Ehrlich and Robert Koch were infectious diseases, sepsis and microbial diseases treatable. Before the time of chemotherapy, only plant extracts

15.1 · History

were taken (Brunel 1951; Walsh and Wencewicz 2014). Tannins and essential oils were effective, although they had a very weak effect by our understanding. Looking back, it can be stated that plants were not a convincing source for antibiotics and microorganisms are far superior to them in structural diversity and activity. A look at ◘ Fig. 15.1 shows that not a single antimicrobial natural substance was isolated from plants.

Since the 1930s, the discovery of new antibiotics has increased rapidly (◘ Fig. 15.1). The diversity was so great that infectious diseases were considered defeated in the 1960s. This overly positive view changed in the 1970s: After the heyday of antibiotics, resistances increasingly occurred, which were due to incorrect or too careless use. Simple and multiple resistant germs became a major problem, especially in hospitals. At the end of the 1990s, a rethink began. On the one hand, the use of antibiotics was reconsidered and adapted, on the other hand, research programs were launched to find new antibiotics. A sensible approach was the search in new genera outside the known *Aspergillus* species and streptomycetes, such as the myxobacteria, among which very successful new antibiotics were found (Cook and Wright 2022).

Many of the antibiotics isolated from microorganisms can be characterized by their structure as polyketides (PK) or non-ribosomal peptides (NRP) (◘ Fig. 15.2), which represent the largest structural families. However, mixed types of PKS/NRPS biosynthesis units have also been found, which synthesize mixed natural substances such as rapamycin.

Both groups are polymers with two different monomeric biological building blocks. Thus, polyketides are built from C2- (acetate) or C3- (propionic acid) building blocks with the help of a polyketide synthase (PKS) and peptides from amino acids with the help of a non-ribosomal peptide synthase (NRPS). The structural polymerization of the building blocks naturally varies, although both biosynthesis pathways follow the basic principle of modularization and "assembly line construction". There are several classes of both enzymes, which differ in their properties, e.g., whether they are modular, work iteratively, or consist of several independent proteins. The most frequently discussed classes are the modular *cis*-acyltransferases

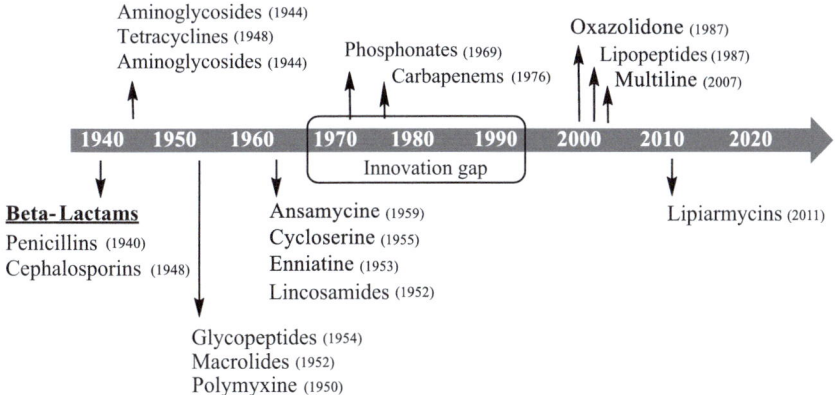

◘ **Fig. 15.1** Historical development of antibiotics. The year indicates the year of market introduction

Actinomycetes		
Lincosamides Chloramphenicol Lipoglycopeptide Carbapenems	Ansamycins Streptogramins Fosfomycins Glycopetide	Tuberactinomycins Aminoglycosides Tetracyclines
Synthesis Azoles Nitrofurans Quinolones Sulfonamides	**Synthesis** Penicillins Cephalosporins Fusidic acid	**Other bacteria** Mupirocin Monobactam Polypeptides Polymyxines

Fig. 15.2 Classification of clinically relevant antibiotics

(AT) of type I of the PKS and of type A of the NRPS. In these classes, each protein module is responsible for the incorporation of a single extension unit into the growing PKS/NRPS chain. Each module of a PKS/NRPS enzyme itself consists of individual catalytic centers, which are called domains and carry out the necessary reactions for the chain extension to take place. In contrast to the types mentioned above, type-II-PKS has also been described, which consists of iteratively working, independent proteins. The minimal type-II-PKS biosynthesis pathway consists of a standalone acyl-carrier-protein (ACP) and two ketosynthases (KSα and KSβ). Both KS proteins are very similar to each other and jointly catalyze the chain initiation and extension steps necessary for polyketide biosynthesis. Also worth mentioning is a final type-III-PKS complex, which consists of several standalone enzymes that catalyze the chain extension without the use of a phosphopantetheine ether. The prosthetic phosphopantethein group is covalently bound to the acyl group via a high-energy thioester bond. Like PKS, NRPS can be iterative or consist of several independent proteins, as will be explained in the following subchapters.

15.2 Non-ribosomal Peptide Biosynthesis

The site of protein biosynthesis is the ribosomes on the endoplasmic reticulum. Short-chain peptides are not assembled by the complex biochemical machinery, and their formation is not subject to strict coding by genes. Non-ribosomal peptide synthases (English: non ribosomal peptide synthases, NRPS) take the place of ribosomes, building short-chain peptides according to an assembly line model. The length and thus the molecular mass determine the difference between peptides and proteins. There is no universally valid definition, but peptides are distinguished from large peptides when they contain no more than 50 amino acids and their molecular mass does not exceed 10,000.

Biosynthesis by non-ribosomal peptide synthases NRPS modules (◼ Fig. 15.3) of type A have a different domain structure than type I PKS modules. To be functional, each module in an NRPS must contain at least one adenylation domain (A), a peptidyl carrier protein (PCP, sometimes also called thiolation domain (T)) and a condensation domain (C). The A domains are responsible for the selection of an

15.2 · Non-ribosomal Peptide Biosynthesis

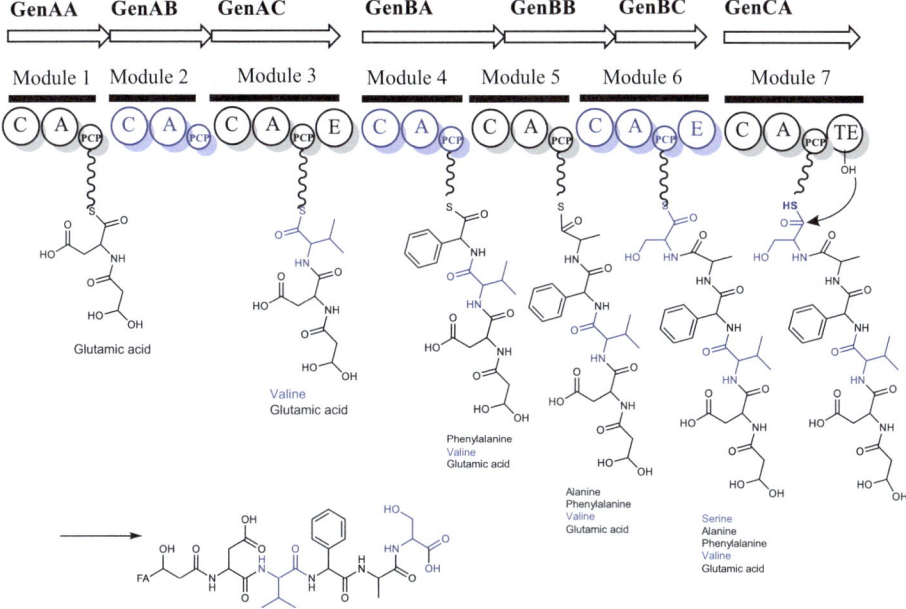

Fig. 15.3 Overview of the modules and the non-ribosomal peptide biosynthesis

extension unit, analogous to the AT domains in PKS enzymes. For this purpose, the A domains catalyze an ATP-dependent adenylation reaction of an amino acid, often with high selectivity. The adenylated amino acid is a high-energy compound with a strong leaving group (AMP), which allows a nucleophilic attack by the thiol of the PCP-bound Ppant group with the release of AMP. The function of the C domains is analogous to that of KS domains, but they catalyze the formation of C-N (peptide) bonds, not the formation of C-C bonds. The PCP-bound NFP chain enters the active center of the C domains, where it is attacked by the α-amino group of the amino acid bound to the PCP domain of the downstream PCP domain. After the final chain extension, the PCP domain is detached from the α-amino group of the amino acid bound to the PCP domain of the downstream PCP domain (Fischbach and Walsh 2006; Jaremko et al. 2020).

Biological Effect Oligopeptide (◼ Table 15.1) are found in nature with very diverse effects due to their chemical diversity. Although this chapter deals with antibiotics, other effect profiles are also well known:
- Metabolic regulation: Peptides can regulate the production or degradation of hormones, enzymes, and other biomolecules. Examples are ciclosporin A or actinomycin D, but also peptides that regulate metabolism like glucagon and insulin or regulate blood pressure like renin and angiotensin.
- Signal transduction: Peptides can transmit signals between cells and tissues. Examples are neuropeptides like endorphins or substance P in pain perception, insulin in the regulation of blood sugar, or oxytocin, which triggers labor as a neuropeptide.

Table 15.1 Cyclosporin of medical importance

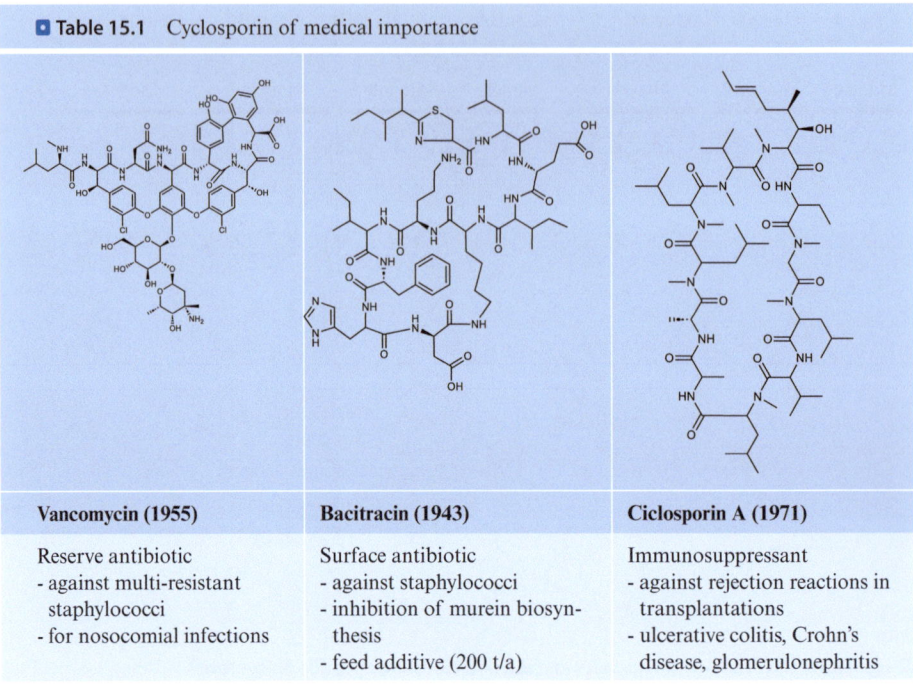

Vancomycin (1955)	Bacitracin (1943)	Ciclosporin A (1971)
Reserve antibiotic - against multi-resistant staphylococci - for nosocomial infections	Surface antibiotic - against staphylococci - inhibition of murein biosynthesis - feed additive (200 t/a)	Immunosuppressant - against rejection reactions in transplantations - ulcerative colitis, Crohn's disease, glomerulonephritis

— Regulation of cell growth and differentiation: Peptides can influence the development and differentiation of cells, like the epidermal growth factor EGF or the fibroblast growth factor FGF.
— Immune system: Peptides play an important role in the immune system by regulating the immune response. The cyclopeptide Ciclosporin A, for example, shows very strong immunosuppressive effects (◻ Table 15.1)

15.3 Penicillins and Cephalosporins

Penicillin belongs to the oldest naturally occurring antibiotics. Despite their early discovery in the 1920s, followed by the discovery of the structurally similar cephalosporin in the 1940s, penicillins, also referred to as β-lactam antibiotics, are indispensable in clinical practice. One of the first penicillins was benzylpenicillin, which was isolated from the fermentation of Penicillium chrysogenum. Due to the turmoil of the Second World War, technical production only began in the 1940s. Various process and later raw material optimizations led to a reduction in the cost per gram to a few cents.

Chemistry Penicillins and cephalosporins have a β-lactam as a common structural feature (◻ Fig. 15.4, blue ring). If this strained ring is followed by a five-membered thiazolidine ring, we recognize penicillins here (◻ Fig. 15.4). If a six-membered dihydrothiazine ring is present, it is cephalosporins. The unusual lactam ring is not stable and can quickly decompose under acidic conditions, such as those in the stomach (◻ Fig. 15.5) The strained β-lactam ring (cyclic amide) is more susceptible

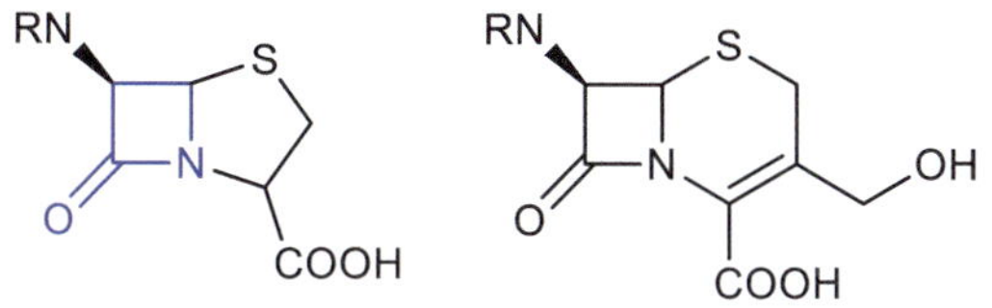

15.3 · Penicillins and Cephalosporins

Fig. 15.4 β-Lactam structure of penicillins, in blue: amide structure in the strained four-ring

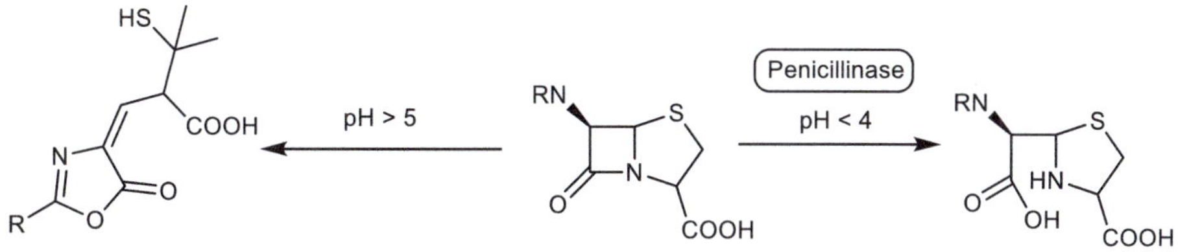

Fig. 15.5 pH-induced degradation reactions of penicillin

to hydrolysis than the unstrained amide function of the side chain, as the normal stabilizing effect of the lone pair of electrons of the neighboring nitrogen atoms is not possible due to geometric constraints. The β-lactam ring is opened by a mechanism involving the side chain carbonyl, leading to the formation of an oxazolidine ring. Penicillinic acid is formed by a nucleophilic attack of the thiazolidine nitrogen on the iminium function, followed by the cleavage of the carboxylate leaving group. Alternatively, the cleavage of the thiol can also lead to the formation of penicillanic acid. At higher pH values (> pH5), benzylpenicillin undergoes a simple β-lactam ring opening and forms penicillanic acid.

The chemical stability of cephalosporins is higher. At acidic pH, the lactam ring remains closed, and bacterial penicillases cannot open it. This stability also has disadvantages, as the antibacterial effect is weaker compared to penicillins.

Biosynthesis of Penicillins The linkage of three amino acids leads to the formation of a tripeptide, which is important for further biosynthesis to penicillins and cephalosporins. Linear condensation of L-cysteine, L-valine, and the non-proteinogenic L-aminoadipic acid by the NRPS complex to δ-(L-α-aminoadipyl)-L-cysteinyl-D-valine (◼ Fig. 15.6). This is cyclized to isopenicillin N, which already has the bicyclic ring system. It is an oxidative reaction that requires molecular oxygen, which must be taken into account in industrial production. The reaction is catalyzed by the iron-containing isopenicillin-N-synthase. The iron binds intermediately to the sulfur of the L-cysteine. In the further course to benzylpenicillin, the L-aminoadipic acid of the isopenicillin N is hydrolyzed and cleaved from the nitrogen of the L-cysteine; it can be replaced by other acids. Phenylethylamine is reduced in the medium to phenylacetic acid, activated with CoA, and bound to the free nitrogen group, so that benzylpenicillin (penicillin G) is formed. However, the intermediate 6-aminopenicillanic acid (6-APA) is of great importance for directed structural changes (◼ Fig. 15.7).

Biological Effect Penicillins and other β-lactam drugs exert their antibacterial effect by binding to proteins (penicillin-binding proteins), peptidase enzymes, which are

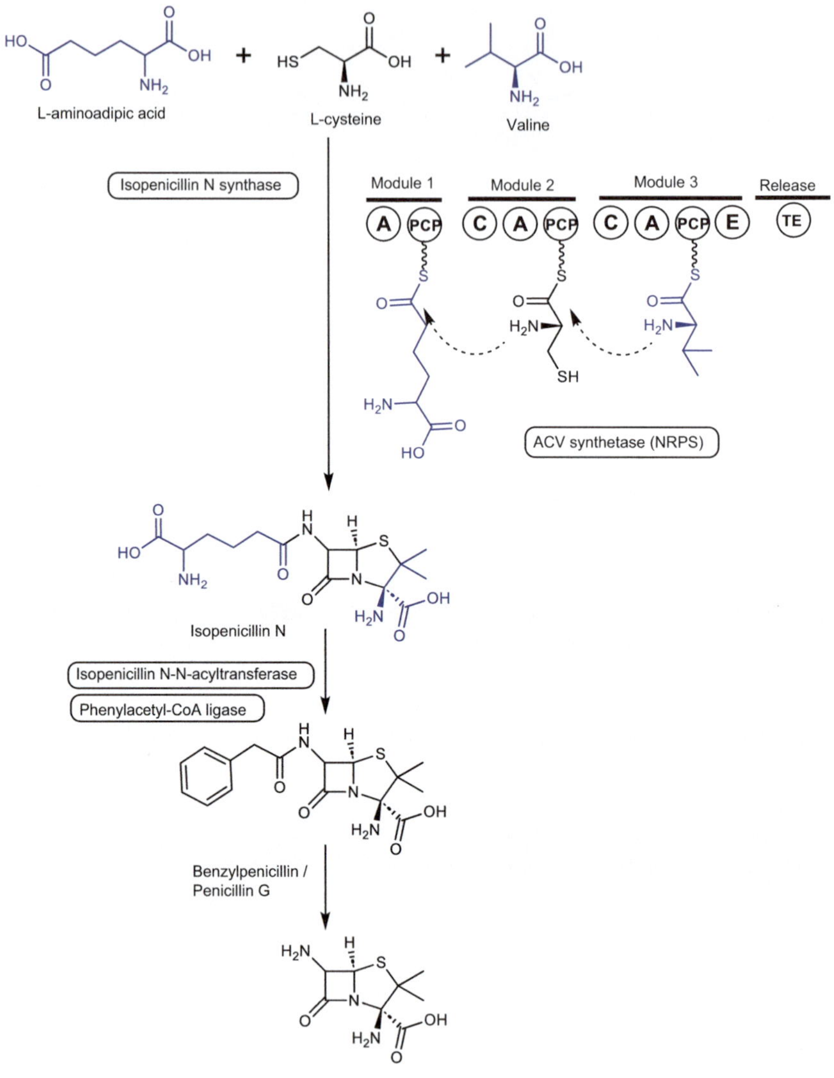

Fig. 15.6 Biosynthesis of aminopenicillanic acid

involved in the late stages of bacterial cell wall biosynthesis. The cross-linking of the peptidoglycan chains that make up the bacterial cell wall is a terminal D-Ala-D-Ala intermediate that is very similar to the penicillin molecule in its transition conformation. As a result, the penicillin occupies the active center of the enzyme and is bound via a serine residue in the active center, leading to irreversible inhibition of the enzyme and cessation of cell wall biosynthesis. A look at the aminopenicillanic acid shows that the lactam ring is responsible for the antibacterial effect by inhibiting the bacterial transpeptidase, the thiazolidine ring is important for binding to the enzyme, and the residues at the 6-amino group determine the pharmacokinetics and spectrum of activity (◘ Fig. 15.8).

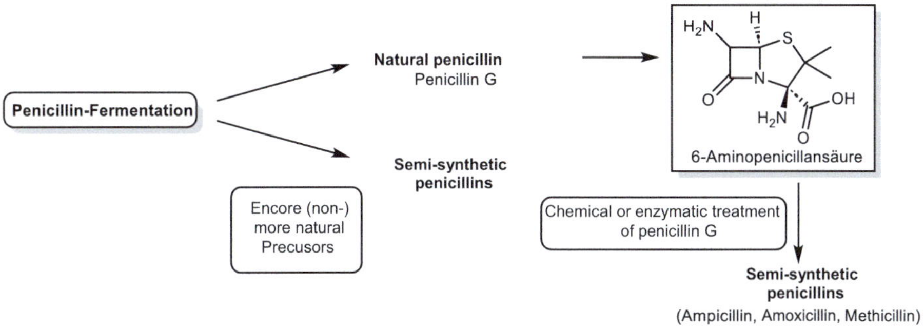

15.3 · Penicillins and Cephalosporins

Fig. 15.7 Bio- and semi-synthesis of penicillins

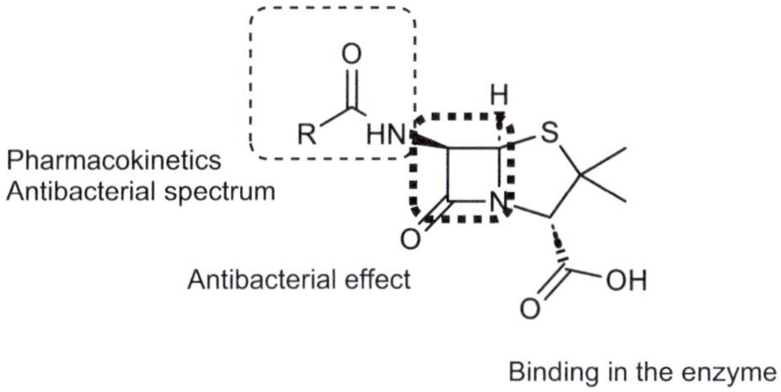

Fig. 15.8 Structure-activity relationship in penicillins

Biosynthesis of Cephalosporins The biosynthesis of cephalosporin C is more complex than that of penicillins, despite similar intermediates (Fig. 15.9). It is well understood in *Cephalosporium aremonium*. It also starts with the three amino acids L-adipic acid, L-cysteine, and L-valine and resembles penicillin biosynthesis up to isopenicillin N. An epimerization step leads to penicillin N, which undergoes ring expansion in an oxidative reaction. An exocyclic methyl group is incorporated into the dihydrothiazine ring and the remaining methyl group is hydroxylated. Finally, this hydroxyl group is acetylated by an acetyltransferase.

The biosynthesis of penicillin and cephalosporin are two similar microbiological and technical processes, but they differ in important aspects (Table 15.2).
— Penicillin acid biosynthesis

The biosynthesis of penicillin acid begins with the precursor molecule L-alanine, which is produced by the cell from glucose. L-alanine is then converted with the precursor molecule phenylalanine to D-penicillin acid.
— Cephalosporin biosynthesis

Cephalosporin biosynthesis begins with the precursor molecule D-cysteine, which is produced by the cell from methionine. D-cysteine is then converted with the precursor molecule L-valine to D-cephalosporanic acid. D-cephalosporanic acid is

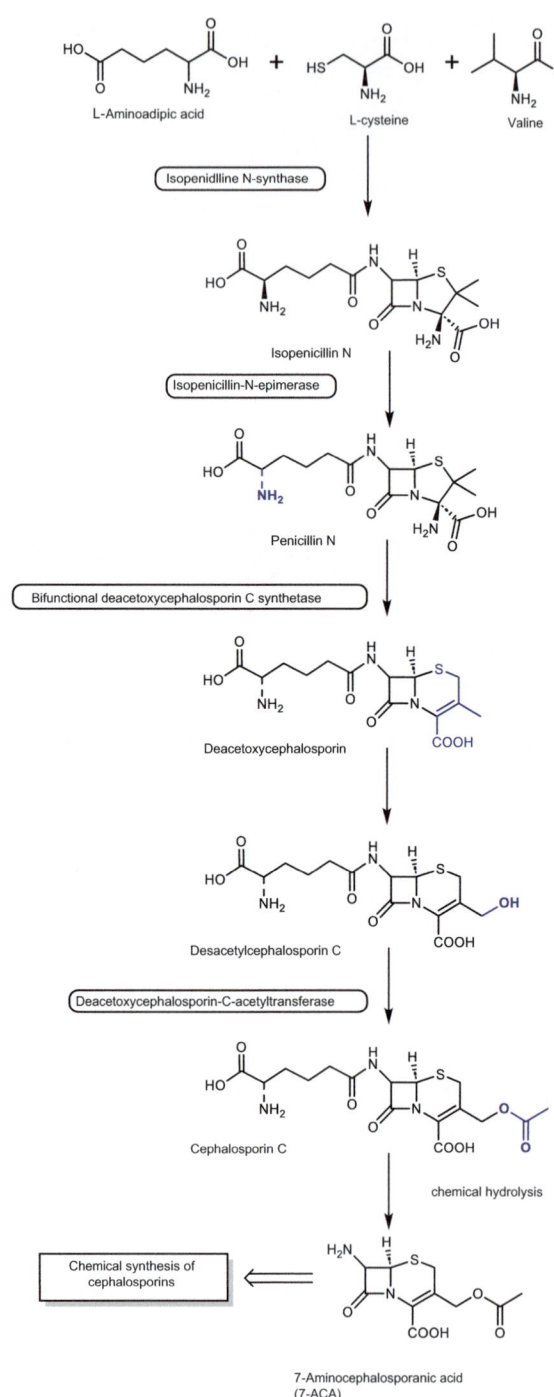

Fig. 15.9 Biosynthesis of 7-aminocephalosporanic acid and derived cephalosporins

Table 15.2 Differences between penicillin and cephalosporin biosynthesis

	Penicillins	**Cephalosporins**
Precursor molecules	D-Alanine and Phenylalanine	D-Cysteine and L-Valine
Intermediate	D-Penicillin acid	D-Cephalosporanic acid
Enzymes	Penicillin synthetase Bacterial epimerization	Cephalosporin synthetase Fungal epimerization
Product	D-Penicillin acid	D-Cephalosporanic acid

the basic structure for a variety of cephalosporin antibiotics, which are produced by modifying the side chain.

15.4 Polyketide Antibiotics

All polyketides derive from the utilization of acetyl-CoA and are subject to the acetate rule. As shown in ◘ Fig. 4.4 in ▶ Chap. 4, according to the acetate rule, a formal linkage of a C2 unit occurs regularly, showing a keto function alternating on every second carbon atom. Through downstream bioorganic reactions, very different natural substances can be produced by chain folding (rearrangements) as well as biochemical reactions on oxygen (reductions, methylations) and on carbon (halogenations). Depending on the number of linked C2 units, one can find, for example, pentaketides (5 × C2), octaketides (8 × C2), and natural substances with many C2-linked units. In principle, the size is almost arbitrary, depending on the performance of the polyketide synthases, which carry out the biochemical synthesis. The most technically interesting polyketides have a number of 8 to 12 linked C2 units, such as the macrolide antibiotics presented here (erythromycin) or the tetracyclines. The chemical structural range of polyketides is large, and flavonoids and stilbenes, which show biosynthetic analogies, have already been presented in the preceding chapters of this book. This chapter focuses on polyketides with antimicrobial activity.

The basic biosynthesis of polyketides shows similarities with fatty acid biosynthesis, although the fatty acid synthase complex and the polyketide synthases show clear differences. If the fatty acids are built up in the synthase complex in a simplified rotating dimeric carousel system, the formation of polyketides follows the assembly line model. In modular structures, which resemble workstations in a technical factory, semi-finished structures are passed from one station to the next, extended by a C2 unit and possibly chemically modified. This assembly line of transfer and chemical modification goes through a species-specifically defined number of reaction modules (= workstations) until the separation of the finished polyketide occurs at the end. Formally, this biosynthesis is the end of polyketide biosynthesis, but postbiosynthetic modifications such as ring closure to macrocycles, methylations or glycosylations, to name just a few, complete the final structure. The number of domains within a type I PKS or type A NRPS module varies from enzyme to enzyme, with the exception of some necessary "core" domains. To be catalytically active, each *cis*-AT-type I PKS module must contain an acyltransferase

(AT), an acyl carrier protein (ACP), and a ketosynthase domain (KS). The ACP of each module serves as an anchor for a phosphopantetheine unit. Each AT domain selects an extension unit and transfers it to the phosphopantetheine group of the adjacent ACP. The most commonly used extension units in PK biosynthesis are acetate (two carbon atoms) and propionate (three carbon atoms), which are typically provided as activated forms of malonyl-CoA and (2S)-methylmalonyl-CoA, respectively. The KS domain catalyzes the synthesis of malonyl-CoA and (2S)-methylmalonyl-CoA, forming the C-C bond by a decarboxylative Claisen condensation. Between the resulting PK chain and the extension unit bound to the ACP of the downstream module, there are flexible phosphopantetheine groups that shuttle back and forth between the various PKS modules of the biosynthetic pathway, with each module increasing the chain size by one extension unit.

15.4.1 Avermectins

Avermectins (◘ Fig. 15.10) are a class of natural substances with strong anthelmintic (antiparasitic) properties. The term "avermectins" is often used for a group of closely related compounds, including abamectin and ivermectin. The soil bacterium *Streptomyces avermitilis* produces these compounds via a complex biosynthetic pathway. The discovery of the avermectins is attributed to the Japanese microbiologist Satoshi Ōmura, who isolated them from soil samples. Avermectins are macrocyclic lactones, which consist of a large and complex ring structure. They are characterized by a 16-membered macrocyclic lactone ring to which two sugar residues are attached. Avermectins exert their anthelmintic effect by binding to glutamate-gated chloride channels in the nervous system of the parasite, leading to increased membrane permeability and thus to the paralysis of the parasite. This mechanism makes Avermectins very effective against a wide range of parasitic organisms, including nematodes and arthropods. In human and veterinary medicine, ivermectin, a synthetic derivative of avermectin, is often used for the treatment of parasitic infections. It is particularly successful in combating diseases such as onchocerciasis (River blindness) and lymphatic filariasis in humans, as well as com-

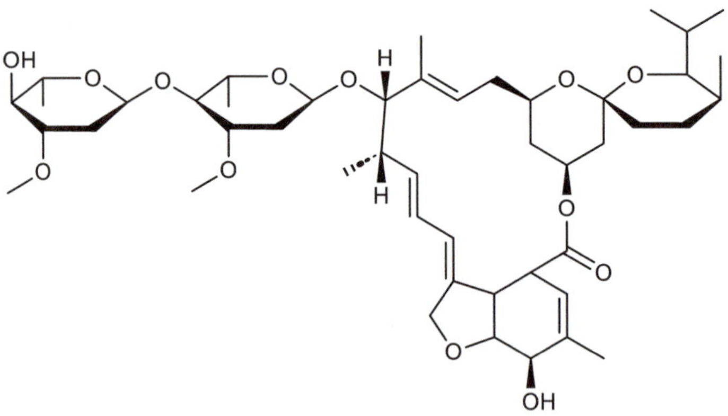

◘ **Fig. 15.10** Avermectin

15.4 · Polyketide Antibiotics

bating various forms of parasitic infestation in animals. The discovery and development of the avermectins had a profound impact on global health. This is particularly true for regions where parasitic infections are endemic. Avermectin has played a crucial role in large-scale health campaigns to control and eradicate the tropical disease river blindness.

15.4.2 Macrolides

15.4.2.1 Erythromycins

Erythromycin The biosynthesis (Zhang et al. 2010) of erythromycin as a macrolide antibiotic can be attributed to polyketide biosynthesis. Characteristic features are the large macrocyclic ring, which gave the class of antibiotics its name, and the ring linkage through a lactone. Erythromycins (◘ Fig. 15.11) are among the oldest antibiotics, first isolated from the strain Streptomyces erythreus in 1948 and marketed as Ilosone® in 1952. The mechanism of action is that erythromycin inhibits the elongation factor in the translocation process of the translation of the mRNA code in the ribosomes. The inhibition of protein biosynthesis can be bactericidal or bacteriostatic depending on the dose and the pathogen.

Chemistry Erythromycin consists of a 24-membered macrolide ring, to which the sugar cladinose is bound at C4 and the amino sugar desosamine at C6 via hydroxyl groups. Due to the lactone function, erythromycins are unstable in acidic solution and tend to ketalize via the keto group at C10 and the hydroxyl group at C7, forming internal tetrahydrofuran rings.

Biosynthesis The basic structure of the lactone ring is built up and structurally modified by the 6-deoxyerythronolide-B-synthase (DEBS) (◘ Fig. 15.12). In summary, it can be stated that the chain extension and the corresponding reduction lead to an enzyme-coupled polyketide, in which one carbonyl group was completely

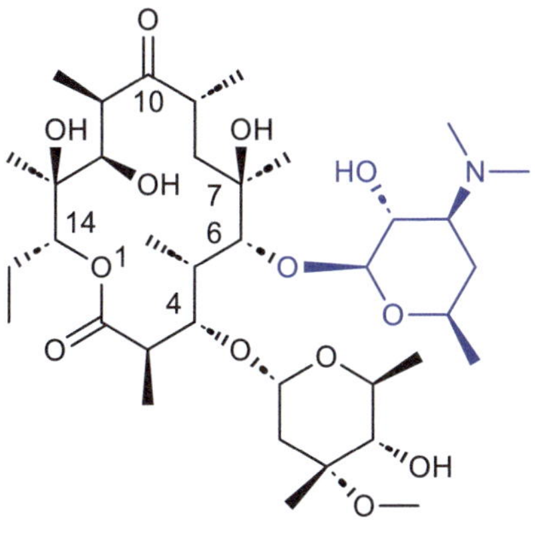

◘ **Fig. 15.11** Erythromycin C

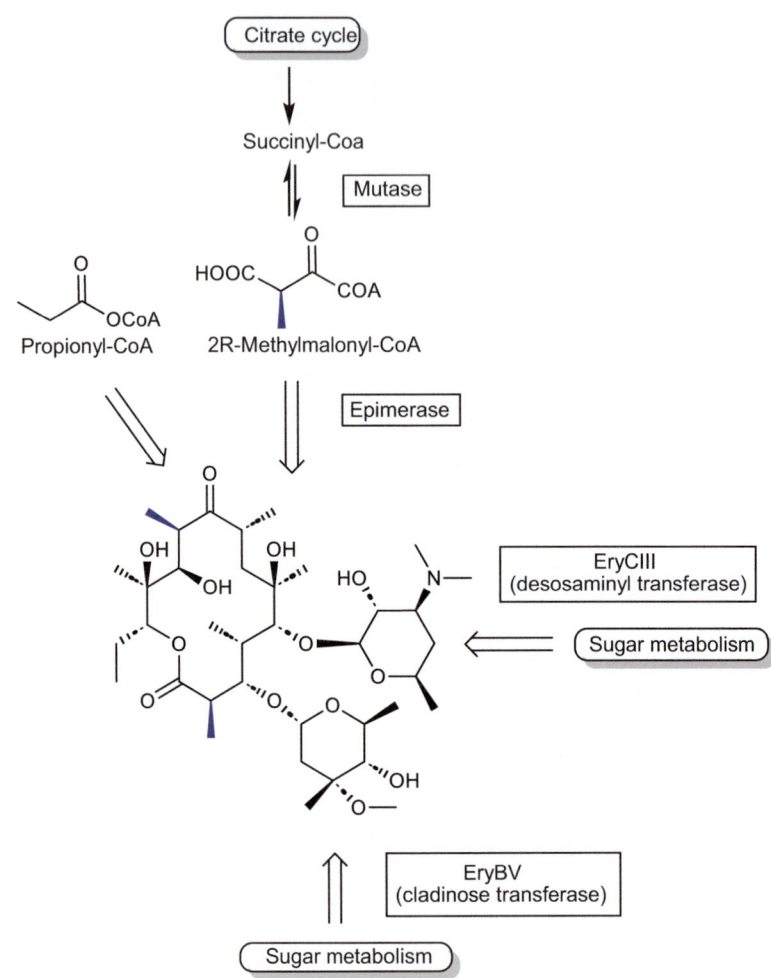

◘ **Fig. 15.12** Simplified biosynthesis of Erythromycin C

reduced, four carbonyl groups were reduced to alcohols, while another carbonyl group was not reduced and remains in the entire sequence.

The starter unit is propionyl-CoA, which is extended to methylmalonyl-CoA. Propionyl-CoA can be found not only as a starter molecule in biosynthesis, as it can be carboxylated to methylmalonyl-CoA with the help of a carboxylase. Structurally relevant is the hydroxylation at position C7. Glycosyltransferases lead to an α-glycosidic bond of cladinose and a β-glycosidic bond with desosamine.

15.4.2.2 Tetracyclines

Tetracyclines are a group of broad-spectrum antibiotics. Their structural hallmark is a tetracyclic system (◘ Fig. 15.13), which is formed from eight C2 units (Malonyl-CoA) and postbiosynthetically modified.

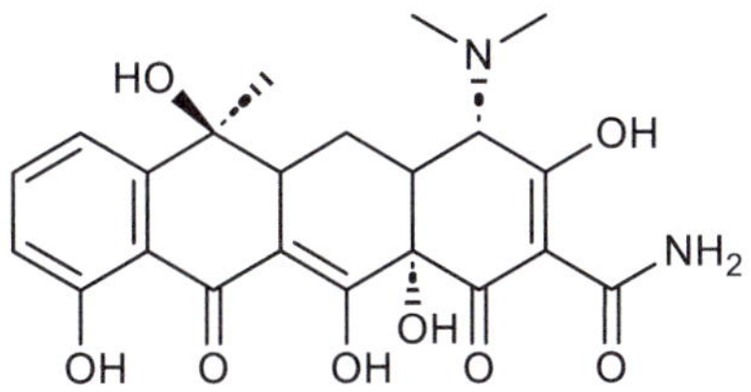

15.4 · Polyketide Antibiotics

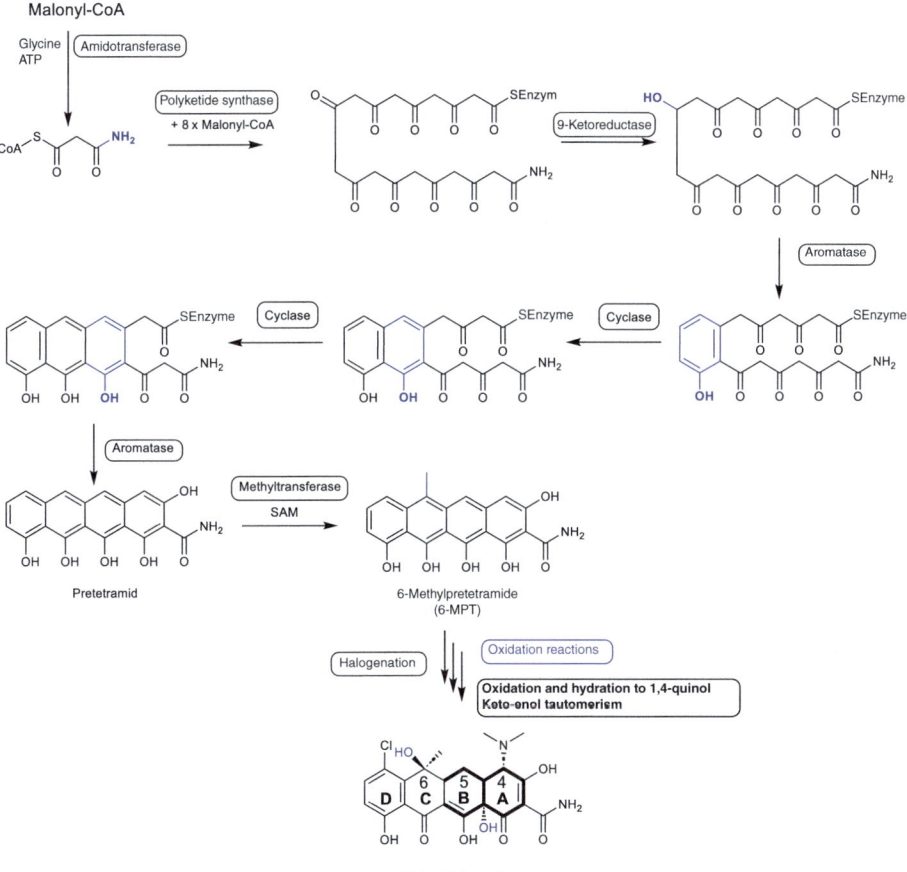

◘ Fig. 15.13 Oxytetracycline

◘ Fig. 15.14 Oxytetracycline biosynthesis in *Streptomyces rimosus* (simplified)

Biosynthesis The basic structure of the tetracyclines is formed by the condensation of eight malonyl-CoA units according to Claisen (◘ Fig. 4.4). The starter unit is malonamyl-CoA, which receives an amino group from glycine at the expense of energy (ATP). This starter unit is important in bacteria, but not in plants. The biosynthetic pathway of tetracyclines was mainly elucidated in *Streptomyces rimosus* for oxytetracycline (◘ Fig. 15.14). In contrast to the above-mentioned macrolide biosynthesis, monofunctional enzymes can be found in tetracycline biosynthesis. The biosynthesis can be divided into three phases. First, with the help of the minimal polyketide synthase complex, a linear octaketide, a poly-β-ketoester, is constructed, which is aromatized to a tetracene (pretetramid) in the second phase with the help of aromatases and cyclases. In the third phase, starting from 4-hydroxy-6-methyl-pretetramid, oxidation reactions to a quinone in ring A and keto-enol rearrangements in rings B and C transform the four-ring system into the tetracycline basic structure. The incorporation of an amino group at ring A is achieved by a transaminase reaction. Further reactions, such as halogenation with a chlorine substituent, can additionally modify the already heavily constructed tetracycline postbiosynthetically. In addition to the well-known shikimate pathway, the polyketide pathway represents an alternative for the biosynthesis of aromatics (Pickens and Tang 2009).

Biotechnology The biosynthesis genes are organized in a biosynthesis cluster. The biosynthesis of oxytetracycline comprises 22 genes, four of which (oxyA–D) code for the minimal polyketide synthase (PKS) (Pickens and Tang 2009). This consists of the keto synthase (KS), the chain length factor CLF (chain length factor) and the acyl carrier protein (ACP). Metabolic engineering experiments involve the genetic modification of the KS and the alteration of the tetracene ring system. Since naturally occurring tetracyclines have been of great commercial importance over the past 60 years, great efforts have been made to improve the strains and optimize the titers. However, most strain improvements were made in the industry, and only a small part was made available to the scientific community. Almost all efforts to improve the strains were carried out using mutagenesis techniques as a "black box" approach, which provides little or no information about the exact position of the mutation in the gene. Randomly introduced mutations can also be advantageous. In one case, a strain was generated by genetic recombination that showed no foaming in the fermentation medium and an increased oxytetracycline production. The oxytetracycline production was attributed to the genetic instability of the oxytetracycline biosynthesis.

Semisynthesis The structure of tetracyclines appears to be conservative, and only a few changes in positions C5 to C6 are successful. Semisynthetic and clinically used tetracyclines are methacycline (◘ Fig. 15.15) and doxycycline (◘ Fig. 15.16), which can be converted into the former by reducing the C6-methylene.

Pharmacology In this class of antibiotics, tetracycline and oxytetracycline were probably the most frequently prescribed antibiotics, which have now been replaced by chlortetracycline and methacycline. Tetracyclines are administered orally or by injection, as ear and eye drops, and applied externally to the skin. Doxycycline is not only an antibiotic, but is also used for malaria prophylaxis in areas with widespread resistance to chloroquine and mefloquine. A new generation of tetracyclines, the glycylcyclines, has been developed to counteract resistances and achieve higher antibiotic efficacy. This new generation includes tigecycline (◘ Fig. 15.17), a glycylamide derivative of minocycline, which was approved in the EU in 2007. It is synthesized by nitrating minocycline, which in turn is obtained from tetracycline by fermentation.

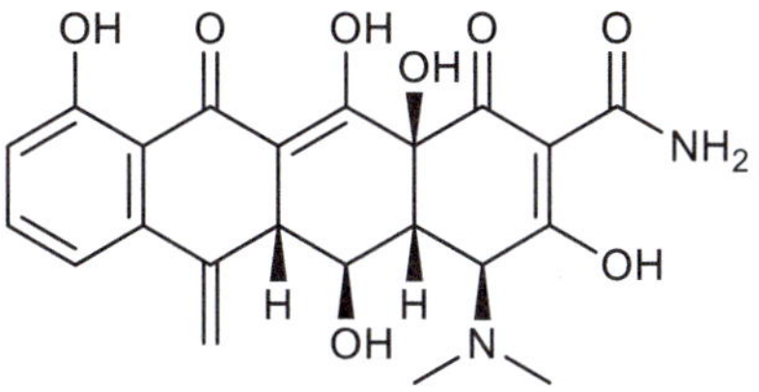

15.4 · Polyketide Antibiotics

◼ **Fig. 15.15** Methacycline

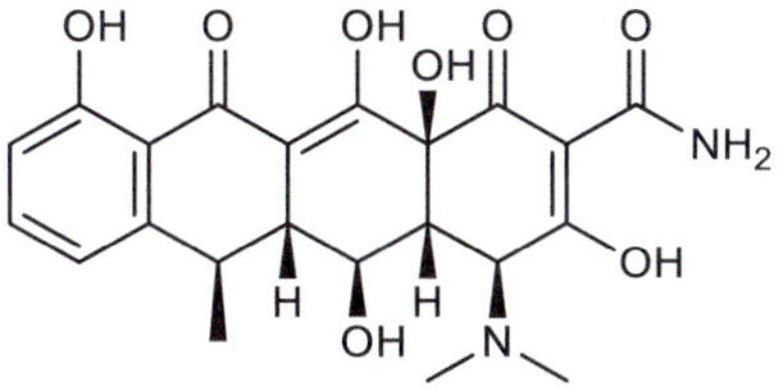

◼ **Fig. 15.16** Doxycycline

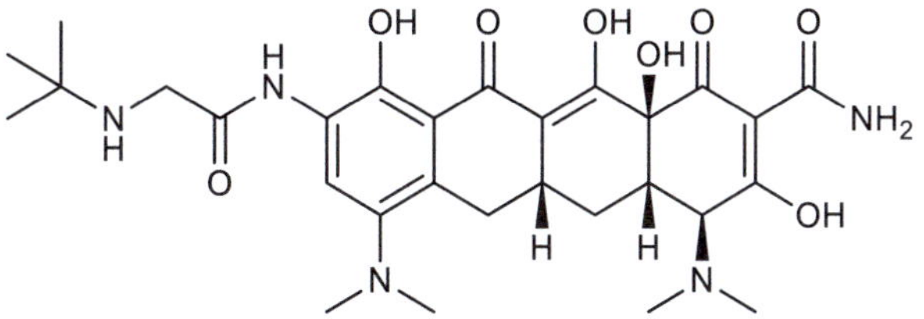

◼ **Fig. 15.17** Tigecycline

❓ Self-check Questions
1. Define the term antibiotics and distinguish it from that of alkaloids.
2. Are there differences in the mechanism of action between synthetic and natural antibiotics?
3. Explain current genetic engineering methods for strain optimization of antibiotic producers.
4. Discuss strategies to identify natural antibiotics in nature.
5. Consider why plants are poor antibiotic producers.

Environmental Biochemistry

Contents

16.1 Biochemistry of Wastewater Treatment – 236

16.2 Biogas and Methanogenesis – 237

© The Author(s), under exclusive license to Springer Fachmedien Wiesbaden GmbH, part of Springer Nature 2025
O. Kayser and N. J. H. Averesch, *Technical Biochemistry*, https://doi.org/10.1007/978-3-658-47121-7_16

> **Learning Objectives**
> - Function of a biogas plant
> - Function of a sewage treatment plant
> - Methanogenesis
> - Ecology of methane-producing microorganisms

The environmental biochemistry is an interdisciplinary field that deals with chemical processes in the environment. It is a subfield of biochemistry and environmental sciences. Environmental biochemistry studies the interactions between chemical substances and biological systems in the environment. It deals with the following questions: How are chemical substances absorbed, stored, distributed, and transformed in the environment? Do chemical substances have short-term and long-term effects on biological systems in the environment? Can chemical substances influence the quality of the environment?

Environmental biochemistry is an important discipline that allows us to understand the interactions between chemistry and the environment. Various topics such as biogeochemistry, ecochemistry, toxicology, and bioremediation can be summarized under this term. This understanding is important for the development of measures to reduce environmental pollution and improve environmental quality. Environmental biochemistry has numerous applications:

- **Environmental protection:** Environmental biochemistry can contribute to the development of new environmental protection measures. For example, it can be used to develop new technologies for cleaning polluted waters or to develop new, less environmentally harmful pesticides.
- **Health protection** for humans: Environmental biochemistry can be used to investigate the effects of environmental pollution on human health. For example, it can help to investigate the effects of pollutants on the development of cancer and other diseases.
- **Nature conservation:** With the help of environmental biochemistry, the effects of environmental changes on ecosystems can be investigated. For example, it can be used to investigate the effects of climate change on biodiversity.

16.1 Biochemistry of Wastewater Treatment

We live on a planet with a constantly increasing population, which is particularly concentrated in urban conurbations. It is estimated that by 2024, about 5 billion people will live in cities, most of which are or will be megacities and are located in so-called emerging countries. This particular situation poses a challenge to municipalities worldwide to not only provide clean drinking water, but also to process and clean service and wastewater in an environmentally friendly manner. The technical challenges are enormous, and the construction of sewage treatment plants is a future task in the densifying cities of Africa and Asia. In addition to the political declaration of intent to want to build sewage treatment plants, however, three essential questions arise in advance of the decision

- **... the availability of resources:** Wastewater treatment requires a number of resources such as water, energy, and chemicals. The availability of these resources can influence the costs of wastewater treatment.
- **... the costs of wastewater treatment:** Wastewater treatment is an expensive process. These costs can be influenced by a number of other factors, such as the size of the plant, the type of wastewater treatment, and local conditions.
- **... the acceptance of the population:** Wastewater treatment can lead to a number of negative effects on the environment, such as odor nuisance and noise. Therefore, the acceptance of the population for wastewater treatment can be a problem.

Technology Wastewater treatment plants are often located at low points in the landscape, so that the wastewater can flow into the treatment plant following the natural force of gravity (◘ Fig. 16.1 and ◘ Table 16.1). The first step in the cleaning process is a mechanical separation with rakes to separate large particles in advance ①. These can be small stones or objects that have accidentally entered the wastewater. What remains is a still polluted suspension with a fat content. The water flows through a sand or grease trap. Since the fat floats on the water surface due to its low density, it is collected with a slowly moving slide ②. The fine-grained sand and smaller particles sink to the bottom of the tank, are compressed by a roller cleaning shield and removed from the water. The remaining sludge, which contains human excrement, is treated in the pre-settling tank ③. As in the previous stage, the sinking particles are collected with a cleaning shield and separated. This is followed by biological cleaning by trillions of bacteria already present in the sludge or wastewater suspension. Oxygen is introduced into the aeration tank by air entry, promoting the degradation of pollutants ④. Here, degraded residues and dead bacteria flocculate and precipitate. In the secondary clarifier ⑤, the residues located at the bottom are separated by a slowly rotating cleaning shield, and the clear water runs off above the tank. The water has sufficient quality to be discharged as cleaned water into natural bodies of water.

The resulting sludge can be used for biogas production. It still consists of 95% water and must be dried or dewatered ⑥. The sludge is pumped into the sludge tower, where anaerobic digestion is aimed for. This produces biogas, which is converted into electricity in local combined heat and power plants ⑦, and heat. The biochemistry of methanogenesis for the formation of biomethane is explained in ▶ Sect. 16.2. The remaining sludge is finally dried and further processed into humus ⑧.

16.2 Biogas and Methanogenesis

Chemistry Biogas is a mixture of the desired methane and other gaseous components, which mainly include carbon dioxide. The methane content is about 60%, the carbon dioxide content about 40%, but gases such as ammonia, steam and hydrogen sulfide are also included. The formation of methane occurs through methanogenesis, the final stage of anaerobic microbial degradation. Methanogenesis is a central part of the Earth's carbon cycle.

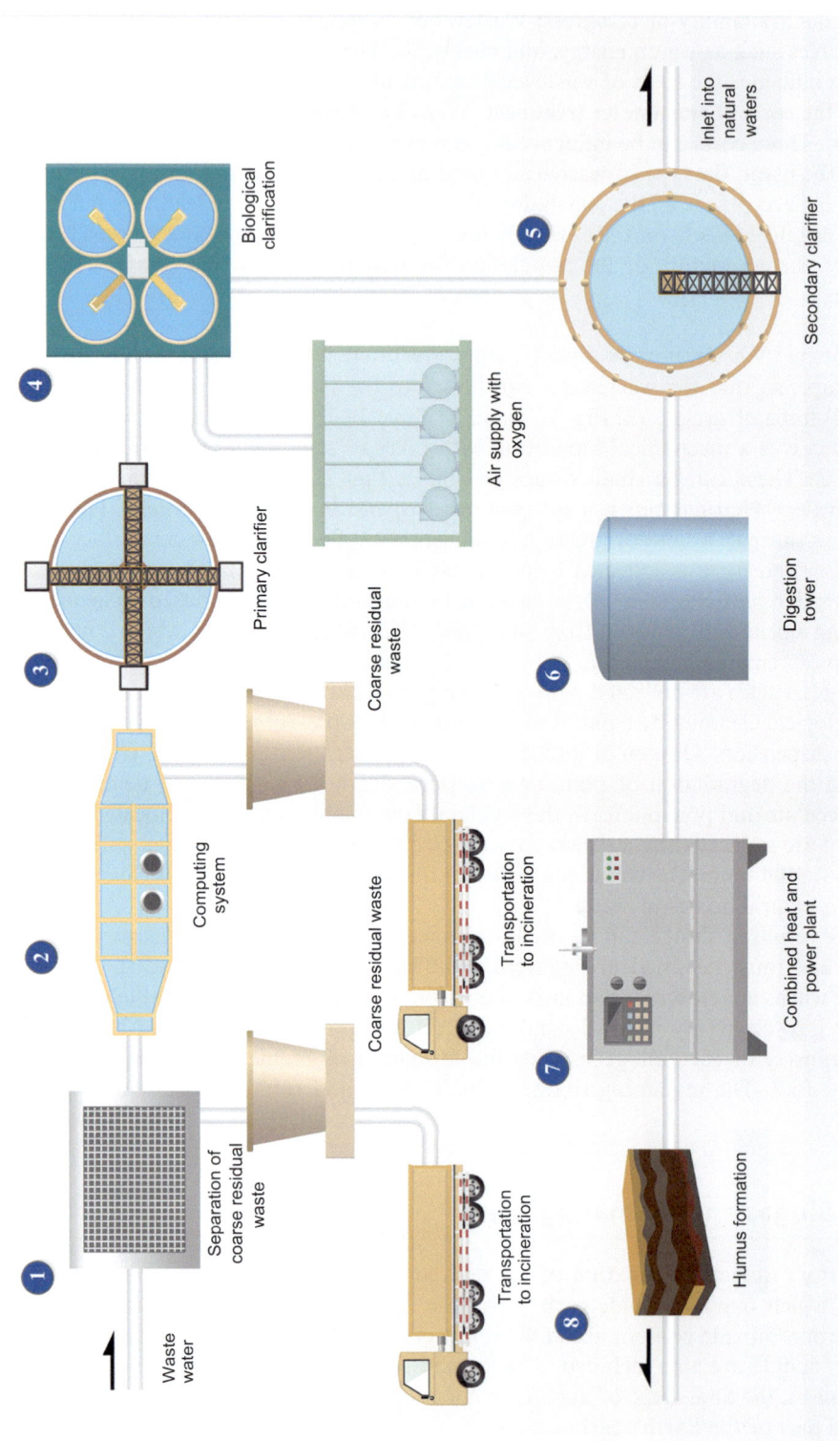

Fig. 16.1 Function and substance transport in a wastewater treatment plant

16.2 · Biogas and Methanogenesis

Table 16.1 Stages of wastewater clarification	
Mechanical stage	Through grate, grid, sand trap, oil and grease separator
Biological stage	Bacteria, algae, protozoa
Chemical stage/Fine cleaning	Pollutants such as phosphates and heavy metals

Biology The metabolic pathway is carried out by methanogens (bacteria and archaea) in the absence of oxygen, which break down carbon dioxide into methane. Methanogenic organisms are widespread in nature, e.g., in swamps, landfills, wastewater treatment plants, and in the gastrointestinal tract of mammals. Efficient methanogenesis occurs in the interplay of an ecological biotope during the degradation process. (◘ Fig. 16.2). Biologically speaking, biogas production is not a classic fermentation, as due to the complex biological composition and the diversity of methanogenic organisms, it cannot be referred to as a single biochemical conversion, such as in alcoholic fermentation. It involves coordinated utilization chains of various bacteria that are mutualistically associated. This association of archaea with mostly reduced metabolism is referred to as syntrophy. Among the archaea, are *Methanobacter*, *Methanococcus* and *Methanopyrus* of importance, which as end consumers in the anaerobic food chain operate methanogenesis with CO_2 as a substrate. Anaerobic bacteria live in the absence of oxygen, but are capable of fully or partially oxidizing glucose in glycolysis to provide acetate for methanogenesis via pyruvate. However, the energy yield under anaerobic conditions is very low, with a maximum of 4 mol ATP per mol glucose, while 36 mol ATP are obtained aerobically via oxidative phosphorylation. Biomass production is also low, as these are degradative processes; it is significantly below the 50% in oxidative processes, at 5% (low load) to 10% (high load). This also reflects the high nutrient requirement in the COD: N: P supply of 800: 5: 1 (anaerobic) to 100: 5: 0.5–1 in aerobic bacteria. Ba-

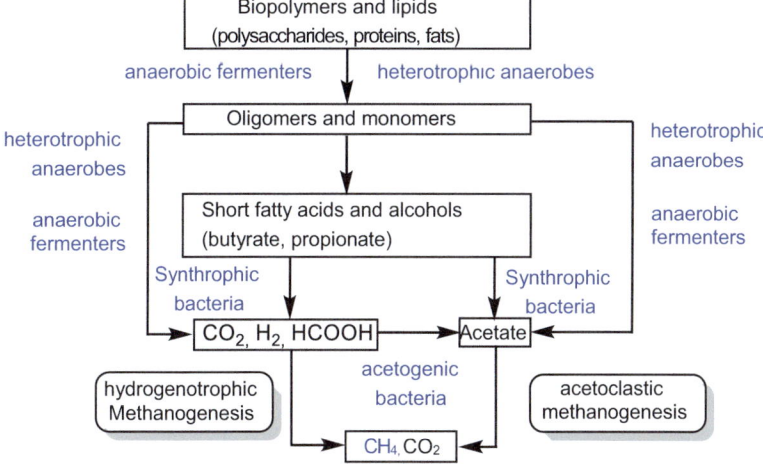

◘ **Fig. 16.2** Ecology and system biochemistry of syntrophic, acetogenic, and anaerobic microorganisms in biogas synthesis

sically, all organic waste can be recycled in a biogas plant. This can be, for example, food waste, pig slurry, solid manure, or green cuttings. The use of starch-containing feed precursors such as maize silage, maize grains, or whey is ethically problematic.

Biosynthesis Methanogenesis (◘ Fig. 16.2) is a complex process that is influenced by a number of factors such as pH, temperature, and substrate concentration. The optimal temperature for the

Methanogenesis is about 37 °C, but it can also occur over a wide temperature range from 10 °C to 80 °C. The pH for methanogenesis is between 6.5 and 8.5. Methanogenesis is an anaerobic biochemical metabolic pathway that occurs in two stages:
- CO_2 reduction with H_2 (30%)
- Degradation of acetic acid (70%)

In the first stage, CO_2 is reduced to methyl groups. In the second stage, the methyl group is reduced to gaseous methane. The reduction of CO_2 to methane is a reductive process in which carbon is reduced from its highest valency (+4) to 0 with the input of energy. The simplified gross equation of methanogenesis is (▶ Reaction 16.1):

Reaction 16.1
$$CO_2 + 4\,H_2 \rightarrow CH_4 + 2\,H_2O$$

Methanogenesis proceeds via two preparatory reaction pathways (◘ Fig. 16.3). It begins with acidogenesis, where organic material is broken down and short-chain organic acids are produced, followed by acetogenesis (◘ Fig. 16.3), which via the acetoclastic pathway of acetate cleavage from sugars and short-chain organic acids directly provides CO_2 and acetic acid. Only now does methanogenesis begin, in which methane (CH_4) is produced. Simplified, methanogenesis starts from carbon dioxide, which is reduced by ferredoxin reductase and bound to methanofuran ① (◘ Fig. 16.5). This reduction requires hydrogen as a reducing agent. The required hydrogen comes from pyruvate, which forms formic acid (HCOOH) with the help

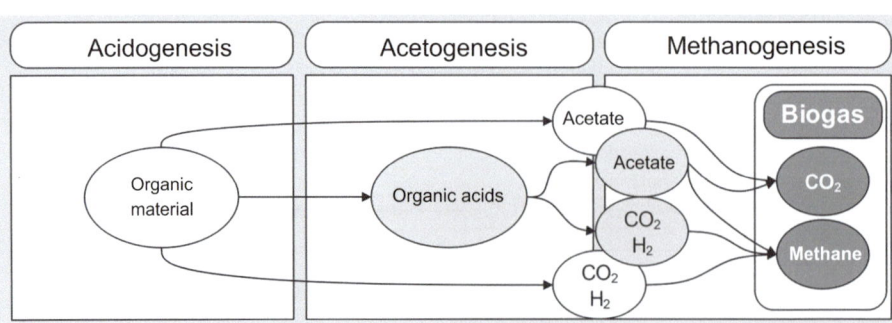

◘ **Fig. 16.3** Acidogenesis, Acetogenesis, and Methanogenesis

16.2 · Biogas and Methanogenesis

of a formate-C-acetyltransferase. With the help of formate dehydrogenase, formic acid is split into hydrogen and carbon dioxide (◘ Fig. 16.4). The transfer of hydrogen for the reduction of oxidized ferredoxin is dependent on sodium (Na^+) and proceeds via a cell wall-bound sodium transporter.

The resulting product is formyl-methanofuran. Methanofuran is a cofactor that is only found in methanogenesis. The goal of this biochemical step is the binding and activation of CO_2 with the help of reduced ferredoxin (Fd). In the second step, the methanofuran is replaced by methanopterin (MP) to formyl-methanopterin ② by a formyltransferase. Methanopterin is structurally similar to the well-known tetrahydrofolate and has similar functions. This activated formyl group is again converted with ferredoxin F420 using tetrahydromethanopterin-cyclohydrolase by splitting off water. The cyclohydrolase is known in tetrahydropterin metabolism and shows different functions. The resulting intermediate products methylene and methyl-methanopterin ③ are reduced in two steps. The hydrogen used here comes only to a small extent from pyruvate and is mainly provided as H_2 in the ferredoxin-420 pool via $NADPH+H^+$. The final step involves the release of the bound methyl group to methane. As is well known in biochemistry, thio compounds are preferred leaving groups, which are also found in methanogenesis. There is a conversion with the sulfur-containing coenzyme-M-SH (CoM-SH) by a thiomethyltransferase. This reaction is linked to a sodium transporter in the bacterial cell wall, which pumps sodium ions out of the cell ④. Coenzyme-M-SH is reduced and replaced by a methyl group. In addition, coenzyme B (CoB) is needed for this reaction, which attacks the thio group of CoM and forms a disulfide (CoM-S-S-CoB) ⑤. This reduction consumes hydrogen, which probably comes from the ferredoxin pool. CoM and CoB are then regenerated and are available for further reactions (◘ Fig. 16.5).

During fermentation in the biogas plant, the bioreactor must be heated with the sludge. The reason for this lies in the anaerobic physiology of the methane producers, which, unlike aerobic organisms, do not generate heat through oxidative phosphorylation. In aerobic organisms, one mole of glucose under ideal conditions provides 2,870 kJ of heat. The same amount of glucose under anaerobic conditions with 5% bacterial growth only provides 131 kJ of heat. However, since heat accelerates the physiology and turnover, the bioreactors are heated with a heater at cold outside temperatures.

Technical Significance The use of biogas should not be confused with the use of pure methane. Biogas is a gas mixture that needs to be processed. The processing of biogas is done by removing the water vapor in a condensate shaft. To achieve a more environmentally friendly combustion and to avoid corrosion, the hydrogen

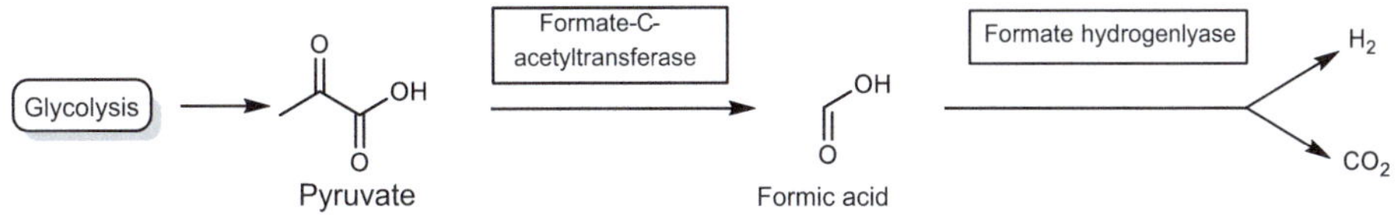

◘ **Fig. 16.4** Energy provision from glycolysis and conversion of formic acid

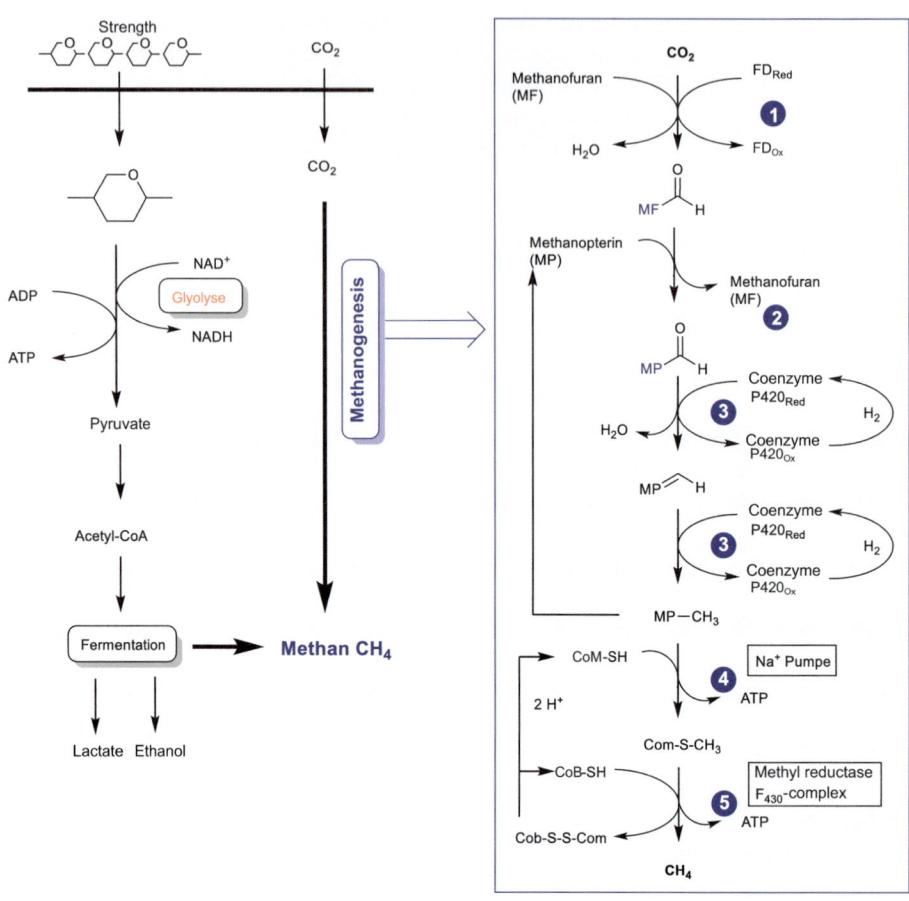

Fig. 16.5 Methanogenesis (simplified representation)

sulfide is removed by desulfurization. The purified methane gas is compressed to ensure an operating pressure for injection or combustion. The combustion can take place in a combined heat and power plant or in an Otto engine.

Self-check Questions

1. Explain why a biogas plant (the fermenter is meant) needs to be heated?
2. Name one microorganism each from the acetogenic and acidogenic phase. Describe optimal culture conditions for both.
3. Can a biogas plant be combined with a sewage treatment plant? Justify your answer.
4. Name and explain the reasons for drying biogas for combustion in engines.
5. Calculate how many biogas plants must be in operation in Germany to replace the natural gas demand. The substrate source is arbitrary.
6. Name the main products of methanogenesis and how they are released. Explain the significance of methane (CH_4) as the end product of this biochemical process.

16.2 · Biogas and Methanogenesis

7. Explain the role that the hydrogen-CO_2 pathway plays in methanogenesis. Explain the biochemical steps of this pathway and the significance of hydrogen (H_2) and carbon dioxide (CO_2) in this process.
8. The fermentation approach is a mixture of various microorganisms. Try to classify them according to their biotransformation.
9. Explain the influence of the H_2 partial pressure (pH_2) on the quality of methanogenesis.

The Future of Bioprocesses Engineering

Contents

17.1 **Artificial Photosynthesis – 247**

17.2 **Natural Substance Biotechnology – 248**

17.3 **Bioinformatics and Artificial Intelligence – 249**
17.3.1 Quantum Computers – 249
17.3.2 Protein Engineering – 251

17.4 **Bioelectricity and Biological Fuel Cells – 252**

© The Author(s), under exclusive license to Springer Fachmedien Wiesbaden GmbH, part of Springer Nature 2025
O. Kayser and N. J. H. Averesch, *Technical Biochemistry*, https://doi.org/10.1007/978-3-658-47121-7_17

> **Learning Objectives**
> - Artificial Photosynthesis
> - Natural Product Biotechnology
> - Bioinformatics and Machine Intelligence
>   - Quantum Computing
>   - Protein Engineering
>   - Artificial Cell as a Factory
> - Bioelectricity
>   - Biological Fuel Cell

The global future market for biotechnologically produced raw materials and end products is a market with growing opportunities and volumes. The worldwide market for bioprocess products is estimated at 71 billion USD in 2022 and is expected to grow by an average of 7–10% per year until 2030 (◘ Fig. 17.1). The most important biobased products in terms of quantity are methanol and ethanol as biofuels and solvents. Already today, ethanol accounts for more than half of all biobased raw materials. It is expected that the issue of dealing with plastic waste and its avoidance will increase the demand for biodegradable plastics in the coming years and thus also significantly increase the demand for various organic acids as building blocks for fiber and plastic production.

What does the future of Industrial Biochemistry look like? The future of Industrial Biochemistry is promising. As the world becomes aware of the need for

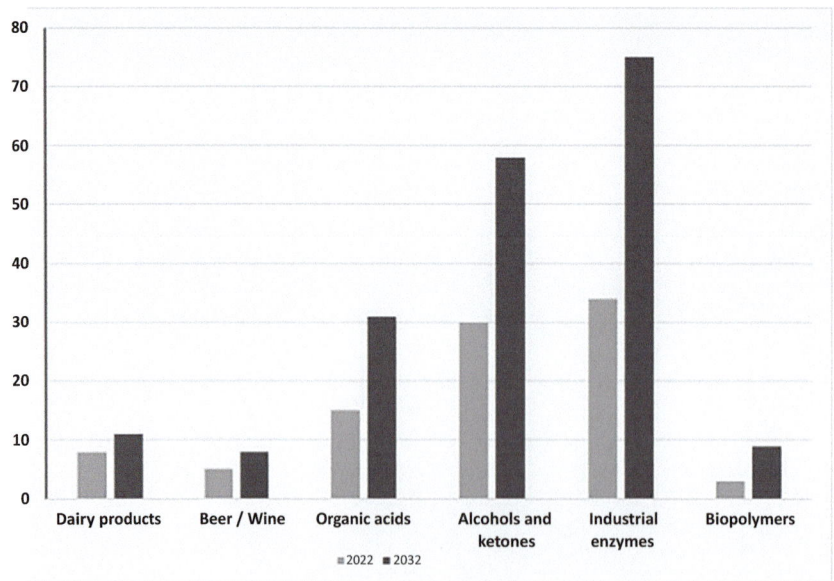

◘ **Fig. 17.1** Global market revenues for chemicals from fermentation processes 2022 and 2032 in billion USD

sustainable and renewable resources, Industrial Biochemistry is gaining importance. It offers a number of advantages over traditional chemical production processes. Primarily, Industrial Biochemistry is more sustainable and uses biological systems more efficiently for limited resources. Waste quantities are greatly reduced and introduced into circular systems. The future of Industrial Biochemistry will be determined by:
- the increasing demand from customers for renewable and sustainable products,
- a greater understanding of biochemical and biotechnical processes through AI in process monitoring, control, and analytics,
- decreasing costs for DNA sequencing and gene editing in Synthetic Biology,
- the development of heterologous photosynthesis in bacteria and the use of CO_2 as a natural resource.

In particular, the biochemical implementation of photosynthetic processes in current production organisms to provide glucose, energy, and reducing agents will greatly change bioprocess engineering, as described in detail in ▶ Chap. 16. If we complain about too high a concentration of carbon dioxide in our atmosphere today, this as yet unused source could become very important for future industrial processes.

In medicine, in addition to the known low molecular weight drugs (molecular weight under 1,000) and proteins, RNA-based drugs are playing an increasingly important role. The clinical breakthrough was achieved with RNA vaccines for the prevention of SARS-CoV2 infections, and in the future, vaccines and antisense drugs against cancer diseases and viral infections will be developed. Since the bioavailability of these new drugs remains problematic even with injection, it remains to be seen what innovative formulations pharmaceutical technology will offer.

Based on the experiences of the two authors, our own considerations and research in recent literature can provide some indications of which areas might be of interest. Of course, it is difficult to look into the biochemical crystal ball, and new technical innovations can immediately revise our assumptions. We have listed selected examples that could influence Technical Biochemistry in the next 10 years.

17.1 Artificial Photosynthesis

Recent approaches and research efforts aim to decouple the photosynthesis from the plant. On the one hand, alternative photosynthetic microorganisms such as cyanobacteria are being sought, on the other hand, ex vivo biohybrid systems are being developed. Many approaches attempt to combine biological elements such as enzymes or organisms with synthetic materials to mimic or improve the process of photosynthesis. The idea of artificial biological photosynthesis is to utilize the efficiency and adaptability of natural photosynthesis while integrating the advantages of synthetic materials. Artificial biological photosynthesis can be realized in various ways:
- Development of **biohybrid systems,** in which living organisms such as algae or cyanobacteria are combined with synthetic materials. The organisms are naturally capable of performing photosynthesis and converting sunlight into

chemical energy, while the synthetic materials support the electron transfer and storage of the generated energy.
- Also exciting is **bio-photovoltaics**. Here biological components such as proteins or enzymes of photosynthesis are used. In reference to plant photosynthesis, photosystems I and II from plants or bacteria, which can absorb light energy and release electrons, are stabilized and made usable for technical applications. An example is biological hydrogen production.
- For technical application, **genetic modification** plays an important role. The genetic modification of organisms such as bacteria or algae is of great interest to improve their ability to photosynthesize and make them produce energy more efficiently. This could enable the development of tailor-made organisms optimized for artificial biological photosynthesis.

Many applications are shaped by the idea of carbon neutrality of our economic actions. An important area of photosynthesis will be sustainable energy generation. Artificial biological photosynthesis could serve as a sustainable method for producing biohydrogen or other fuels (biofuels) that can be used as clean energy sources. A second, so far little considered area could be environmental remediation. By using microorganisms in biohybrid systems, pollutants in the environment could be degraded while simultaneously generating energy. But where are the problems and limits? The current technology has a low efficiency that needs to be increased. The efficiency of artificial biological photosynthesis is currently much lower than that of natural photosynthesis or conventional photovoltaics. Also, proteins are unstable biochemical tools, their low stability is due to the constant renewal in the cell. The stability of biological components in synthetic systems is a challenge as they must react sensitively to environmental conditions.

17.2 Natural Substance Biotechnology

Biotransformations in the cell or in in-vitro systems are becoming increasingly important. Today, more than 150 bioprocesses are known that are entirely or partially controlled by microorganisms or pure enzymes. This group includes not only natural, unaltered cells and catalysts, but also those that have been specifically altered through genetic modification and protein design. Despite the impressive number of bioprocesses, which is expected to double every 10 years, only a few stand out that reach a volume of 5,000 t/year or more in the industry (◘ Table 17.1). The most important enzymatic biotransformations occur in the food and feed industry, but also in the pharmaceutical industry (◘ Table 17.2).

The importance of the bio-based industry will increase over the next twenty years, and a shift away from fossil raw materials will be inevitable if the availability of natural resources allows the expansion of the green industry. The cultivation and production of plant-based raw materials can be critical, as forecasts show that with an expected world population of 10 billion people in 2050, an additional cultivation area the size of Brazil will be needed. This is unrealistic, and biotechnological processes must close this demand gap. Therefore, an even stronger increase in enzymatic and cell-based bioprocesses can be expected, which will not be possible without genetic modification of the platform organisms and their biochemical metabolic pathways.

17.3 · Bioinformatics and Artificial Intelligence

□ **Table 17.1** Biotechnological processes with a production volume of more than 10,000 t per year since 2010 (own research)

Product	Quantity (in t in 2023)	Production organism
Astaxanthin		
- biotechnical	500	*Haematococcus pluvialis*
- synthetic	10,000	
Ethanol		
- from glucose/starch	50,000,000	*Saccharomyces cerevisiae*
- from lignocellulose	1,000,000	*Zymomonas nobilis*
Biodiesel	38,200,000	Rapeseed, provides raw material
Biogas (Biomethane)	30,000,000 m³	Mixture consisting of archaea
Biowater	330,000	*Clostridium acetobutylicum*
Amino acids		
- Tryptophan	5,000	*E. coli*, genetically modified
- Lysine	6,000	*C. glutamicum*
- Cysteine	100,000	*C. glutamicum*
- Glutamine	10,000	*C. glutamicum*
Vitamins		
- Ascorbic acid	30,000	*Pantoea agglomerans*
- Vitamin B_2	10,000	*Ashbya gossypii*
- Vitamin B_{12}	30	*Propionobacterium freudenreichii, Bacillus megaterium*
Laundry enzymes	200,000	*Bacillus sp.*
Biopolymers		
- Polylactate	1,500,000	*Lactobacter sp.*
Organic acid		
- Acetic acid	6,600,000	*Acetobacter sp.*
- Lactic acid	40,000	*Lactobacter sp.*

17.3 Bioinformatics and Artificial Intelligence

17.3.1 Quantum Computers

The application of quantum computers in biochemistry can become a fascinating and promising field of research, as quantum computers have the potential to solve complex and computationally intensive problems in biochemistry that are very difficult or even impossible to handle with classical computers. An important area of application is molecule simulations. Quantum computers can simulate molecules at the atomic level and precisely take into account quantum mechanical effects that occur in biochemical processes and bioorganic reactions. This will benefit drug research, as the identification of potential drug candidates requires the screening of large molecule databases in silico. Quantum computers can accelerate this process by predicting the probability with which certain molecules penetrate into biological

◘ **Table 17.2** Important enzymes in biotechnological industrial processes

Enzyme EC-Number	Strain	Product
Acetolactate Decarboxylase 4.1.1.5	*Bacillus brevis*	No direct product
Alcohol Dehydrogenase 1.1.1.1	*Actinetobacter calcoaceticus*	No direct product
Aldehyde Reductase 1.1.1.2	*E. coli*	No direct product
Amidase 3.5.1.4	*Comamonas acidovorans* *Klebsiella oxytoca* *Klebsiella terrigena*	No direct product
Aminopeptidase 3.4.11.1	*Bacillus* sp.	Aspartame
Aminoacylase 3.5.1.14	*Bacillus* sp.	L-Tryptophan
Alpha-Amylase 3.2.1.3	*Bacillus licheniformis*	Glucose
L-Aspartase 4.3.1.1	*Brevibacterium flavum* *E. coli*	Aspartame
Cyclodextrin-Glucosyltransferase 2.4.1.19	*Bacillus* spp.	Cyclodextrins
Lipase 3.1.1.3	*Bacillus subtilis* *Pseudomonas* sp. *Burkholderia cepacia*	Metoprolol Diltiazem Ibuprofen
Penicillin-Acylase 3.5.1.11	*Bacillus subtilis*	Amoxicillin Ampicillin
Subtilisin 3.4.21.62	*Bacillus lentus* *Bacillus licheniformis*	Aspartate

target structures. This allows for more accurate predictions of molecule structures, interactions, and biochemical reactions, which will be of great benefit for the development of drugs. In the development of drugs, the interactions between molecules and proteins such as enzymes and receptors in the body must be accurately predicted. Quantum computers can perhaps more accurately predict binding energies and structural changes in the interaction of drug molecules with target proteins. This speeds up the development process and reduces costs in the laboratory, animal testing, and clinical trials.

In biochemistry, catalysis, i.e., the acceleration of chemical reactions by enzymes, plays an important role. Quantum computers can help to better understand the mechanisms of enzymatic reactions and optimize the development of enzymes for industrial and medical applications.

However, there are also limits and doubts about this promising vision of the future. The technical challenges are great, as quantum computers are still in the experimental stage and are prone to errors. The technology must be significantly improved before it can be used on a large scale in biochemistry. A very important argument is the high costs. Not only the acquisition, but also the construction and maintenance of quantum computers are extremely expensive. It is unclear whether the investments will pay off in the long term for many biochemical research institutions and companies. It is being discussed whether quantum computers actually offer superior performance in all biochemical applications. In some cases, conventional high-performance computers may still be competitive.

In summary, quantum computers have the potential to advance biochemical research in many ways by enabling complex calculations and simulations. However, many technical challenges still need to be overcome, and the cost-benefit ratio must be carefully weighed. The future role of quantum computers in biochemistry will likely depend on the development of this technology and the specific application.

17.3.2 Protein Engineering

AlphaFold is an AI system developed by the UK company DeepMind (▶ https://alphafold.ebi.ac.uk) to predict the three-dimensional structure of proteins. It combines deep neural networks with biological knowledge to accurately model protein folding. This makes it possible to predict the molecular structure of proteins with high accuracy. AlphaFold can revolutionize protein research by accelerating the deciphering of disease mechanisms and the development of drugs. It has the potential to deepen our understanding of biology and medicine and accelerate the development of new therapies. AlphaFold has been praised for its impressive accuracy in planning protein folding. The company and its algorithms are also of great importance for biotechnology as they enable the accurate prediction of catalysis enzymes. This in turn accelerates the development of tailored proteins and applications in various areas of biotechnology. Some examples already show today how AlphaFold influences biotechnology:

- Enzymes and Biocatalysis: By accurately predicting protein structures, AlphaFold helps to develop enzymes for biotechnological processes. These enzymes can help to accelerate or make chemical reactions more selective. This can be beneficial in the production of chemicals, drugs, or food.
- Biopharmaceuticals: With AlphaFold, the structure of therapeutic proteins and antibodies can be precisely predicted. This helps to develop biopharmaceutical drugs that are used to treat, for example, cancer, autoimmune diseases, and other diseases.
- Vaccine development: The prediction of protein structures helps in the identification of antigens and epitopes on pathogens. This is crucial for the development of vaccines to trigger effective immune responses against viruses, bacteria, or other pathogens.
- Food production: In the food industry, proteins developed with the help of AlphaFold can be used to improve the texture, taste, or nutrient composition of foods.
- Material science: In material science, the precision control of protein structures is used to produce biological materials with tailored properties, for example, biodegradable plastics or materials with special mechanical properties.

In all these application areas, AlphaFold enables targeted design of proteins and biological molecules. This can improve the efficiency, effectiveness, and sustainability of biotechnological processes. This makes a significant contribution to the further development and application of biotechnology in various industries.

17.4 Bioelectricity and Biological Fuel Cells

Biological fuel cells are devices for converting chemical energy from biological, mostly microbiological processes, into electrical energy. They convert organic substances such as biomass or organic waste into electrical energy with the help of microorganisms, mostly bacteria. This technology enables sustainable and environmentally friendly energy production and waste recycling. In the field of renewable energies, specially adapted proteins used in biological fuel cell systems can improve energy extraction from organic material. This technology has the potential to become a sustainable energy source and offers numerous applications in both environmental technology and medicine.

The functioning of biological fuel cells is based on the biochemical metabolism of microorganisms. The basic functioning of such a fuel cell proceeds in the following steps: The organic fuel is oxidized. The oxidation of, for example, glucose takes place at the anode of the fuel cell. Here, the glucose is broken down by bacteria, releasing electrons. As in the electron transport chain, the released electrons are transported by the bacteria via an electrically conductive path to the cathode. This electron transport creates an electric current. At the cathode of the fuel cell, the reduction of an electron acceptor, usually oxygen, takes place. The electrons bind to the electron acceptor and form water or another compound, depending on the type of fuel cell. The electron flow from the anode via an external circuit to the cathode generates electrical energy, which can be used to power devices or stored in an energy storage device.

Biofuel cells have the advantage of being sustainable energy sources. They use organic fuels, which are abundantly available in nature, e.g., waste. Fuel cells can efficiently break down organic waste and generate energy, contributing to waste prevention. They are also environmentally friendly, as biological fuel cells are based on renewable resources in addition to waste recycling. But there are also challenges and limitations here. Current biological fuel cells are not very efficient. They often have a lower energy yield than conventional fuel cells. Efficiency can be limited by factors such as low power density and voltage losses. The use of microorganisms carries the risk of contamination and requires strict controls to ensure the stability and efficiency of the fuel cell. The performance of biological fuel cells can depend on environmental conditions such as temperature, pH value, and nutrient concentration, and scaling up biological fuel cells for commercial applications can bring technical challenges.

Supplementary Information

Answers to the Self-Control Questions – 254

Important Databases – 267

Glossary – 269

References – 271

© The Editor(s) (if applicable) and The Author(s), under exclusive license to Springer Fachmedien Wiesbaden GmbH, part of Springer Nature 2025
O. Kayser and N. J. H. Averesch, *Technical Biochemistry*, https://doi.org/10.1007/978-3-658-47121-7

Answers to the Self-Control Questions

▶ Chap. 3 The Basis of Biochemical Reactions—Primary Metabolism

Question 1: **Ethnomedicine** is the interdisciplinary field between ethnology and medicine. Ethnomedicine researches the culture-specific healing methods and applications of plants, animals, and/or minerals. **Phytochemistry** is a subfield of organic chemistry or natural product chemistry and deals with the analysis, identification, and structural description of natural substances from plants. **Physiology** is the science and the subfield of biology that deals with the physical and biochemical processes in the cell or in tissues. In contrast to biochemistry and anatomy, physiology distinguishes itself by considering the relationships in the entire organism.

Question 2: see book chapter 3.
Primary metabolism: Acetic acid, Phenylalanine, Valine, Lactic acid, Citric acid.
Secondary metabolism: Morphine, Podophyllotoxin, Flavonoids, Papaverine. See also: Metabolism Secondary Plant Substances.

Question 3: a. Living beings can perceive information (stimuli) from their environment and react to it (irritability). b. Living beings are capable of reproducing and multiplying (reproduction and multiplication). c. Living beings possess a (own!) metabolism for the construction and maintenance of their body and its functions. d. Living beings grow and develop (growth and development). e. Living beings can move themselves or at least show movements within their body (or within their cells) (movement, mobility or motility).

Question 4: Viruses only meet some of these criteria. For example, they can only multiply through hosts and show neither their own growth nor do they have their own metabolism. In contrast to the cell, viruses are traditionally not considered living beings.

Question 5: Biogas production, Erythromycin, Tetracycline or Penicillin production, Citric acid production, Vanillin biosynthesis, Lysine biosynthesis, Glutamate biosynthesis.

Question 6: About 0.5% to a chimpanzee and about 50% to a banana.

▶ Chap. 4 Bioorganic Reactions and Building Blocks of Biosynthesis

Question 1: A classification of biosynthetic building blocks can be made by dividing them according to the number of carbon or nitrogen atoms and the structural features formed by these. (see also Question 2).

Question 2: C1 (e.g. methyl group), C2 (e.g. acetyl-CoA), C5 (e.g. isoprene), C6C1 (e.g. shikimic acid), C6C3 (e.g. phenylpropane), C6C2N (e.g. aromatic amino acids), Indol C2N (e.g. tryptophan), C4N (e.g. pyrrolizidine alkaloids), C5N (e.g. lysine).

Question 3: C_6C_3: Vanillin, cinnamic acid; C_6C_2N: Mescaline, adrenaline.

Question 4: Nucleophilic substitution and electrophilic addition are two types of reactions in organic chemistry. In nucleophilic substitution, it is important that the nucleophile is an electron donor that exchanges (= substitutes) a residue. In electrophilic addition, an unsaturated carbon (e.g. C=C) reacts with various classes of substances. In a reaction with SAM, no transfer of the methyl group to a carbon takes place, which is why an electrophilic addition, among other things, is ruled out. What about coenzyme A? If there is an electron donor in nucleophilic substitution, there must also be an electron acceptor. The thiol of the coenzyme can be seen as an acceptor that attacks an electron donor such as a hydroxyl or amino group. Truthfully, one should also say that this S_N2 reaction is fundamentally possible in C-methylations, but very rarely occurs in nature.

Question 5: Tautomerism is a special form of isomerism. Tautomeric compounds are compounds with the same empirical formula, but differ in their structural formula due to "intramolecular migration" of a small part, e.g. a proton. Both compounds are in a chemical equilibrium with each other, the rate constants for the forward and reverse reactions are equal.

Question 6: Ethanol is converted in the liver by alcohol dehydrogenase to CO_2 and fatty acids. Acetaldehyde, a harmful compound (hazard identification: Xn; H-statement 351: May cause cancer), is formed as an intermediate. Acetaldehyde is then further oxidized by acetaldehyde dehydrogenase to acetate. Subsequently, the degradation takes place with the enzyme cytochrome P450E1 down to CO_2 and water.

Question 7: The synthetic metabolic pathway consists of a 3-dehydroshikimate dehydratase (3DSD) from the mold *Podospora pauciseta*, an O-methyltransferase (OMT) from *Homo sapiens*, and an aromatic carboxyl reductase (ACAR) from a bacterium of the genus *Nocardia*. The ACAR enzyme must be activated by phosphopantetheinylation, which was achieved by a phosphopantetheinyl transferase from *Corynebacterium glutamicum*. The glycosylation to vanillin-β-D-glucoside was carried out using an Arabidopsis thaliana family 1 UDP-glycosyl transferase (UGT). This was done because vanillin is very toxic to microorganisms (for yeast 20 mg/L), glycosylated vanillin is not.

▶ Chap. 7 The basis of all biochemical reactions—primary metabolism

Question 1: Opposite biochemical reactions occur in both metabolic pathways. In the light reaction, ATP and NADPH are formed. Both energy carriers or reduction equivalents are converted or degraded in the dark reaction.

Question 2: The end product is glucose. However, this monosaccharide is not the transport form. This is created by linking the monosaccharides glucose and fructose to form the disaccharide sucrose. The reason for this is the reduction of osmotic pressure.

Question 3: An aldose is a sugar that carries a terminal carbonyl group. Examples of aldoses are D-(+)-glyceraldehyde, D-(+)-glucose, D-(-)-ribose, D-(-)-erythrose, or D-(+)-xylose. Ketoses are sugars that carry a carbonyl group on the penultimate C-atom. Examples of ketoses are D-fructose, D-ribulose, or D-xylulose. Also compare in the book ▶ Chap. 8 (Sugar).

Question 4: Net equation of photosynthesis: $6\ CO_2 + 6\ H_2O \rightarrow C_6H_{12}O_6 + 6\ O_2$, gross equation of photosynthesis: $12\ H_2O + 12\ NADP^+ + 18\ ADP + 18\ Pi \rightarrow 6\ O_2 + 12\ NADPH + 12\ H^+ + 18\ ATP$. Photosynthesis is a complex biochemical reaction cascade that usually cannot be summarized in a net or gross equation. This is because it is a complex process with about 20 individual steps that are catalyzed by different enzymes (e.g., enzyme 1 splits off a proton, enzyme 2 stabilizes a charge, enzyme 3 transfers oxygen, etc.). Under this assumption, it is chemically incorrect to describe all these reactions with one equation. For the sake of clarity, however, all the chemical substances entering the reaction (left side) and all the substances leaving the reaction (right side) are often summarized, resulting in the gross equation. It is possible, for example, that H_2O enters the reaction but is also released as a product. Stoichiometrically, this makes no sense, which is why this is not captured in a net equation. For the sake of completeness, however, the water is still included in the gross equation.

Question 5: The Z-scheme is an energy diagram that shows the electron transport in the light-dependent reactions of photosynthesis in plants and cyanobacteria. The y-axis shows the ability of the involved molecules to transfer electrons to the molecule immediately to the right, thereby acting as a reducing agent. Molecules that are higher than their right partner readily give up electrons, as this occurs along the energy hill (English, "downhill reaction"). However, the chlorophylls of the reaction centers of the photosystems must first be excited by light energy (English, "uphill reaction").

Question 6: The goal of the light-dependent reaction is to store energy in the form of ATP and the reduction equivalent NADPH for the subsequent dark reaction. Using photosystems II and I, a portion of the spectral range is utilized to chemically bind light energy. Comparatively, it can be noted that ATP and NADPH are formed in the light reaction and consumed in the dark reaction to build energy-rich structures like glucose from energy-poor CO_2.

Question 7: Naturally, every reaction step is essential. However, characteristic for the Calvin cycle is the fixation of CO_2 by the enzyme RuBisCO to ribulose-1,5-diphosphate.

Question 8: The redox potentials at pH 7.2 are E = −0.28 V for NADH and E = −0.37 V for NADPH. Due to the presence of the phosphate group in NADPH, a stronger binding of the negative charge occurs, which explains the lower redox potential. The ratio of NAD^+ : NADH is kept constant in the cell at about 30:1. This allows the use of NAD^+ for the oxidation of organic molecules. The ratio of $NADP^+$ to NADPH is reversed at 1:50. This, in turn, allows the cell to use NADPH as a reducing agent.

Question 9: All green plants carry out photosynthesis to build carbohydrates. In this process, carbon dioxide (CO_2) is fixed in the "dark reaction" and built into carbohydrates. Most plants (C3 plants) operate a mechanism described as C3 metabolism, in which carbon dioxide passively enters the cells through the stomata and is fixed as a substrate during the day in the Calvin cycle. A variation of this mechanism is found in C4 plants, which actively (and thus with energy consumption) increase the CO_2 concentration for fixation, resulting in a C4 body. This involves a spatial separation (two cell types, mesophyll cells and bundle sheath cells) for the pre-fixation and metabolization of carbon dioxide. This allows the plants to partially close their stomata, as they are not limited by the simple diffusion of carbon dioxide into the cells like C3 plants. When the stomata are partially closed, this also reduces the evaporation of water from the plant. Therefore, C4 plants are predominantly found in dry and sunny locations. The further steps of CO_2 fixation in the Calvin cycle correspond to those of C3 plants. To survive in arid regions, CAM plants have mechanisms that temporally separate the steps of CO_2 fixation from those of the Calvin cycle. This allows the stomata to remain closed during the daytime heat to minimize water loss. They are then opened for CO_2 uptake during the cooler night. While carbon dioxide is fixed into malate at night and stored in the vacuoles, it is released the following day and processed by RuBisCO, the key enzyme of the "dark reaction", similar to a C3 plant. These biochemical reactions take place in a single cell.

- ▶ Chap. 8 Metabolism of Sugars

Question 1: see ▶ Sect. 8.1

Question 2: The citric acid cycle (◘ Fig. 8.5) provides the redox equivalents in the form of NADH and $FADH_2$, which then serve in the respiratory chain (◘ Fig. 8.10) to build up the proton gradient. This ultimately serves to drive the ATP synthase.

Question 3: Most citric acid is produced by *Aspergillus* at the end of its logarithmic growth phase. This is due to two metabolic processes: On the one hand, pyruvate carboxylase forms oxaloacetate from pyruvate and CO_2 in anaplerotic reactions, which is converted to citric acid in the citric acid cycle. On the other hand, malate dehydrogenase in the cytosol converts oxaloacetate to malate, which is transported into the mitochondrion by an antiporter while simultaneously transporting citrate into the cytosol. See ◘ Fig. 8.7.

Answers to the Self-Control Questions

Question 4: The preparatory phase of glycolysis begins with the phosphorylation of glucose to glucose-6-phosphate. This allows the cell to accumulate the compound internally, as it cannot cross the cell membrane unlike glucose. Furthermore, this reduces the glucose concentration inside the cell, as the external and internal glucose concentrations are in equilibrium and this equilibrium is disrupted by the phosphorylation inside. The formed glucose-6-phosphate is then isomerized to fructose-6-phosphate by the glucose-6-phosphate isomerase. The enzyme phosphofructokinase then helps to phosphorylate fructose-6-phosphate by converting a molecule of ATP to ADP and producing fructose-6-bisphosphate. The aldolase is responsible for splitting fructose-6-bisphosphate into dihydroxyacetone phosphate (DHAP), which is subsequently isomerized to glyceraldehyde-3-phosphate (GAP) by the triose phosphate isomerase. At the beginning of the pay-off phase, glyceraldehyde-3-phosphate is oxidized to 1,3-bisphosphoglycerate by the glyceraldehyde-3-phosphate dehydrogenase (GAPDH). During this oxidation, an equivalent of NAD^+ is reduced to NADPH. Thermodynamically, the equilibrium lies on the side of GAP, but the product is converted faster than the reverse reaction can occur. By converting an equivalent of ADP to ATP, the phosphoglycerate kinase converts the 1,3-bisphosphoglycerate to 3-phosphoglycerate. The phosphoglycerate mutase converts the 3-phosphoglycerate to 2-phosphoglycerate, which reacts with the help of the enzyme enolase by splitting off water to phosphoenolpyruvate (PEP). This compound has a high group transfer potential. The pyruvate kinase is thus able to catalyze the reaction of PEP to pyruvate with the gain of an ATP from ADP.

Question 5: The name is derived from pentoses (C5-sugars). A distinction is made between the oxidative and non-oxidative part (◘ Fig. 8.2). In organisms capable of photosynthesis, the non-oxidative part overlaps with the Calvin cycle.

Question 6: Coenzyme A is capable, with its thiol function (-SH), of forming thioester bonds with carboxyl groups. This creates reactive intermediate compounds ("active ester analog"). This enables the conversion of compounds with carboxylic acid functions (as acyl-CoA in the metabolism of fats or as acetyl-CoA in the metabolism of carbohydrates and proteins) at an acceptable speed.

Question 7: Phosphoenolpyruvate (PEP) is the connecting link between glycolysis and the citric acid cycle, thus being a central metabolite. Furthermore, PEP is one of the two initial building blocks of the shikimate pathway.

Question 8: The electron transport chain refers to the biological process in which electrons are formally transferred from a donor to one or more acceptors. This creates an electrochemical proton gradient, which facilitates the synthesis of ADP to ATP with the help of the enzyme ATP synthase, provided the components are in close proximity, e.g., in a biomembrane.

Question 9: see ◘ Figs. 8.10 and 8.11.

Question 10: Other coenzymes (or also cofactors) include, for example, FADH, ATP, Coenzyme A, Coenzyme Q (Ubiquinone), Coenzyme F (Tetrahydrofolate), and Pyridoxal phosphate.

Question 11: Vitamins cannot be produced by the body itself (or not in sufficient quantity), whereas enzymes as subsequent compounds can indeed be produced by the body from vitamins. Hydrophilic vitamins: Vitamin C, B_1, B_2, B_3, B_5, B_6, B_7, B_9, B_{12}. Lipophilic vitamins: Vitamin E, D, K, A.

Question 12: The function of ATP synthase is based on the principle of a molecular motor, which consists of a rotor and a stator. ADP and P are anchored in the complex

▶ Chap. 9 Metabolism of Amino Acids

Question 1: Characteristic for amino acids are the amino and carboxyl functions, which are directly adjacent. Through these functional groups, amino acids can be linked into chains: In a condensation reaction, i.e. with the elimination of water, an amide is formed (◘ Fig. 9.1). The resulting bond is also called a peptide bond in amino acids. Peptides are short chains of up to 50 amino acids. From about 100 linked amino acids, we speak of proteins.

Question 2: Differentiation of amino acids is based on their distinct chemical properties: **Position** of the nearest amino group in relation to the carboxyl group in the amide bond; here, counting starts from the carbon of the carboxyl group (= 1-position). α-amino acids carry their amino group in the 2-position, β-amino acids in the 3-position, and so on. **Properties** of additional functional groups: If the amino acids carry additional functional groups, a distinction can be made based on these (◘ Fig. 9.3).

If an amino acid carries an additional carboxylic acid function (e.g., glutamic acid, aspartic acid), it is classified as **acidic**. If it carries an amino function (e.g., lysine) or other basic groups (arginine, histidine), it is classified as **basic**. Furthermore, amino acids that possess an **aromatic system** (phenylalanine, tryptophan, tyrosine) can be differentiated. **Solubility** or poor solubility in water: thus, hydrophobic amino acids, which are not soluble in water due to the presence of aliphatic carbon atoms (valine, leucine, isoleucine), and hydrophilic amino acids, which possess polar functional groups and are soluble in water for this reason. Another distinction can be made according to the general **size** of the amino acids. For example, glycine represents the smallest proteinogenic amino acid. In addition, amino acids can be differentiated according to the **configuration** of their chirality center (or their chirality centers), with glycine occupying a special position as the only non-chiral amino acid. In the case of chirality, the amino acids can be divided into L- and D- (or, according to another nomenclature, into (R) and (S) amino acids).

Question 3: By combining different functional groups and carbon chain lengths, an infinite number of amino acids can be conceived. However, amino acids can generally be divided into two different categories: proteinogenic and non-proteinogenic. Specifically, the proteinogenic standard amino acids have a fixed number of 20, whereas the number of non-proteinogenic amino acids is unlimited. The proteinogenic standard amino acids can be distinguished based on the possibility of their own biosynthesis in an organism (such as humans). Generally, isoleucine, leucine, lysine, methionine, phenylalanine, threonine, tryptophan, and valine are referred to as essential, although there are exceptions: semi-essential, i.e., conditionally vital, are amino acids that can only be formed from other amino acids in the body (for example, tyrosine can only be synthesized from the essential amino acid phenylalanine and is therefore often also counted among the essential amino acids): arginine, histidine, asparagine, cysteine, glutamine, glycine, proline, tyrosine. Non-essential are alanine, aspartate (aspartic acid), glutamate (glutamic acid), and serine.

Question 4: The amino acids are derived from glycolysis (serine, glycine, alanine, valine, leucine), the citric acid cycle (aspartate, methionine, lysine, threonine, proline, glutamate) and the shikimate pathway (phenylalanine, tyrosine, tryptophan), see also Fig., p. 82.

Question 5: see Biosyntheses in ▶ Chap. 9, ◘ Fig. 9.4 generally and specifically for example Glutamate in ◘ Fig. 9.5.

Question 6: In the production of an amino acid through transamination, a comparatively "cheap amino acid" can be "refined," so to speak, by enzymatically catalyzed transfer of an amino group. The necessary process-technical effort is less than with a complete (bio-)synthesis of the amino acid. With clever coupling of the enzymatic steps, an almost complete conversion is possible.

▶ Chap. 10 Fatty Acid Biosynthesis and ABE Metabolism

Question 1: Palmitic acid is the most common fatty acid in nature. The general formula for saturated fatty acids is $H_3C\text{-}(CH_2)_n\text{-}COOH$.

Question 2: Chemically, unsaturated fatty acids differ from their saturated analogs by the presence of at least one double bond (C=C). Higher unsaturated polycarboxylic acids usually have two or three double bonds. Fatty acids with more than three double bonds are less common, but are very important as essential fatty acids for the human body.

Question 3: Course of fatty acid biosynthesis and the degradation, which is referred to as β-oxidation.

Question 4: Lipids and surfactants show strong structural similarities. Both can identify a head and a tail. Both are therefore amphiphilic, but to different degrees: surfactants are much more polar, as they have a lipophilic part (fatty acid residue) as well as a strongly hydrophilic part (phosphate group or choline residue). This makes them surface-active, i.e., they behave physically like dishwashing detergent with soap effect. Lipids do not have this property. Since the lipophilic part predominates, they do not form emulsions in aqueous solutions.

Question 5:

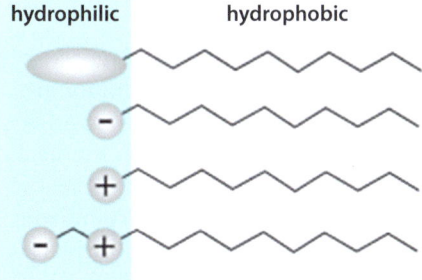

Question 6: The three groups of fatty acid derivatives are prostaglandins, thromboxanes, and leukotrienes.

Question 7: Isobutanol is formed via the 2-ketovalerate pathway and butanol via the ABE pathway. The precursor of isobutanol is formally an amino acid (◘ Fig. 10.4), that of butanol is acetyl-CoA (◘ Fig. 10.1).

Question 8: The distinction between first and second generation biofuels is based on the raw materials and production methods. First generation biofuels are mainly made from edible plant oils and food crops such as corn, sugarcane, soybeans, and rapeseed oil. Second generation biofuels are based on non-edible plants and waste materials, such as straw, wood waste, algae, *Miscanthus* and *Jatropha*. These raw materials do not directly compete with food production.

▶ Chap. 11 Secondary metabolism and important biotechnological pathways

Question 1: As "hydrates of carbon," the name "carbohydrates" derives from the empirical ratio of C to H to O of 1:2:1, which would correspond to a sum formula of $C_n(H_2O)_n$. Carbohydrates are among the most important sources and storage substances of energy in almost all living organisms.

Question 2: Glucose must be activated to make it available for further metabolism (in glycolysis). This is done by transferring a phosphate, where one mole of ATP is converted to ADP and the glucose is phosphorylated to glucose-6-phosphate.

Question 3: Monosaccharides typically have between three and nine carbon atoms. Hexoses with six carbon atoms are referred to as "true sugars".

Question 4: The D-(+)-isomers of carbohydrates are dominant in nature.

Question 5: Dextrose, also called D-(+)-glucose, which in ring form is also referred to as α-D-glucopyranose, for example. Of course, it can also be drawn in its beta form.

Question 6: While up to 38 moles of ATP per mole of sugar can be obtained aerobically, alcoholic fermentation only yields 2 moles of ATP per mole of sugar.

Question 7: Compounds of two to six sugar units, which are linked together via a glycosidic bond, a so-called hemiacetal.

Question 8: Important polysaccharides: Starch, glycogen, chitin, and cellulose. Starch and glycogen are storage substances. Chitin and cellulose are structural materials in the cell walls of insects and plants. DNA (and/or RNA) is found in all living beings.

Question 9: Humans do not possess cellulase and are therefore unable to break down cellulose to its basic building block D-glucose (see also yellow flesh). Many bacteria, on the other hand, possess cellulases. Animals, such as ruminants, which largely rely on cellulose-containing food, often do not have endogenous cellulases and are therefore dependent on (prokaryotic) endosymbionts, which convert the cellulose into fatty acids via C2 units. The same applies to insect species such as termites. Humans can break down cellulose-containing food to a certain extent into short-chain fatty acids (C4 to C8) with the help of anaerobic bacteria. Cellulose, as well as hemicelluloses, pectin, and lignin, represent an important component of human nutrition as dietary fiber.

Question 10: The polymerization of monomeric sugars into long-chain sugar chains is based on a condensation reaction. The reverse breakdown of sugars back to their monomers is chemically a hydrolysis, which is found in all saccharides and makes them biocompatible under the assumption of hydrolytic enzymes.

Question 11: The biological activity of heparins can be explained by the sulfate and amino groups typical for this class of substances. These functional groups specifically affect factors of the blood coagulation cascade and inhibit them.

Question 12: Chitin is the polysaccharide that gives hardness to the shell and exoskeleton of insects, among other things. Unlike cellulose, chitin has an acetamide group, i.e., a hydroxyl group is replaced by an amino group that is esterified with acetic acid. Chitosan, a polyglucosamine that has lost the acetic acid through hydrolysis, is obtained from chitin, among other things. The hydrolysis can be carried out chemically with caustic soda or enzymatically.

▶ Chap. 12 Phenolic Natural Substances

Question 1: Both glycolysis itself and the pentose phosphate pathway branching off from glycolysis feed the shikimate pathway.

Question 2: With the help of the shikimate pathway, it is possible to obtain aromatic natural substances both with and without nitrogen. In principle, products of the shikimate pathway have an aromatic structure. Since these are simple aromatics with hydroxyl groups, they are also considered phenolic natural substances, whose OH groups are typically found in *para*-or *meta*-position. Aromatics of the shikimate pathway can often be distinguished from those of polyketide biosynthesis by the alternating oxygenation (keto or hydroxy group) occurring here.

Question 3: The most important biosynthesis steps to tryptophan are given in the book in ◘ Fig. 9.11. The essential difference from the classic shikimate pathway is that the introduction of an amino group occurs at anthranilate and ribose is involved in the construction of the pyrrole ring.

Question 4: Benzoic acid, vanillin, gallic acid, cinnamic acid, quercetin, lysergic acid, morphine, dopamine.

Question 5: Both in nature and in the synthetic biosynthesis pathway, vanillin is obtained via the shikimate pathway. In natural biosynthesis, vanillin is obtained via phenylalanine starting from *para*-coumaric acid via the ferulic acid pathway or the benzoate pathway. The synthetic biosynthesis pathway diverges much earlier from the shikimate pathway (◘ Fig. 9.10): Starting from 3-dehydroshikimate, vanillin is built up in only three steps.

Question 6: Ammonium lyases are capable of replacing the amino group of phenylalanine with a keto group, which is then successively reduced either to a hydroxyl group or to a C-C double bond.

Question 7: See best question/answer 7 from Chap. "Building Blocks of Biosynthesis", ▶ Sect. 4.1. Sometimes the reversal of metabolic pathways is also possible, as shown in the example of β-oxidation. To create a non-natural metabolic pathway, enzymes from different organisms are often combined, sometimes even newly found in a natural organism and heterogeneously expressed in a technically favorable host organism. This can be very difficult, as the functionality of the heterogeneous enzymes is not trivial. A prerequisite for success is in any case that thermodynamics allows the individual reactions as exergonic reactions.

▶ Chap. 13 Terpenes

Question 1: Unlike terpenes, terpenoids also contain functional groups such as hydroxyl groups and are built from a multiple of the C5 monomer. This monomer is isopentenyl diphosphate (IPP).

Question 2: Terpenoids can be accessed both via the mevalonate pathway and the MEP (methylerythritol phosphate) pathway.

Question 3: Terpenes are classified according to the number of their carbon atoms:

Monoterpenoids:	C10 or 2 × C5
Sesquiterpenoids:	C15 or 3 × C5
Diterpenoids:	C20 or 2 × C10
Triterpenoids:	C30 or 2 × C15
Carotenoids:	C40 or 2 × C20
Polyterpenoids:	n × C5

Question 4: see ◻ Table 13.2.

Question 5: Natural rubber has, on average, about 10 times the carbon number of gutta-percha (carbon number = C5n with n = ± 5000 for natural rubber and n = ± 500 for gutta-percha). In addition, the double bonds of natural rubber always have the *trans* configuration as opposed to the *cis* configuration of gutta-percha.

Question 6: The reason is as simple as it is problematic: To date, biosynthesis is only known up to dihydroartemisinic acid; what follows is unknown. Possibly a non-enzymatic chemical conversion occurs, as postulated by some authors.

Question 7: Steroids are a class of terpenes, which share the structural feature of three six-membered and one five-membered ring. Sterane thus represents the basic structure of steroids (◻ Fig. 13.15). While sexual steroids, glucosteroids, and bile salts share this basic structure, they differ in their further functionalization. For example, bile salts always carry a propyl carboxylic acid substituent in the 1-position of the five-ring (in some cases, the carboxylic acid is further functionalized, e.g., to an amide). Sexual steroids, such as estrogen or androgen, are very potent hormones in the body, having significant effects on sexual reproduction and muscle building. Glucosteroids are important steroids that play a major role in physiology, psychology, and chronic stress responses. Bile salts are surfactants in the mammalian organism, which are released into the gastrointestinal tract (small intestine and large intestine) via the bile and strongly influence digestion.

Question 8: The sterically specific oxidation is an enzymatic reaction that was rather accidentally found in the laboratory. An overview is given in ◻ Figs. 13.18 and 13.19.

▶ Chap. 14 Alkaloids

Question 1: In the course of his discovery of the alkaloid "morphine", F. W. Sertürner found that the compound reacted "like alkali", that is, it was basic. This is the origin of the word "alkaloid".

Question 2: Historically, it has been very difficult to find a "lowest common denominator", so the current definition broadly refers to nitrogenous natural substances with a strong physiological effect (with the exception of antibiotics). Examples of the chemical structural diversity of alkaloids are given in the book in ◻ Fig. 14.1.

Answers to the Self-Control Questions

Question 3: Catecholamines are some of the simplest alkaloids, which are based on phenols with an amine. Important representatives are dopamine and adrenaline. The biosynthesis begins with tyrosine from the shikimate pathway and is mainly characterized by changes in the side chain such as decarboxylation, hydroxylation, and N-methylation.

Question 4: The biosynthesis of morphine derivatives is complex and begins with a dimerization of two tyrosine molecules to (S)-reticulin. The steps of morphine biosynthesis are well documented in the book in ◘ Fig. 14.4 for thebaine. It is important to note that codeine is the end product of biosynthesis and the medically important drugs etorphine, buprenorphine, and naloxone are derived from the intermediate product thebaine.

Question 5: Ajmaline, Galanthamine, Morphine.

Question 6: The biosynthesis of morphine derivatives is complex and begins with a dimerization of two tyrosine molecules to (S)-reticulin. The steps of morphine biosynthesis are well documented in the book in ◘ Figs. 14.3 and 14.4. It is important to note that codeine is the end product of biosynthesis. The biosynthetic intermediate thebaine is used for the synthesis of medically important drugs such as etorphine, buprenorphine, and naloxone.

Question 7: See Scopolamine and Cocaine.

Question 8: See ◘ Fig. 14.6. The intersection point is the Secologanin.

Question 9: Ajmaline.

Ajmaline is a complexly structured alkaloid. Determining the building blocks can sometimes be a challenge even for experts. This example was chosen to illustrate the complexity of biosynthesis reactions. When considering the basic structure, a tryptophan based on a C_6C_2N building block, shown here in blue, is certainly quickly recognizable.

The second part is harder to identify: Upon closer inspection, however, it is noticeable that there are 10 carbon atoms, which are interrupted by nitrogen and oxygen. If you look at this building block more closely, you will notice that it is secologanin, which is made up of a C10 or two C5 building blocks (red and green). The trick is that during the rearrangement of geraniol via loganin, a bond in the C5 building block (green) is broken, so that a carbon atom (green) is added to an existing carbon building block (red). This is further oxidized to an aldehyde to finally react in a Schiff base reaction with the exocyclic nitrogen of tryptophan. Neutral bonds are shown in black.

Geraniol → Loganin → Secologan

Galantamine

Composed of phenylalanine and tyrosine
  Emetin

Composed of twice tyrosine and secologanin

▶ Chap. 15 Antibiotics

Question 1: An antibiotic (Greek ἀντί-/anti- "against" and βίος/bios "life") in its original sense is a naturally formed low-molecular metabolic product of fungi or bacteria, which inhibits the growth of other microorganisms or kills them. An antibiotic in a broader sense is also an antimicrobial substance that does not occur in nature and is produced semi-synthetically, fully synthetically, or genetically, but is not a disinfectant.

Question 2: No, synthetic antibiotics are chemically produced compounds that typically do not occur in nature. They are specifically developed to combat a wide range of pathogenic microorganisms. Their mechanisms of action can be diverse, but generally aim to inhibit the growth or reproduction of bacteria or to kill them directly. Some of the most common mechanisms of action of synthetic antibiotics are: inhibition of cell wall synthesis, inhibition of protein synthesis, disruption of DNA replication, and inhibition of metabolite synthesis.

Question 3: **Recombinant DNA Technology:** Recombinant DNA technology allows for the targeted manipulation of the genomes of microorganisms, including antibiotic producers, to increase the production of antibiotics. This includes the introduction, deactivation, or enhancement of genes involved in antibiotic synthesis. For example, genes coding for the production of antibiotic precursors can be overexpressed. **Genome-wide Analyses:** Modern genome sequencing technologies allow researchers to analyze the complete genome of antibiotic producers. This enables the identification of genes associated with antibiotic production, as well as regulatory elements that influence gene expression. **Metabolomics and Proteomics:** The analysis of the metabolome (totality of metabolic products) and proteome (totality of proteins) of microorganisms aids in the identification of metabolic pathways and enzymes involved in antibiotic production. This allows for targeted optimization of these metabolic pathways. **Synthetic Biology:** Synthetic biology enables the construction of custom strains by using modular genetic elements to introduce new metabolic pathways or modify existing ones. This allows for the targeted production of antibiotic variants. **Metabolic Engineering:** This approach aims to specifically alter the metabolism of microorganisms to increase the production of antibiotics. This can be achieved through targeted modification of metabolic pathways and flow regulation.

Question 4: Sampling from environmental sources: One of the first strategies is to collect samples from various natural environments, including soil, water, sediments, plants, microorganisms, and animals. These samples can contain a variety of potential antibiotic producers.

Question 5: **Evolutionary Differences:** Plants and microorganisms have very different evolutionary origins and lifestyles. Microorganisms, especially bacteria and fungi, are more closely related to pathogens and have developed antibiotics over the course of evolution to protect themselves from competing microorganisms. Plants have a different evolutionary history and have specialized more on chemical defense mechanisms against herbivores and pathogenic insects. **Specificity:** Plants usually produce bioactive compounds that target specific interactions with herbivores or insects. These compounds are often not broad enough in their effectiveness to act as antibiotics against a variety of pathogens. Microorganisms, on the other hand, have often developed antibiotics that can combat a wider range of bacteria or fungi. **Production Capacity:** Plants have limited resources for the production of bioactive compounds. Most of their secondary metabolites are focused on self-protection and adaptation to their environment. Therefore, the production of antibiotics in sufficient quantities for medical purposes could be difficult in plants.

- ▶ Chap. 16 Environmental Biochemistry

Question 1: In anaerobic fermentation, organic substrates such as biomass or waste are broken down by microorganisms under oxygen-free conditions. This process is temperature-dependent, and the microorganisms involved in fermentation do not have a respiratory chain that thermodynamically gives off heat under anaerobic conditions.

Question 2: Clostridium acetobutylicum is an anaerobic, gram-positive bacterial strain that plays a key role in the acetogenic phase of the fermentation process. *C. thermocellum* is an anaerobic, thermophilic gram-positive bacterial strain that plays an important role in the acidogenic phase of anaerobic fermentation.

Question 3: Yes, that makes sense, as the sludge can be broken down into biogas.

Question 4: Biogas contains water, which can damage the engine when burned.

Question 5: The answer to this question depends on several factors, including the amount of biomass available for biogas production, the efficiency of biogas plants, and the technical prerequisites for feeding biogas into the natural gas grid. According to the Federal Association of the German Biogas Industry (BDBW), up to 100,000 bio gas plants could be built in Germany to cover the entire natural gas demand. This corresponds to a tenfold increase in the current number of plants.

Question 6: The main products of methanogenesis are methane (CH_4) and carbon dioxide (CO_2). These products are released by the biochemical reactions of methanogenesis: Methane (CH_4) is formed as the end product of methanogenesis and is released by the methanogens. It is a colorless and odorless gas that is of great importance as a renewable energy carrier and as a potent greenhouse gas. Carbon dioxide (CO_2) is formed in the biochemical reactions of the hydrogen-CO_2 pathway before it is released as a byproduct. Carbon dioxide is an important intermediate product and contributes to carbon binding in this process. The importance of methane as the end product of methanogenesis lies in its role as a renewable energy carrier and in its effects on the carbon cycle and climate change. Methane can be used as a fuel to generate energy and is used in biogas plants and as a source of natural gas. At the same time, methane is a potent greenhouse gas, the release of which into the atmosphere influences climate change. Therefore, the research and understanding of methanogenesis are of great ecological and energy economic importance.

Question 7: The hydrogen-CO_2 pathway, also known as hydrogenotrophic methanogenesis, is one of the main ways in which methanogens produce methane. In this pathway, hydrogen (H_2) and carbon dioxide (CO_2) are the main substrates and play a crucial role: Step 1: Formation of formate ($HCOO^-$): First, CO_2 reacts with hydrogen (H_2) involving an enzyme called formate dehydrogenase. This forms formate ($HCOO^-$), which serves as an intermediate product. Step 2: Formation of methane (CH_4): In a further step, formate is converted by the enzyme formate-methane-lyase into methane (CH_4) and carbon dioxide (CO_2). Here, methane is released as the end product. The hydrogen-CO_2 pathway is important because it allows methanogens to produce methane using hydrogen and carbon dioxide as substrates. This pathway is widespread in anaerobic environments, such as the digestive tract of animals or sediments.

Question 8: (1) Hydrolytic microorganisms (yeasts, bacteria), (2) acidogenic bacteria (clostridia), (3) acetogenic bacteria (*Streptomonas*), (4) methanogenic bacteria (*Methanococcus*)

Question 9: The H_2 partial pressure must be kept at a low level to ensure the thermodynamic conditions for a conversion of volatile acids and alcohols to acetate.

Important Databases

- **1000 Plant Genome Project**
Database for genome research of 1,000 plant genomes.

- **1001 Genomes**
Database for research in 1001 sequence variations of *Arabidopsis thaliana*.

- **BioCarta**
Database of metabolic pathways with structural formulas and explanations.

- **Brenda**
Probably the most comprehensive database on enzymes.

- **Entrez Genome Database**
Database for the most relevant genomes in biology and medicine.

- **EcoCyc**
Collection of model organisms, their metabolic pathways, enzymes, genes, and substrates. Specialized in *E. coli*.

- **ExPASy**
SIB Bioinformatics offers a vast portal for research around the topic of life sciences, genomics, phylogeny, transcriptomics, population genetics, and much more. Particularly interesting website, as a lot of software is also offered here.

- **GeneMANIA**
Database to predict and graphically represent interactions and functions of genes.

- **Hazardous Substances Data Bank (HSDB)**
Database for researching toxic substances.

- **Human Metabolome Database (HMDB)**
Database for researching human metabolomes.

- **JournalSeek**
Free database that offers all freely available journals with full-text search.

- **Kyoto Encyclopedia of Genes and Genomes (KEGG)**
The most popular biochemistry database: All known biochemical metabolic pathways in various organisms are recorded here. A list of additional databases in biochemistry can be found here.

- **Medicinal Plant Genomics Resources**
Database on genomes of important medicinal plants.

- **METLIN**

Collection of metabolite information and tandem mass spectrometry data. The provided data is useful in the application of metabolomics.

- **Natural Products Alert (Napralert)**

Database for researching low molecular natural substances.

- **MetaCyc**

Database on metabolic pathways, natural product biosynthesis, metabolic reactions, etc.

- **PDB—Protein Data Bank (PDB)**

Provides comprehensive information about proteins, their chemical, biological, and physical properties, as well as proteomes. A further list of databases for protein biochemistry can be found here.

- **PhylomeDB**

This database specializes in phylogenetics. By collecting relevant genes, it offers a so-called phylome, which allows predictions about orthology and paralogy.

- **PubMed**

Database for searching for (medical) publications and medically relevant data such as genes, chemical substances, organisms, etc.

- **Super Natural Database (SuperNatural)**

Database for researching low molecular natural substances.

- **UniProt**

Database for searching protein sequences and protein functionalities. Another list of databases can be found here.

- **Zinc12**

List of databases from commercial providers of natural substances.

Glossary

Acetyl group Functional group in chemistry that is derived from a carboxylic acid.

Aldehyde Chemical compound that is an alcohol from which water (H_2O) has been removed. Functional group on a hypothetical residue R: R-CHO. Aldehydes are very reactive in natural product chemistry.

Alkaloid A group of natural chemical substances that contain at least one nitrogen atom, have a strong effect on humans, but are not antibiotics.

Alcohol Group of organic molecules that carry at least one hydroxyl group (OH).

Amine Chemical compound that contains at least one nitrogen as derivatives of ammonia. Cyclic amines are considered heterocycles and are very commonly found in alkaloids.

Anesthesia State of insensitivity in pharmacology during surgical procedures, which is generally also referred to as numbness.

Androgens Chemical group of male sex hormones.

Drug substance An active ingredient that has been approved by an authority for medical use in humans or animals.

Base A molecule that can absorb hydrogen ions when dissolved in water. Bases have an alkaline pH value above 7.

Chromatography A technique in chemistry for the separation of a mixture of substances using a stationary phase (solid that does not dissolve in the mobile phase) and a mobile phase (liquid, gas).

Inflammation Body's own response to harmful stimuli, leading to warmth, redness, swelling, pain.

Ester A group of chemical substances that are formed in a chemical reaction between an alcohol and an acid, with the elimination of water. Example: Acetylsalicylic acid, Heroin.

Ethyne A chemical gaseous substance, also called acetylene.

Fermentation Microbiological conversion of a substrate by cells or enzymes in biology and biotechnology.

Fat A group of chemical substances in which glycerin is esterified with three fatty acids. Fats can be solid (butter) or liquid (olive oil) at room temperature.

Progestogens Group of pregnancy hormones that serve the initiation and maintenance of pregnancies.

Glucocorticoids Chemically modified steroid hormones that have an effect on glucose metabolism.

Hydrolysis Splitting of a chemical compound, in which water is consumed.

Ketone Chemical compound that has a carbonyl group as a non-terminal functional group (R-CO-R). Ketones are very reactive.

Condensation Chemical reaction in which two substances are combined with the elimination of water. For example, an ester from an alcohol and an acid.

Constitution Indicates which and how many atoms are present in a molecule and how they are connected to each other.

Cancer A collective term for diseases characterized by uncontrolled growth of cells and the seeding of metastases into healthy tissue.

Natural substance Chemical compounds that are formed by an organism to fulfill a biological function.

Neurotransmitter A signaling molecule that is released at the synapse of a nerve cell to transmit an electrical signal.

Metabolism The chemical processes that occur within a living organism in order to maintain life.

Molecule Group of atoms that are connected with each other.

Molecular mass Mass of $6.022140857 \times 10^{23}$ molecules of a compound.

Oncogene An unusually activated gene that can turn a cell into a tumor cell.

Opiates Alkaloids of opium, which as natural or semi-synthetic substances have an effect on opiate receptors in humans.

Opioids Natural or synthetic substances that have an effect on opioid receptors in humans.

Estrogen Group of female sex hormones. The newer term is Estrogen.

Oxidation Chemical reaction in which electrons are given off and in which oxygen is very often involved.

Pharmacy Science of the production and testing of drugs.

Pharmacognosy Science in pharmacy and the study of plant, fungal, and animal drugs, medicines, and toxins.

Pharmacology Sub-discipline of medicine that deals with the effect and efficacy of chemical substances and drugs in the human and animal body.

Phytochemistry Analytical science that chemically investigates the occurrence of constituents in plants.

Plant drug Dried whole plant or its parts such as leaves or roots.

pH value Chemical term for the acidity and alkalinity of a liquid. A liquid with a pH value of 7 is neutral (e.g., distilled water), values less than 7 indicate acids, and those greater than 7 indicate bases.

Reduction Chemical reaction in which electrons are gained and in which the oxidation state is reduced.

Acid, organic Chemical compound with a carboxyl group that transfers protons to another reaction partner. Functional group with a hypothetical residue R: R-COOH.

Metabolism Chemical transformations in the cell of substances such as food. These chemical conversions are also called biochemical reactions, which metabolize energy, heat, carbon dioxide, and degradation products from sugar and fat in the body.

Synthesis, chemical Chemical term for the production of a chemical substance from at least two chemical starting materials.

Terpenes Largest known group of natural substances, predominantly composed of carbon, followed by oxygen. The general structure consists of isoprene units.

Effect Biological effect of a chemical substance in a living biological system from the cell to the organism

Efficacy Effect of a drug in humans.

Active ingredient Chemical substance with a biological effect.

References

General Literature

Introduction

Sahm, H., Antranikian, G., Stahmann, K.-P., Takors, R. (2013) Industrielle Mikrobiologie. Springer Spektrum. 1. edn * [Der Klassiker, der eine Neuauflage braucht]

Berg, M.J., Tymozcko, J.L., Gatto Jr., G.J., Stryer, L. (2018) Stryer Biochemie. Springer Spektrum Berlin, Heidelberg. 8. edn [Das Standardwerk in der Biochemie]

Kurreck, J., Engels, J.W., Lottspeich, F. (2022) Bioanalytik. Springer Spektrum Berlin, Heidelberg. 4. edn [Das Standardwerk für die Analytik biologischer Proben]

Görtz, H.-D., Brümmer, F. (2012) Biologie für Ingenieure. Springer Spektrum Berlin, Heidelberg. [Ein schönes Buch für den schnellen Einstieg]

Madigan, M.T., Bender, K.S., Buckley, D.H., Sattley, W.M., Stahl, .D.A. (2020) Brock Mikrobiologie kompakt, Pearson Studium. 15. edn

Koolman, J., Röhm, K.-J. (2019) Taschenatlas Biochemie des Menschen, Thieme Verlag, Stuttgart. 5. edn

Primary Metabolism

Heldt, H.W., Piechulla, B. (2014) Pflanzenbiochemie. Springer Spektrum Berlin, Heidelberg. [Das Standardwerk für die Pflanzenbiochemie]

McMurry, J., Begley, T. (2006) Organische Chemie der biologischen Stoffwechselwege. Spektrum Akademischer Verlag. [Ein gelungener Versuch, die Organische Chemie und die Biochemie zusammenzubringen, um Mechanismen der Enzymkatalyse zu erklären.]

Secondary Metabolism

Hänsel, R, Sticher (Editoren) (2010) Pharmakognosie—Phytopharmazie. Springer Berlin, Heidelberg. 9. edn [Der Klassiker über die Chemie und die biologische Wirkung sekundärer Naturstoffe]

Breitmaier, E. (2008) Alkaloide. Vieweg-Teubner, Stuttgart, Wiesbaden. 3. edn [Ein älteres Buch, das aber einen guten Überblick über die Chemie der Alkaloide gibt]

Giannis, A. (2023) Naturstoffe im Dienst der Medizin. Springer, Berlin, Heidelberg. [Ein Buch, das ausgewählte Naturstoffe, ihre Pharmakologie und gesellschaftliche Bedeutung einordnet]

Kayser, O. (2024) Von den Molukken zu Molekülen. Springer, Berlin, Heidelberg. [Eher ein Geschichtsbuch, das die Geschichte ausgewählter Naturstoffe beschreibt]

Other Areas of Technical Biochemistry

Steinhilber, D., Schubert-Zsilavecz, M., Roth, H.J. Medizinische Chemie, (Deutscher Apotheker Verlag, Stuttgart. 2. edn. [Ein sehr umfassendes Werk, das den aktuellen Stand der Medizinischen Chemie angibt]

Stolz, A. (2018) Extremophile Mikroorganismen: von der Anpassung zur Anwendung. Springer, Spektrum, Berlin, Heidelberg. [Ein Buch, welches das mikrobielle Leben im physiologischen Grenzbereich gut beschreibt]

Wink, M., van Wyk, B.e., Wink, C. (2008) Handbuch der giftigen und psychoaktiven Pflanzen. Wissenschaftliche Verlagsgesellschaft mbH, Stuttgart. [Ein zusammenfassendes Kompendium der wichtigsten tierischen und pflanzlichen Gifte]

Nuhn, P. (2006) Naturstoffchemie. S. Hirzel Verlag Stuttgart. 4. edn. [Ein Buch, das trotz seines Alters einen sehr guten Überblick über die Chemie der meisten pflanzlichen Naturstoffe gibt.]

References to Literature Citations in the Book

Abdelaal, A. S., & Yazdani, S. S. (2022). Engineering *E. coli* to synthesize butanol. *Biochemical Society Transactions*, *50*(2), 867–876. ▶ https://doi.org/10.1042/BST20211009

Atsumi, S., Hanai, T., & Liao, J. C. (2008). *Non-fermentative pathways for synthesis of branched-chain higher alcohols as biofuels. 451.* ▶ https://doi.org/10.1038/nature06450

Atsumi, S., Li, Z., & Liao, J. C. (2009). Acetolactate synthase from *Bacillus subtilis* serves as a 2-ketoisovalerate decarboxylase for isobutanol biosynthesis in *Escherichia coli*. *Applied and Environmental Microbiology*, *75*(19), 6306–6311. ▶ https://doi.org/10.1128/AEM.01160-09

Ault, A. (2004). The monosodium glutamate story: The commercial production of MSG and other amino acids. *Journal of Chemical Education*, *81*(3), 347–355. ▶ https://doi.org/10.1021/ED081P347

Bang, H. B., Choi, I. H., Jang, J. H., & Jeong, K. J. (2021). Engineering of *Escherichia coli* for the economic production L-phenylalanine in large-scale bioreactors. *Biotechnology and Bioprocess Engineering*, *26*(3), 468–475. ▶ https://doi.org/10.1007/S12257-020-0313-1/METRICS

Baritugo, K. A., Kim, H. T., David, Y., Choi, J. il, Hong, S. H., Jeong, K. J., Choi, J. H., Joo, J. C., & Park, S. J. (2018). Metabolic engineering of *Corynebacterium glutamicum* for fermentative production of chemicals in biorefinery. *Applied Microbiology and Biotechnology*, *102*(9), 3915–3937. ▶ https://doi.org/10.1007/S00253-018-8896-6

Becker, A., Katzen, F., Puè, A., & Ielpi, L. (1998). Xanthan gum biosynthesis and application: A biochemical genetic perspective. *Applied Microbiology and Biotechnology*, *50*, 145–152.

Brown, S., & Pummill, P. (2008). Recombinant production of hyaluronic acid. *Current Pharmaceutical Biotechnology*, *9*(4), 239–241. ▶ https://doi.org/10.2174/138920108785161488

Brunel, J. (1951). Antibiosis from Pasteur to Fleming. *Journal of the History of Medicine and Allied Sciences*, *6*(3), 287–301. ▶ https://doi.org/10.1093/JHMAS/VI.SUMMER.287

Buchholz, J., Schwentner, A., Brunnenkan, B., Gabris, C., Grimm, S., Gerstmeir, R., Takors, R., Eikmanns, B. J., & Blombacha, B. (2013). Platform engineering of *Corynebacterium glutamicum* with reduced pyruvate dehydrogenase complex activity for improved production of l-lysine, l-valine, and 2-ketoisovalerate. *Applied and Environmental Microbiology*, *79*(18), 5566–5575. ▶ https://doi.org/10.1128/AEM.01741-13

Cai, M., Liu, Z., Zhao, Z., Wu, H., Xu, M., & Rao, Z. (2023a). Microbial production of L-methionine and its precursors using systems metabolic engineering. In *Biotechnology Advances* (Bd. 69). Elsevier Inc. ▶ https://doi.org/10.1016/j.biotechadv.2023.108260

Cai, M., Liu, Z., Zhao, Z., Wu, H., Xu, M., & Rao, Z. (2023b). Microbial production of L-methionine and its precursors using systems metabolic engineering. *Biotechnology Advances*, *69*. ▶ https://doi.org/10.1016/j.biotechadv.2023.108260

Chen, Y., Daviet, L., Schalk, M., Siewers, V., & Nielsen, J. (2013). Establishing a platform cell factory through engineering of yeast acetyl-CoA metabolism. *Metabolic Engineering*. ▶ https://doi.org/10.1016/j.ymben.2012.11.002

Chen, Z., & Elowitz, M. B. (2021). Programmable protein circuit design. *Cell*, *184*(9), 2284–2301. ▶ https://doi.org/10.1016/J.CELL.2021.03.007

Cook, M. A., & Wright, G. D. (2022). The past, present, and future of antibiotics. *Science Translational Medicine*, *14*(657). ▶ https://doi.org/10.1126/SCITRANSLMED.ABO7793/ASSET/7B9DBBD9-64BC-4FF2-92CC-E257DFE0E092/ASSETS/IMAGES/LARGE/SCITRANSLMED.ABO7793-F3.JPG

Coutinho, E. M. (2002). Gossypol: A contraceptive for men. *Contraception*, *65*(4), 259–263. ▶ https://doi.org/10.1016/S0010-7824(02)00294-9

Damerow, Peter. 2012. „Sumerian Beer: The Origins of Brewing Technology in Ancient Mesopotamia." Cuneiform Digital Library Journal 2012 (2). ▶ https://cdli.mpiwg-berlin.mpg.de/articles/cdlj/2012-2.

Dewick, P. M. (2009). *Medicinal natural products : A biosynthetic approach* (Vol. 2). ▶ https://onlinelibrary.wiley.com/doi/book/10.1002/9780470742761

Dicker, K. T., Gurski, L. A., Pradhan-Bhatt, S., Witt, R. L., Farach-Carson, M. C., & Jia, X. (2014). Hyaluronan: A simple polysaccharide with diverse biological functions. *Acta Biomaterialia*, *10*(4), 1558–1570. ▶ https://doi.org/10.1016/j.actbio.2013.12.019

Fischbach, M. A., & Walsh, C. T. (2006). Assembly-line enzymology for polyketide and nonribosomal peptide antibiotics: Logic, machinery, and mechanisms. *Chemical Reviews*, *106*(8), 3468–3496. ▶ https://doi.org/10.1021/CR0503097

Gruchattka, E., Hädicke, O., Klamt, S., Schütz, V., & Kayser, O. (2013). In silico profiling of *Escherichia coli* and *Saccharomyces cerevisiae* as terpenoid factories. *Microbial Cell Factories*, *12*(1). ▶ https://doi.org/10.1186/1475-2859-12-84

References

Hofmann, A. (1978). Historical view on ergot alkaloids. *Pharmacology*, *16*(Suppl. 1), 1–11. ▶ https://doi.org/10.1159/000136803

Hosseini, M., & Pereira, D. M. (2023). The chemical space of terpenes: Insights from data science and AI. *Pharmaceuticals*, *16*(2), 202. ▶ https://doi.org/10.3390/PH16020202/S1

Ibrahim, G. G., Yan, J., Xu, L., Yang, M., & Yan, Y. (2021). Resveratrol production in yeast hosts: Current status and perspectives. *Biomolecules*, *11*(6). ▶ https://doi.org/10.3390/BIOM11060830

Jaremko, M. J., Davis, T. D., Corpuz, J. C., & Burkart, M. D. (2020). Type II non-ribosomal peptide synthetase proteins: Structure, mechanism, and protein-protein interactions. *Natural Product Reports*, *37*(3), 355–379. ▶ https://doi.org/10.1039/C9NP00047J

KAYSER, OLIVER. (2024). *Von den Molukken zu Molekülen: Wie Naturstoffe Geschichte schreiben*. Springer-Verlag Berlin.

Lee, S. Y., Kim, J. M., Song, H., Lee, J. W., Kim, T. Y., & Jang, Y.-S. (2008). From genome sequence to integrated bioprocess for succinic acid production by *Mannheimia succiniciproducens*. *Applied Microbiology and Biotechnology*, *79*(1), 11–22. ▶ https://doi.org/10.1007/s00253-008-1424-3

Li, S., Li, F., Zhu, X., Liao, Q., Chang, J. S., & Ho, S. H. (2022). Biohydrogen production from microalgae for environmental sustainability. *Chemosphere*, *291*(Pt 1). ▶ https://doi.org/10.1016/J.CHEMOSPHERE.2021.132717

Liu, B., Wang, H., Du, Z., Li, G., & Ye, H. (2011). Metabolic engineering of artemisinin biosynthesis in *Artemisia annua* L. In *Plant Cell Reports* (Bd. 30, Nummer 5, S. 689–694). ▶ https://doi.org/10.1007/s00299-010-0967-9

Liu, X., Xie, H., Roussou, S., Miao, R., & Lindblad, P. (2021). Engineering cyanobacteria for photosynthetic butanol production. *Photosynthesis: Biotechnological Applications with Microalgae*, 43–66. ▶ https://doi.org/10.1515/9783110716979-002

Men, X., Wang, F., Chen, G. Q., Zhang, H. B., & Xian, M. (2019). Biosynthesis of natural rubber: Current state and perspectives. *International Journal of Molecular Sciences*, *20*(1). ▶ https://doi.org/10.3390/IJMS20010050

Mindt M, Walter T, Kugler P, Wendisch VF (2020). *Biotechnology Journal* 15(7):1900451.

Nagarajan, D., Lee, D.-J., Kondo, A., & Chang, J.-S. (2016). Recent insights into biohydrogen production by microalgae—From biophotolysis to dark fermentation. *Bioresource Technology*, *227*, 373–387. ▶ https://doi.org/10.1016/j.biortech.2016.12.104

Nair, K. P. P. (2010). Rubber (Hevea brasiliensis). In *The Agronomy and Economy of Important Tree Crops of the Developing World*. ▶ https://doi.org/10.1016/b978-0-12-384677-8.00008-4

Pang, X. C., Zhang, L., & Du, G. H. (2018). Podophyllotoxin. In *Natural Small Molecule Drugs from Plants*. ▶ https://doi.org/10.1007/978-981-10-8022-7_90

Pickens, L. B., & Tang, Y. (2009). Decoding and engineering tetracycline biosynthesis. *Metabolic Engineering*, *11*(2), 69. ▶ https://doi.org/10.1016/J.YMBEN.2008.10.001

Proskauer, C. (1958). Development and use of the rubber glove in surgery and gynecology. *Journal of the History of Medicine and Allied Sciences*, *XIII*(3), 373–381. ▶ https://doi.org/10.1093/JHMAS/XIII.3.373

Rahmat, E., & Kang, Y. (2020). Yeast metabolic engineering for the production of pharmaceutically important secondary metabolites. *Applied Microbiology and Biotechnology*, *104*(11), 4659–4674. ▶ https://doi.org/10.1007/s00253-020-10587-y

Renneberg, R. (2007). Biotech history: Yew trees, paclitaxel synthesis and fungi. *Biotechnology Journal*, *2*(10), 1207–1209. ▶ https://doi.org/10.1002/biot.200790106

Rudo, A., Siehl, H. U., Zeller, K. P., Berger, S., & Sicker, D. (2015). Diosgenin aus Yams als Hormonvorstufe: Von der Pflanze für das „Tier" Mensch. *Chemie in Unserer Zeit*, *49*(6), 372–384. ▶ https://doi.org/10.1002/ciuz.201500723

Salehi, M., Bahmankar, M., Naghavi, M. R., & Cornish, K. (2022). Rubber and latex extraction processes for Taraxacum kok-saghyz. *Industrial Crops and Products*, *178*, 114562. ▶ https://doi.org/10.1016/J.INDCROP.2022.114562

Saranraj, P., & Naidu, M. A. (2013). Hyaluronic acid production and its applications—A review. *International Journal of Pharmaceutical & Biological Archives*, *4*(5), 853–859. ▶ http://www.ijpba.info/ijpba/index.php/ijpba/article/viewFile/1126/795

Savile, C. K., Janey, J. M., Mundorff, E. C., Moore, J. C., Tam, S., Jarvis, W. R., Colbeck, J. C., Krebber, A., Fleitz, F. J., Brands, J., Devine, P. N., Huisman, G. W., & Hughes, G. J. (2010). Biocatalytic asymmetric synthesis of chiral amines from ketones applied to sitagliptin manufacture. *Science*, *329*(5989), 305–309. ▶ https://doi.org/10.1126/science.1188934

Shah, Z., Gohar, U. F., Jamshed, I., Mushtaq, A., Mukhtar, H., Zia-Ui-haq, M., Toma, S. I., Manea, R., Moga, M., & Popovici, B. (2021). Podophyllotoxin: History, recent advances and future prospects. *Biomolecules*, *11*(4), 603. ▶ https://doi.org/10.3390/BIOM11040603

Sheng, Q., Wu, X. Y., Xu, X., Tan, X., Li, Z., & Zhang, B. (2021). Production of l-glutamate family amino acids in *Corynebacterium glutamicum*: Physiological mechanism, genetic modulation, and prospects. *Synthetic and Systems Biotechnology*, *6*(4), 302. ▶ https://doi.org/10.1016/J.SYNBIO.2021.09.005

Smith, K. M., & Liao, J. C. (2011). An evolutionary strategy for isobutanol production strain development in *Escherichia coli*. *Metabolic Engineering*, *13*(6), 674–681. ▶ https://doi.org/10.1016/J.YMBEN.2011.08.004

Stehle, F., Degenhardt, F., Zirpel, B., & Kayser, O. (2017). Heterologe Biosynthese der Tetrahydrocannabinolsäure. Biotechnological synthesis of tetrahydrocannabinolic acid. *PHARMAKON, 5*(2), 142–147. ▶ https://doi.org/10.1691/pn.201700016

Steiger, M. G., Blumhoff, M. L., Mattanovich, D., & Sauer, M. (2013). Biochemistry of microbial itaconic acid production. *Frontiers in Microbiology, 4*. ▶ https://doi.org/10.3389/fmicb.2013.00023

Stolz, A. (2017). Extremophile Mikroorganismen. In *Extremophile Mikroorganismen*. ▶ https://doi.org/10.1007/978-3-662-55595-8

Sugahara, K., Schwartz, N. B., & Dorfman, A. (1979). Biosynthesis of hyaluronic acid by *Streptococcus*. *Journal of Biological Chemistry, 254*(14), 6252–6261.

Thomas, F., Schmidt, C., & Kayser, O. (2020). Bioengineering studies and pathway modeling of the heterologous biosynthesis of tetrahydrocannabinolic acid in yeast. In *Applied Microbiology and Biotechnology*. ▶ https://doi.org/10.1007/s00253-020-10798-3

Turconi, J., Griolet, F., Guevel, R., Oddon, G., Villa, R., Geatti, A., Hvala, M., Rossen, K., Göller, R., & Burgard, A. (2014). Semisynthetic artemisinin, the chemical path to industrial production. *Organic Process Research and Development, 18*(3), 417–422. ▶ https://doi.org/10.1021/op4003196

Walsh, C. T., & Wencewicz, T. A. (2014). Prospects for new antibiotics: A molecule-centered perspective. *Journal of Antibiotics, 67*(1), 7–22. ▶ https://doi.org/10.1038/JA.2013.49

Wink, M. (1998). A short history of alkaloids. *Alkaloids*, 11–44. ▶ https://doi.org/10.1007/978-1-4757-2905-4_2

Xu, S., & Li, Y. (2020). Yeast as a promising heterologous host for steroid bioproduction. *Journal of Industrial Microbiology & Biotechnology, 47*(9–10), 829–843. ▶ https://doi.org/10.1007/S10295-020-02291-7

Zha, J., Wu, X., Gong, G., & Koffas, M. A. G. (2019). Pathway enzyme engineering for flavonoid production in recombinant microbes. *Metabolic Engineering Communications, 9*. ▶ https://doi.org/10.1016/J.MEC.2019.E00104

Zhang, H., Wang, Y., Wu, J., Skalina, K., & Pfeifer, B. A. (2010). Complete biosynthesis of erythromycin A and designed analogs using *E. coli* as a heterologous host. *Chemistry and Biology, 17*(11), 1232–1240. ▶ https://doi.org/10.1016/J.CHEMBIOL.2010.09.013

Zhang, X., Sun, Z., Bian, J., Gao, Y., Zhang, D., Xu, G., Zhang, X., Li, H., Shi, J., & Xu, Z. (2022). Rational metabolic engineering combined with biosensor-mediated adaptive laboratory evolution for L-cysteine overproduction from glycerol in *Escherichia coli*. *Fermentation, 8*(7), 299. ▶ https://doi.org/10.3390/FERMENTATION8070299/S1

Zhong, J.-J. (2002). Plant cell culture for production of paclitaxel and other taxanes. *Journal of Bioscience and Bioengineering, 94*(6), 591–599. ▶ https://doi.org/10.1016/S1389-1723(02)80200-6

MIX
Papier aus verantwortungsvollen Quellen
Paper from responsible sources
FSC® C105338

If you have any concerns about our products,
you can contact us on
ProductSafety@springernature.com

In case Publisher is established outside the EU,
the EU authorized representative is:
**Springer Nature Customer Service Center GmbH
Europaplatz 3, 69115 Heidelberg, Germany**

Printed by Libri Plureos GmbH
in Hamburg, Germany